Personality and Individual Differences

A Natural Science Approach

PERSPECTIVES ON INDIVIDUAL DIFFERENCES

CECIL R. REYNOLDS, *Texas A&M University, College Station*
ROBERT T. BROWN, *University of North Carolina, Wilmington*

PERSPECTIVES ON BIAS IN MENTAL TESTING
Edited by Cecil R. Reynolds and Robert T. Brown

PERSONALITY AND INDIVIDUAL DIFFERENCES
A Natural Science Approach
Hans J. Eysenck and Michael W. Eysenck

DETERMINANTS OF SUBSTANCE ABUSE
Biological, Psychological, and Environmental Factors
Edited by Mark Galizio and Stephen A. Maisto

A Continuation Order Plan is available for this series. A continuation order will bring delivery of each new volume immediately upon publication. Volumes are billed only upon actual shipment. For further information please contact the publisher.

Personality and Individual Differences

A Natural Science Approach

Hans J. Eysenck

Institute of Psychiatry
University of London
London, England

and

Michael W. Eysenck

Birkbeck College
University of London
London, England

Plenum Press • New York and London

Library of Congress Cataloging in Publication Data

Eysenck, H. J. (Hans Jurgen), 1916–
  Personality and individual differences.

  (Perspectives on individual differences)
  Bibliography: p.
  Includes indexes.
  1. Personality. 2. Individuality. I. Eysenck, Michael W. II. Title. III. Series.
BF698.E948   1985   155.2   84-24851
  ISBN 0-306-41844-4

First Printing — February 1985
Second Printing — April 1987

© 1985 Hans J. Eysenck and Michael W. Eysenck

Plenum Press is a Division of Plenum Publishing Corporation
233 Spring Street, New York, N.Y. 10013

Printed in the United States of America

Preface

This book presents an introduction to the study of personality and individual differences, but it is not a textbook in the usual sense. As we shall point out in some detail later, typically textbooks on personality and individual differences either deal with statistical and psychometric problems, methodology, and the technical issues of measurement, or else they present the different theories of personality associated with various authors such as Maslow, Cattell, Freud, Jung, Murray, Rogers, Rotter, or to whomever the various eponymous chapters may be dedicated. The theories are presented, together with a brief mention of some empirical studies, but the student is not enlightened as to the weight to be given to the supporting evidence, nor is any comparison attempted between the different theories, formulating judgments regarding completeness, criteria adopted, or validity in terms of experimental proof.

It is small wonder that philosophers of science have concluded that the social sciences, unlike the "hard" sciences, suffer from the lack of a *paradigm* (Kuhn, 1970); this defect is more noticeable, perhaps, in the study of personality and individual differences than in any other part of psychology (except perhaps in clinical and abnormal psychology, where an equal lack of consensus rules). Is it really true that there are in this field no paradigms, that is, comprehensive theories based on empirical studies giving rise to testable deductions and able to show many instances of laboratory verification of such predictions? It is our argument that such a paradigm does exist in this field, and our effort is to trace the history of this paradigm, present it in sufficient detail to make it intelligible, and discuss the relevant experimental evidence so as to make the reader capable of formulating at least a preliminary judgment as to its adequacy. Thus, our effort has not been to present the reader with a general survey of what many different types of psychologists have had to say about personality; instead, we are trying to organize a large amount of material around a few fundamental concepts and to show that these concepts, in both their descriptive and causal aspects, embrace a sufficiently large area of what is usually called personality to be considered a true paradigm. It is from this point of view that the book is written, and it is from this point of view that it should be judged.

There are many other ways in which this book differs from the traditional textbook. We have paid particular attention to the growth of the concepts involved and their historical development from the days of the ancient Greeks, through philosophers like Immanuel Kant and psychologists like Wilhelm Wundt, to the pioneers of modern work. As Boring never tired of pointing out, the historical orientation gives an added degree of profundity to the study of a particular topic, and much can be learned from the mistakes as well as the insights of earlier workers.

Another difference from the typical textbook has been our effort to relate the work of different workers in the field to the paradigm and thus to show that there is a surprising degree of agreement on the fundamental dimensions of personality among many different authors whose concepts would, at first sight, seem to have little in common, either with each other or with the paradigm. Fortunately, a good deal of psychometric work is available to relate the different conceptions of Cattell, Guilford, Comrey, Gough, Rotter, the authors of the MMPI, and many others in this way, and we believe most readers will agree that there is a commonality to all this work that emerges most strikingly from these comparisons.

We have also taken issue with a number of theoretical views that, although popular, we believe are quite mistaken. The belief, for instance, that personality traits do not possess any great degree of consistency and that behavior is highly situation-specific is widespread but mistaken; we give a detailed discussion of the evidence, and we believe that it demonstrates considerable consistency of behavior, not only at any given time but also longitudinally, over 20, 30, or more years of a person's life. The data on which this judgment is based deserve serious consideration from anyone interested in the topic of personality and individual differences.

Another direction in which our treatment differs from a typical textbook is in the emphasis given to genetic factors. For a long time scientists believed that heredity played little part in the genesis of personality traits and individual differences generally, but in recent years a large amount of work has been done to demonstrate the falsity of this assumption, and it is now generally agreed that at least half of the total variance in personality traits is due to genetic causes, and indeed their

contribution may be even larger than that. The point is an important
one, not only because it lends support to the doctrine of consistency of
human conduct, but also because it suggests that in looking for causal
aspects of personality we should take a careful look at physiological, neurological, and hormonal factors as being most likely to mediate the
genetic determinants of behavior.

Last but not least, our treatment differs from that of the usual textbook in emphasizing particularly laboratory studies designed to test
deductions from theories of personality, both physiological and psychological, and in using these laboratory tests as measures of personality.
This, we believe, is an unusual approach to the investigation of personality and one which we believe links it with experimental psychology, a
link that to us appears to be most important in creating a unified science
of psychology.

Personality, as we look at it, has two major aspects: temperament
and intelligence. Most textbooks of personality deal with temperament
only, but we have devoted one chapter to a discussion of intelligence,
partly for the sake of completeness, but also in part because recent work
has completely revolutionized this field and has demonstrated, as has
been the case for temperament, that psychophysiological and laboratory
investigations can throw much light on the concept of intelligence, taking us well beyond the traditional notions of the IQ. Indeed, it may be
possible to say that a new paradigm is in the process of being born in
this field, and hence its inclusion in this book seemed most desirable.

The reader must not assume that because we believe that paradigms
are beginning to emerge in the study of temperament and ability our
major problems have therefore been solved. Paradigms, as Kuhn has
pointed out, always confront anomalies that require exploration and
solutions along the lines of the puzzle-solving process characteristic of
"normal science"; one of our aims has been to indicate some of these
anomalies and the best ways of dealing with them. No doubt the theories
here delineated will be changed profoundly in the course of this puzzle-
solving process, but that, of course, is the fate of all scientific theories. If
the paradigm here outlined has one advantage over most theories of personality dealt with in textbooks, it is that it is eminently testable and
subject to falsification. Most personality theories, particularly those
deriving from psychoanalysis, are impervious to this proof; it is this
quality that makes them so widely acceptable, but it is this same quality
that also means that they are not part of science. If there is no empirical
result that the theory cannot explain after the event, then the theory is
not in fact making any prediction and is not saying anything that is not
tautological. Insofar as the theories here discussed are wrong, they are

capable of being shown to be wrong; scientifically, this is a proud claim that can be made by few theories in the field of personality and individual differences!

Although our main concern has been with psychometric and causal theories associated with our main concepts, we have also been interested in the degree to which the theory can be integrated with, and is relevant to, important social actions and activities such as antisocial and criminal behavior, emotional and neurotic behavior, and cross-cultural differences in behavior. Relations between these topics and personality are not dealt with in any detail, because that would have made the book unduly long; we are simply indicating the kind of deductions that can be made from personality theory onto these various types of behavior, and we quote some of the most relevant studies to indicate that the predictions made can be tested and often give correct results. In that sense, they contribute to the nomological network that constitutes the essence of the theory in question.

We have tried to make the book readable and intelligible to undergraduates without any specialized knowledge in either the psychometric or the experimental fields which we treat. Where we found it necessary to introduce complex modes of statistical treatment, such as factor analysis, we have tried to introduce it briefly by stressing the logic, rather than the mathematics, of the approach in question; we believe that for most readers this will be much more important and relevant and will give them an understanding that a more directly quantitative treatment could not have done. References throughout are given to more advanced treatments of the questions involved, and indeed we have purposely given far more references than would normally be required. The reason for this is plain. In making the claim that we are presenting here a true scientific paradigm, we will surely be confronted by critics who doubt the admissibility of this claim; and to contravert them it is necessary to direct their attention to the large body of evidence which favors our interpretation. Most readers will not need to consult the books and articles mentioned in the bibliography, but if and when they wish to pursue a particular topic further, or assure themselves that what we have said in the text is in fact correct, they will be able to do so by following up the list of references given.

We have benefited from the advice of friends and colleagues, not only in writing the book, but also in carrying out much of the research mentioned in it; the list is too long to reproduce, and it would be invidious to single out individuals for mention. We are grateful to them all, as well as to critics who have drawn attention to weaknesses in theory construction or experimental testing or have suggested alternative theories or solutions to specific problems, and who have successfully managed to keep us from sinking into the slough of complacency—a danger from which scientists are perhaps not immune! In any case, we value their contribution, although of course whatever errors may remain are strictly our own.

HANS J. EYSENCK

Institute of Psychiatry
University of London

MICHAEL W. EYSENCK

Birkbeck College
University of London

Acknowledgments

The author wishes to thank the following for permission to reprint copyrighted materials in this book.

From *Theory of Personality and Individual Differences: Factors Systems, and Processes*, by J. R. Royce and S. Powell, 1983, Englewood Cliffs, NJ: Prentice-Hall. Reprinted by permission.

From "Still Stable after All These Years: Personality as a Key to some Issues in Adulthood and Old Age," by P. T. Costa and R. R. McCrae, 1980. In *Life Span Development and Behavior*, Volume 3, P. B. Baltes and O. G. Brim, eds., New York: Academic Press. Reprinted by permission.

From "Eysenck's Personality Dimensions: A Model for the MMPI," by J. A. Wakefield, B. H. L. Yom, P. E. Bradley, E. B. Doughtie, J. A. Cox, and I. A. Kraft, 1974. In *British Journal of Social and Clinical Psychology, 13:* 413–420. Reprinted by permission.

From *The Biological Basis of Personality*, by H. J. Eysenck, 1967, Springfield, IL: Charles C Thomas. Reprinted by permission.

From "Interaction of Lack of Sleep with Knowledge of Results, Repeated Testing and Individual Differences," by R. T. Wilkinson, 1961. In *Journal of Experimental Psychology, 62:* 263–271. Reprinted by permission.

From "Interaction of Noise with Knowledge of Results and Sleep Deprivation," by R. T. Wilkinson, 1963. In *Journal of Experimental Psychology, 66:* 332–337. Reprinted by permission.

From "The Psychophysiological Nature of Introversion-Extraversion," by J. A. Gray, 1970. In *Behaviour Research and Therapy, 8:* 249–266. Adapted by permission.

From "Extraversion and Pupillary Response to Affective and Taboo Words," by R. M. Stelmack and N. Mandelzys, 1975. In *Psychophysiology, 13:* 536–540. Reprinted by permission.

From "Drug Tolerance and Personality: Some Implications for Eysenck's Theory," by G. S. Claridge, J. R. Donald, and P. M. Birchall, 1981. In *Personality and Individual Differences, 2:* 153–166. Reprinted by permission.

From "Konditionierung, Introversion-Extraversion und die Stärke des Nervensystems," by H. J. Eysenck and A. B. Levey, 1967. In *Zeitschrift für Psycholgie, 174:* 96–106. Reprinted by permission.

From *Readings in Extraversion-Introversion: 3. Bearings on Basic Psychological Processes,* by H. J. Eysenck, 1971, Oxford: Pergamon. Reprinted by permission.

From "Extraversion Arousal and Paired-Associate Recall." by E. Howarth and H. J. Eysenck, 1968. In *Journal of Experimental Research in Personality, 3:* 114–116. Reprinted by permission.

From "The Interactive Effect of Personality, Time of Day, and Caffeine: A Test of the Arousal Model," by W. Revelle, M. S. Humphreys, L. Simon, and K. Gilliland, 1980. In *Journal of Experimental Psychology: General, 109,* 1–31. Reprinted by permission.

From "Effects of Ego Threat and Threat of Pain on State Anxiety," by W. F. Hodges, 1968. In *Journal of Personality and Social Psychology, 8,* 364–372. Reprinted by permission.

Form "Mental Load, Effort, and Individual Differences," by S. Dornic, 1977. In *Reports from the Department of Psychology,* No. 509, University of Stockholm. Reprinted by permission.

From "The Effects of Manifest Anxiety on the Academic Achievement of College Students," by C. D. Spielberger, 1962. In *Mental Hygiene, 46:* 420–426.

From *Accident Proneness* by L. Shaw and H. Sichel, 1970, Oxford: Pergamon. Reprinted by permission.

From *Psychoticism as a Dimension of Personality,* by H. J. Eysenck and S. B. G. Eysenck, 1976, London: Hodder & Stoughton. Reprinted by permission.

From *The Measurement of Personality,* by H. J. Eysenck, 1976, Lancaster: Medical & Technical Publishers. Reprinted by permission.

From *Sensation Seeking: Beyond the Optimal Level of Arousal* by M. Zuckerman, 1979, Hillsdale, NJ: Lawrence Erlbaum. Reprinted by permission.

From "Impulsiveness and Venturesomeness in a Detention Center Population," by S. B. G. Eysenck and B. J. McGurk, 1980. In *Psychological Reports, 47:* 1299–1306. Reprinted by permission.

From *Sex and Personality*, by H. J. Eysenck, 1976. London: Open Books. Reprinted by permission.

From *Personality Structure and Measurement*, by H. J. Eysenck and S. B. G. Eysenck, 1969, London: Curtis Brown Ltd. Reprinted by permission.

From "The Place of Impulsiveness in a Dimensional System of Personality Description," by S. B. G. Eysenck and H. J. Eysenck, 1977. In *British Journal of Social and Clinical Psychology, 16:* 57–68. Reprinted by permission.

From "Will the Real Factor of Extraversion-Introversion Please Stand Up! A Reply to Eysenck," by J. P. Guilford, 1977. In *Psychological Bulletin, 84:* 412–416. Reprinted by permission.

From "Intelligenz informations psychologische Grundlagen," by S. Lehrl, 1983. In *Enzyklopädie Naturwissenschaft und Technik*, Landsberg, West Germany: Moderne Industrie. Reprinted by permission.

From "Effects of Intensity of Visual Stimulation on Auditory Sensitivity in Relation to Personality," by T. Shigehisa and J. R. Symons, 1973. In *British Journal of Psychology, 64:* 205–213. Reprinted by permission.

Contents

PART THREE: EPILOGUE

Chapter Twelve

PART ONE

DESCRIPTIVE

CHAPTER ONE

The Scientific Description of Personality

PERSONALITY AND TAXONOMY: THE PROBLEM OF CLASSIFICATION

At the age of 99, the Greek philosopher Theophrastus wrote a book on personality entitled *Characters*, in the Proem to which he put the question that has motivated all efforts to study personality and individual differences in general ever since: "Why is it that while all Greece lies under the same sky and all the Greeks are educated alike, nevertheless we are all different with respect to personality?" Theophrastus himself was in the tradition of the literary description of personality and intuitive understanding; as Roback (1931) points out:

> It is thanks to these writers of antiquity and their imitators that we can say with a high degree of confidence that human nature, though ages and oceans apart, is about the same wherever found, i.e., the same differences among individuals will be discovered whether they be ancient Greeks or 20th Century Americans. (p. 9)

Opposed to this literary or *idiographic* method, we can already discern among the ancient Greeks a more scientific one, which would nowadays be called *nomothetic*. This was pioneered by Hippocrates and later canonized by Galen, a Roman physician who lived in the second century A.D. It is to these men and to the many others who worked in this field that we owe the doctrine of the four temperaments—phlegmatic, sanguine, choleric, and melancholic. The highly successful typology thus established all those years ago was based on careful observation and provided a paradigm for scientific investigation that has lasted over 2,000 years and may still have something to teach us.

3

When faced with the puzzle of individual differences and with personality generally, we encounter two rather different types of questions. The first of these is static, descriptive, and noncausal; it concerns the descriptive analysis of those types of behavior we include under such terms as *personality, character*, and *temperament*. The second question is concerned with the more dynamic, causal problem of *why* a particular individual behaves in the way he does, shows certain traits of personality rather than others, or demonstrates one kind of ability rather than another. The first type of question leads to a investigation of the *taxonomy* of human behavior, whereas the second leads to an investigation of the *dynamics* of human behavior. Taxonomy inevitably precedes dynamics, and thus the first part of this book will be largely devoted to a discussion of personality description, or taxonomy. The term *taxonomy* in essence refers to classification, and no scientific study of any field is possible without some prior degree of classification of the multivarious material presented to the scientists.

This point is often denied by psychologists of the *idiographic* persuasion. Their case has been best put by Allport (1937), who presents the argument in the strongest possible manner. Earlier German protagonists like Heidegger or Windelband, writing in a prose that is both exuberant and practically unintelligible, have addressed the question in philosophical rather than psychological terms, but Allport's writings are well within the psychological tradition and deserve careful study.

What the idiographic psychologist says, in essence, is that all human beings are unique and consequently cannot be placed on any particular point of a trait or ability continuum. The way in which any particular trait or ability will manifest itself in a given individual will be determined by its congruence with, or opposition to, other traits and abilities, presenting a unique configuration and thus making it impossible for meaningful predictions to be made from isolated measures of traits or abilities. In addition, such traits or abilities are artificial abstractions from reality and do not singly or in combination reproduce the unique living reality that characterizes a particular person's existence.

These arguments are perfectly correct as far as they go and should certainly be taken seriously by psychologists interested in individual differences. However, they tend to go too far, as we shall try to show presently. Taken to the extreme, they lead to a complete nihilism with respect to the possibility of scientific study of personality, and indeed of physics, chemistry, and astronomy as well! Everything that exists is unique, in the same way that a given person is unique; no star or planet is precisely like any other; no atom is precisely identical with any other (even though the differences may only be in position and speed and direction of movement), and no physical or chemical structures are ever identical. Position may seem to be a relatively unimportant aspect of a particular object, but clearly this is not so. Nuclear fusion and fission are caused essentially by changes in the position and velocity of elementary particles. In a similar way, whether the given cells in a chicken embryo develop into one bodily part of another depends on their position in the body of the embryo (MacKenzie, 1982). Two cars, produced in the same works and apparently identical copies of the same model, may differ grossly from each other, as anyone who has ever bought a "lemon" will be able to testify! Even quite small differences in tire pressure may make one model of a given car show understeer and another one oversteer. If the existence of uniqueness is fatal to the scientific study of personality, it must be equally fatal to the scientific study of other organisms, or of physical elements and their combinations.

The opposite pole to the Allport-type idiographic psychologist is the experimental psychologist, who seems to be working on the hypothesis that all human beings are essentially identical, that general laws can be found by the study of small and unrepresentative samples of the population (e.g., sophomores), and that individual differences may be safely disregarded. Cronbach (1957) and H. J. Eysenck (1967a) have argued strongly against this point of view, quoting a rich body of evidence to show that a large proportion of the total variance shown in most psychological experiments is in fact due to individual differences and appears as error variance when such individual differences are not taken into account. We will have ample opportunity to return to this argument.

One of the major arguments against the hypothesis that all organisms are unique, profoundly different from each other, and hence incapable of being studied by the ordinary methods of science appeals to the simple fact that the existence of *differences* implies the existence of *similarities* and that both differences and similarities must be along certain measurable dimensions. How can we say that all individuals differ from each other unless we can quantify these differences and organize them in terms of certain traits, abilities, or other similar concepts? The idiographic psychologist is certainly right in suggesting that these concepts are artificial, but this can hardly be regarded as a drawback. All scientific concepts are artifacts, created by the human mind in order to impart order to an unruly universe and to facilitate understanding and prediction. Concepts like heat, gravitation, and magnetism do not *exist* in the same way that tables and chairs or pigs and flat-earthers exist (although even there the philosopher might inject some caution in the use of the term *exist* in relation to any observable entities); these are all scientific concepts, meaningful within the context of a scientific theory but artificial and likely to be abandoned when other more inclusive and more promising concepts appear on the scene. In this way psychology is no different from any other science, and the criticism, although true as a statement, is not in fact a criticism at all; it is simply an acknowledgment of the nature of scientific concepts.

Thus, to say that individuals differ from each other implies a *direction* or dimension along which these differences may appear. Individual A may differ from Individual B with respect to such fairly obvious dimensions as age and sex, height and weight, size of nose or length of foot. Differences may also exist with respect to intelligence, extraversion, musical ability, amorousness, persistence, and a host of other variables; we cannot say in the abstract that A differs from B unless we can specify the nature of these differences, and that must be done with respect to certain concepts such as those mentioned. Equally, if it is true that two people may *differ* with respect to any of these dimensions, it must also be possible to say that they are *similar* to each other; if I can differ from someone else with respect to intelligence, or beauty, or libido, then it must also be possible to say that I have an intelligence, a degree of beauty, or a measure of libido which is similar to that of some other person.

This immediately leads us into quantification. If we want to make any factual statement about differences and similarities, then we must have a measure of the particular dimensions along which we wish to consider people as being similar or dissimilar to varying degrees. If we wish to make statements about differences in intelligence, then we must have a measure of intelligence that can be applied to every single individual, giving quantitative results that can be translated into statements of differences and similarities. Common speech does not usually insist on refined measurement but implicitly accepts the arguments put forward so far. We say that Mary is more beautiful than Jane, that John is more intelligent than Philip, that Glen has more libido than Frank, or that Michael and David are equally strong or agile or fast in running on the track.

From the simple point of view of description, we may say: "To the scientist, the unique individual is simply the point of intersection of a number of quantitative variables" (H. J. Eysenck, 1952b, p. 18). There are some 340,000 discriminable color experiences, each of which is absolutely unique and distinguishable from any other. From the point of view of descriptive science, they can all be considered as points of intersection of three quantitative variables, namely, hue, tint, and saturation. A combination of perfectly general, descriptive variables is sufficient to allow any individual to be differentiated from any other by specifying his position on each of these variables in a quantitative form. As Guilford (1936) has pointed out, many psychologists "seem unable to see that one individual can differ quantitatively from another in many variables, common variables though they may be, and still have a unique personality" (p. 675).

Let us for the sake of argument assume that human beings are distinguishable along just ten dimensions and that there are ten steps along each dimension. Let us further assume some kind of rectangular distribution and the independence of these dimensions. By the various combinations of these ten dimensions, there would be generated 10,000,000,000 different human beings, distinguished from each other and unique in every way. This would probably be in excess of the total number of human beings that ever lived, including the present generation, and of course there are far more than ten traits along which a human being can be distinguished—and far more steps along each trait. Uniqueness is by no means ruled out by a descriptive system of this kind. The idiographic psychologist indulges in sciamachy!

For many people such a descriptive system is too static to be of much interest; they compare it disadvantageously with colorful idiographic or literary productions such as biographies and autobiographies, descriptive of individual people, or with the tragedies of Shakespeare, or with the works of Proust or Dickens. But this is hardly fair. Description and taxonomy are only the first steps in scientific analysis; as we shall see, they provide a scaffolding that enables us to go on to causal analysis, motivational studies, psychophysiological investigations, and a comparative analysis of genetic and environmental causes, and so forth. The dual nature of the scientific enterprise should never be forgotten; it is nonsensical to criticize one part of the exercise for not having the virtues of the other, and vice versa. No dynamic analysis is possible without a descriptive framework, and the concepts provided within this framework are the stepping stones to a more dynamic analysis and understanding. Historically, this has been true of all sciences; it is not obvious why psychology should be an exception. Literary masterpieces are intellectually and emotionally appealing but cannot be generalized or made the basis of scientific advances.

We shall not pursue here the debate between idiographic and nomothetic psychologists; H. J. Eysenck (1952b) has argued the case in some detail, and there is little point in repeating the arguments. From one point of view the debate is a philosophical one and of little practical interest to psychologists. Essentially, the idiographic approach has not proved fruitful; even Allport, in spite of his advocacy, has always used nomothetic methods for his actual empirical investigations, and neither

Heidegger nor Windelband nor any other idiographic philosopher or psychologist has ever produced data or results that proved acceptable to his peers. Perhaps all that needs to be said is that obviously nomothetic studies are possible and will continue to be pursued. Whether they will ultimately be successful in unraveling the secrets of human behavior, in particular those of personality and individual differences, is a question that only a prophet could answer. All we can do is work along these lines and see to what extent success will attend our efforts. It is perfectly possible, of course, that the philosophers are right and that much of human behavior will elude our grasp. We cannot discover whether this is true until we try. Let us just remember that the prophecies of philosophers have not always been found to be correct in the light of history. Thus Hegel predicted that quite obviously the seven planets known in his time were the only ones that could exist and that no other planets would ever be found; it was not long after this that Neptune was discovered.

Modern personality theory, with its types and traits and abilities, is situated somewhere between idiographic psychology stressing *uniqueness* and experimental psychology stressing the *identity* of human beings. What one must say, essentially, is that although human beings clearly do differ from each other, they differ along certain dimensions, and their differences and similarities can therefore be quantified and measured. Measurements of traits and abilities lead us to certain *type* constructs, such as extraversion–introversion, or verbal ability; this simply means that for certain purposes we can point to groups of people similar with respect to a given trait or ability and contrast these with groups of other people not sharing this trait or ability or showing its opposite. The rules of this mathematical game are fairly well understood, and in the first part of this book we will be concerned very much with an investigation of the degree to which this approach has led us to an acceptable paradigm, in the Kuhnian sense (Kuhn, 1970). It is often objected that the social sciences in general, and psychology in particular, have failed to come up with acceptable paradigms. We believe that this is an exaggeration and that certain aspects of personality theory present us with exactly the kind of paradigm that Kuhn has suggested as forming the basis of a true science.

Kuhn, as is well known, perceived a repetitive pattern in the history of science consisting of two markedly different enterprises: *normal science* and *revolution.* Normal science refers to the accumulation of knowledge within a widely accepted global orientation, called by Kuhn a *paradigm.* This, Kuhn suggests, consists of a strong network of commitments—conceptual, theoretical, instrumental, and methodological and quasimetaphysical. Such paradigms serve as a source of the methods, problem field, and standard of solution accepted by any mature scientific community at any given time. Normal science, according to Kuhn, consists of empirical and theoretical efforts within the paradigm to gather additional relevant information and improve the match between theoretical prediction and data. Such a paradigm possesses two related social functions. In the first place, it attracts groups of workers interested in the particular field in question, and second, it provides them with a number of unanswered questions that keep them busy experimenting and theorizing. The paradigm thus provides an intellectual structure that integrates important facts but at the same time generates numerous research opportunities. In this sense, as the rest of this book will make clear, there does exist a personality theory at the moment which fulfills this function and which may thus be regarded as a paradigm in the Kuhnian sense. We shall return to this question in the Epilogue.

Before turning to a consideration of the scientific meaning of such terms as *traits* and *abilities*, let us first define the concept of personality. Allport (1937) has given a most valuable summary of the many uses of the term in many different contexts, but we will not follow him in this but rather try and give a definition that is widely accepted by psychologists concerned with this field. Accordingly, we shall define *personality* as

> a more or less stable and enduring organization of a person's character, temperament, intellect, and physique, which determines his unique adjustment to the environment. *Character* denotes a person's more or less stable and enduring system of conative behavior *(will); temperament*, his more or less stable and enduring system of affective behavior *(emotion); intellect*, his more or less stable and enduring system of cognitive behavior *(intelligence); physique*, his more or less stable and enduring system of bodily configuration and neuroendocrine endowment. (H. J. Eysenck, 1970c, p. 2)

Such a definition, which emphasizes stable and enduring characteristics of the individual, goes counter to the doctrine of *specificity*, which has been very popular in the United States for many years and is often identified with the behaviorist movement. Thorndike (1903) was among the first to adopt this doctrine of specificity when he maintained that "there are no broad, general traits of personality, no general and consistent forms of conduct, which, if they existed, would make for consistency of behaviour and stability of personality, but only independent and specific stimulus–response bonds or habits" (Thorndike, 1903, p. 29). This "sarbond" theory, as McDougall (1921) tauntingly called it, was very popular for many years, but there are probably very few psychologists nowadays who would adopt it. Indeed, even if S–R bonds were the building stones of behavior, it is not immediately obvious why, once established, they should not furnish us with precisely the type of stable and enduring systems of behavior that are postulated in our definition of personality.

Let us assume that during childhood a given infant is rewarded for behaving in an orderly fashion and punished for failing to do so. Learning theory, whether of the Skinnerian or Hullian type, would lead us to predict that a gradual buildup of habit strength would occur for behaviors of the desired kind, so that in adulthood such a person would tend always to behave in an orderly fashion. Thus, even adopting the Thorndikian type of "law of effect" would not necessarily lead us to believe in specificity of human conduct, and Skinner's stress on the "history of reinforcement" suggests that he himself would accept this argument.

In recent years it has been Mischel (1969, 1977; Mischel & Peake, 1982) who has taken up the argument relating to specificity, suggesting that although trait theory predicts behavioral consistency, it is behavioral inconsistency that is typically observed. He writes:

> I am more and more convinced, however, hopefully by data as well as on theoretical grounds, that the observed inconsistencies so regularly found in studies of non-cognitive personality dimensions often reflect the state of nature and not merely the noise of measurement. (p. 1014)

The basis for this assertion was the partial review of the relevant literature by Mischel (1968), who concluded that measures of consistency in personality rarely produced correlations as high as .30. This estimate harks back to the review of some 350 questionnaire studies by Ellis (1946), who concluded that

> group administered pencil-and-paper personality questionnaires are of dubious value in distinguishing between groups of adjusted and maladjusted individuals and that they are of much less value in the diagnosis of individual adjustment or personality traits. (p. 438)

In a review of the validity of personality inventories in military practice, Ellis and Conrad (1948) came to a slightly different conclusion:

> Military applications of personality inventories have yielded enough favourable results to command attention. In contrast, personality inventories in civilian practice have generally proved disappointing. (p. 424)

In these evaluations Ellis has taken an unusually severe criterion of validity. He claims that he will in his review

> usually evaluate the reported coefficients of correlations in terms of the conventional estimations given them in the consideration of psychological and educational tests. Thus we shall say that rs from zero to .19 indicate negative validity; from .20 to .39 mainly negative validity; from .40 to .69 questionable positive validity; from .70 to .79 mainly positive validity; and from .80 to 1.00 positive validity. (p. 386)

This, of course, is absurd from the psychometric point of view; if any questionnaire gave correlations with the criterion of .80 to 1.00, one could only conclude that the author of the paper in question was either cheating or else had miscalculated his coefficients! Few questionnaires have reliabilities in excess of .80, and the correlation with the criterion cannot exceed the square root of the product of the reliability of each scale. To consider a correlation with the criterion of .69 "questionable" is to apply standards that are quite unusual in the psychometric field. Even so, the studies quoted by Ellis in his Table 2 indicate that some 35% give validation coefficients in excess of .70, whereas only about 40% give validation coefficients lower than .40! These results would seem rather promising to most psychologists, particularly in view of the fact that they were obtained in civilian work and that according to Ellis and Conrad military application of questionnaire results have even higher validation coefficients.

In any case, it would seem inadmissible to take these coefficients at their face value. They validate questionnaires in terms of criteria which are themselves imperfect, and consequently even a perfect measuring instrument could not be expected to give very high correlations with such imperfect criteria. All we can deduce from figures such as those given by Ellis is that there are high agreements between some questionnaires and some external criteria.

As we shall see later on, there is also good evidence that genetic factors play a large part in the determination of individual differences in personality and intellect, and these would certainly favor some kind of stable and enduring traits and abilities. We shall discuss the evidence relating to the influence of genetic and environmental factors on personality later on; here let us merely note that it is not necessary to adopt a hereditarian position in order to accept some such definition as that given above. A lengthy critique of the specificity position is given by H. J. Eysenck (1970c), and even earlier Allport (1937) provided a brilliant theoretical refutation of the specifist contention. He attacked in particular the notion of "identical elements" as accounting for similarities of behavior by showing that the very notion of "elements" is completely ambiguous in the writings of those who support the Thorndikian view, and that the alleged identity of these elements is merely an *a posteriori* justification of the observed phenomena, without any value in predicting and without any possibility of verification. (Thorndike hypothesized that correlations between nonidentical situations could be found because these shared "identical elements." The nature of these "identical elements" was never properly explicated or submitted to experimental tests.)

TYPE AND TRAIT THEORIES: THE MODERN VIEW

In looking for a model for our description of personality organization, we find two claimants in the field, two concepts that have for a long time been used by those who have theorized about the mechanics of consistent and congruent behavior—the concepts, namely, of *trait* and *type*. The former of these found a particularly warm champion in Stern (1921), who wrote:

> We have the right and the obligation to develop a concept of trait as a definitive doctrine; for in all activity of the person, there is besides a variable portion, likewise a constant purposive portion, and this latter we isolate in the concept of trait. (p. 14)

And how are these traits to be discovered? According to Allport (1937), who has done much to popularize this concept in the Anglo-American countries:

> Traits . . . are discovered not by deductive reasoning, not by fiat, not by naming, and are themselves never directly observed. They are discovered in the individual life—the only place where they can be discovered—only through an inference (or interpretation) made necessary by the demonstrable consistency of the separate observable acts of behaviour. (p. 128)

And again:

> Traits are not directly observable; they are inferred (as any kind of determining tendency is inferred). Without such an inference, the stability and consistency of personal behaviour could not possibly be explained. Any specific action is a product of innumerable determinants, not only of traits but of momentary pressures and specialized influences. But it is the repeated occurrence of actions having the *same significance* (equivalence of response) following upon a definable range of stimuli having the same personal significance (equivalence of stimuli), that makes necessary the postulation of traits as states of Being. Traits are not at all times active, but they are persistent even when latent, and are distinguished by low thresholds of arousal. (p. 129)

It will be clear from these quotations that the notion of *trait* is intimately connected with the notion of *correlation*. Stability, consistency, repeated occurrence of actions—all these terms, when translated into more rigorous and operationally definable language, refer to covariation of a number of behavioral acts. Such covariation may refer to correlations between tests, correlations between persons, or even to correlations between different occasions of measurement within the same person.

The term *type* in modern personality theory refers to a concept superordinate to that of *trait*. Traits are often intercorrelated, and these intercorrelations give rise to a type. Thus, type concepts, like extraversion-introversion or neuroticism-stability, are postulated because in each case there are a number of traits that are found to be correlated, giving rise to a higher-order concept.

The concept of *type* has often been criticized because it is erroneously believed that it postulates categorically distinct groups of people, or a bimodal or multimodal distribution of scores on a continuum. This was indeed the position held by the ancient Greeks, and also by Immanuel Kant, as we shall see presently, but it is not a view embraced by any type theorist since the First World War. Thus Jung (1923), one of the best known type psychologists, has the following to say:

> A trait may be defined as a mechanism. . . . Type never denotes more than the relative dominance of the one mechanism. . . . It follows that there can never be a pure type in the sense that the one mechanism is completely dominant to the exclusion of the other. (p. 23)

This quotation could be multiplied many times, but it will suffice to show that Jung was very far from conceiving of all human beings as being either extraverted or introverted. Rather, he considered that most of them were characterized by a balance between the extravertive and the introvertive mechanisms; a relatively small number he considered to be unbalanced and characterized by the more or less marked dominance of one function or the other. Nothing is farther from his thoughts than the hypothesis of discontinuity; stress is laid again and again on the notion of complete continuity and balance. Admittedly, his description is in terms of *ideal types*, that is, of completely introverted or extraverted individuals, but he emphasizes repeatedly that these are abstractions, in the same sense that Newton's laws of motion are idealized abstractions not to be found in actual experiments.

What, then, is at the basis of his concept of type? We may answer this question by quoting a passage from Kretschmer (1948), who appears to hold a view of typology similar to that of Jung and who has discussed this concept with admirable lucidity. According to Kretschmer:

> The concept of type is the most important fundamental concept of all biology. Nature . . . does not work with sharp contrasts and precise definitions, which derive from our own thought and our own need for comprehension. In nature, fluid transitions are the rule, but it would not be true to say that, in this infinite sea of fluid empirical forms, nothing clear and objective could be seen; quite the contrary. In certain fields, groupings arise which we encounter again and again; when we study them objectively, we realize that we are dealing here with focal-points of frequently occurring groups of characteristics, concentrations of correlated traits. . . . What is essential in biology, as in clinical medicine, is not a single correlation but groups of corre-

lations; only those lead to the innermost connections. It is daily experience in the field of typology, which can be deduced quite easily from the general theory, that in dealing with groups of characteristics one obtains higher correlations than with single characteristics. . . . What we call, mathematically, focal-points of statistical correlations, we call, in more descriptive prose, constitutional types. . . . A true type can be recognised by the fact that it leads to ever more connections of biological importance. Where there are many and ever-new correlations with fundamental biological factors . . . we are dealing with focal-points of the greatest importance. (p. 12)

A type is defined, then, as a group of correlated traits, just as a trait was defined as a group of correlated behavioral acts or action tendencies. According to this view, then, the difference between the concepts of *trait* and *type* lies not in the continuity or lack of continuity of the hypothesized variable, nor in its form of distribution, but in the greater *inclusiveness* of the type concept. As an example, let us consider the three major type concepts or dimensions which we shall consider in this book and which appear to have emerged from many if not all the major investigations carried out in this field.

These three concepts, types, or dimensions have been variously named, as we shall see later on, but they will be referred to in this book under the names of *psychoticism* as opposed to impulse control, *extraversion* as opposed to introversion, and *neuroticism* as opposed to stability. Figures 1, 2, and 3 show how these type concepts are based on observed correlations between traits; it is these traits in their interrelationships which define the type concept. *P*, *E*, and *N* are all continuously and more or less normally distributed; there is no question of bimodal or trimodal distributions, or of actual mutually exclusive classifications. It is true that the term *type* used to carry the interpretation given to it by Kant, but there are no modern typologists who

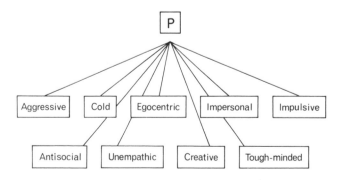

FIGURE 1. Traits making up the type concept of psychoticism.

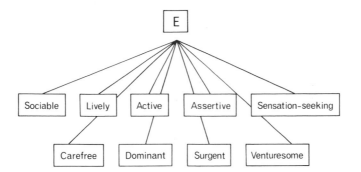

FIGURE 2. Traits making up the type concept of extraversion.

would accept discontinuity or multimodal distribution of the underlying concept. In what follows, therefore, we shall use the term *type* to correspond to what in factor analysis would be called second-order factors, or superfactors, restricting the term *trait* to what in factor analysis would be called primary factors. These technical terms will be defined more precisely in the next section.

Before turning to a brief discussion of factor analysis, let us here note a popular misunderstanding and criticism of the scientific study of personality, types and traits, abilities, and the like. These concepts are contrasted with physical entities such as heat, magnetism, or gravitation, and it is suggested that the psychological entities are more ephemeral, vague, and insubstantial than concepts in physics. This is not so; as already noted, concepts in science are always man-made and hence insubstantial and ephemeral. Some of the difficulties which people expe-

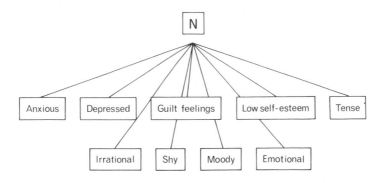

FIGURE 3. Traits making up the type concept of neuroticism.

rience in talking about personality or intelligence derive from the fact that these concepts have three correlated but somewhat different meanings. Let us clarify the situation by looking at intelligence (Figure 4). As is customary, we can distinguish intelligence A, intelligence B, and intelligence C. Intelligence A is the *biological substratum*, determined very largely genetically, which disposes a given individual to develop high or low degrees of success in problem solving, learning, and other cognitive tasks. Intelligence B is the *social manifestation* of intelligence, modified by cultural and educational factors, socioeconomic status, and the like; this is the kind of intelligence to which the man in the street usually refers when he uses the term. Intelligence C is intelligence *as measured by current IQ tests;* it correlates with intelligence A and intelligence B, but clearly not perfectly with either. Indeed, different measures of intelligence C may show different correlations. Thus g_c, or crystallized ability as Cattell (1982) has called it, correlates more highly with intelligence B than g_f, or fluid intelligence, which has a stronger genetic component, and hence correlates more strongly with intelligence A. Finally, g_p, or the psychophysiological measurement of intelligence by means of evoked potentials (H. J. Eysenck, 1982c), would appear to be an even better measure of intelligence A and to be located closer to it than g_f and g_c. It is obvious that discussions about intelligence without specification of which intelligence we are talking about will be misleading and not very fruitful. Similarly, we can postulate personality A, referring to the underlying biological and genetic component of personality; personality B, as expressed in everyday life and experienced by other people; and personality C, which would be the questionnaire or other measures of personality adopted for a particular investigation.

It might be said that the very existence of these three different types of intelligence or personality indicates the inferiority of psychological

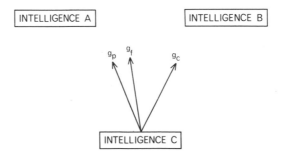

FIGURE 4. Correlations between intelligence A, intelligence B, and intelligence C.

concepts to physical ones, but this is not so. Let us consider heat. Here, too, we have heat A, which is essentially the speed of movement of atoms or molecules constituting the substance involved. Next we have heat B, which is the heat experienced by people exposed to given conditions; this is partly determined by the actual temperature, but also by factors such as humidity, the chill factor (wind speed), alcoholic and food intake, thickness of fat deposits, exercise, emotion, fever, and so forth. A well-known experiment consists of asking the subject to immerse his right hand in a bowl of hot water and his left hand in a bowl of cold water; he is then instructed to move both hands to a bowl in the middle which is filled with lukewarm water; this appears cold to the right hand, hot to the left hand, demonstrating conclusively that heat B is not identical with heat A! Heat C is the measurement of heat by means of different types of thermometers, and we shall show later that these do not always give the same readings but show consistent differences depending on the type of thermometer adopted (Baker, Ryder, & Baker, 1975).

Nor would it be true to say that there is an agreed theory of heat in physics, but no agreed theory of personality (or intelligence) in psychology. In physics we have the thermodynamic and the kinetic theories of heat. Thermodynamics deals with unimaginable concepts of a purely quantitative kind: *temperature*, measured on a thermometer; *pressure*, measured as the force per unit area; and *volume*, measured by the size of the container. Nothing is said in the laws of thermodynamics about the nature of heat. Bernoulli, in his famous treatise on hydraulics, postulated that all "elastic fluids" such as air consist of small particles that are in constant irregular motion and constantly collide with each other and with the walls of the container. This was the foundation stone of the kinetic theory of heat, which results in a picture of events that is eminently visualizable and gives to many people a feeling of greater understanding and a better and more thorough explanation than do the laws of thermodynamics. Nevertheless, even today many phenomena which submit easily to thermodynamic solutions are quite intractable to kinetic interpretations (H. J. Eysenck, 1970b). This would appear to be a physical distinction similar to that between intervening variables and hypothetical constructs in psychology (MacCorquodale & Meehl, 1948.)

Traits are essentially *dispositional* factors that regularly and persistently determine our conduct in many different types of situations. The contrast is often made between *traits* and *states or moods*, which may be defined as singular occurrences. A generally sociable person may on a particular occasion behave in an unsociable manner; a generally fearless person may on occasion show anxiety. This distinction was first brought out in psychological theorizing by Allport and Odbert (1936).

They defined traits as "broad patterns of determining tendency that confer upon personality such consistency as it displays. Consistent and stable moods are indicative of an individual's adjustment to his environment" (p. 26). States were defined as "present activity, temporary states of mind and mood" (p. 26). An obvious example of the state–trait distinction can be found in the work of Spielberger (1971, 1980) on trait and state measurements of anxiety and anger, and in the many mood checklists available at present.

The distinction is an obvious one, often made by the man in the street, who distinguishes between habitual behavior and atypical behavior. It was first made by Marcus Tullius Cicero, 2,000 years ago, in his *Tusculan Disputations*, written in the year 45 B.C. after he had completed the *De Finibus* and before he began the *De Natura Deorum*. In this work Cicero is simply reporting the present state of philosophical and psychological speculation among Greek and Roman philosophers; he is not putting forward original notions. He points out that some men are more *prone* to some diseases, other men to others, "and so we say of certain people that they are liable to catch cold, certain others to attacks of cholic, not because they are suffering at the moment, but because they frequently do so" (1971, p. 355). He then goes on to state explicitly the state–trait theory, using trait *(angor)* and state anxiety *(anxietas)* as an example, as well as *irascibility* as opposed to *anger:*

> In the same way some men are prone to fear, others to another disorder, in
> consequence of which in some cases we speak of an *anxious temper*, and
> hence of anxious people, in other cases of *irascibility* which is different from
> *anger*. It is one thing to be irascible, another thing to be angry, just as an
> anxious temper is different from feeling anxiety; for not all men who are at
> times anxious are of an anxious temper, nor are those who have an anxious
> temper always feeling anxious, just as for instance there is a difference
> between intoxication and habitual drunkenness, and it one thing to be a gal-
> lant, and another thing to be in love. (p. 356)

Having thus laid the foundation for such attempts as Spielberger's to embody trait and state concepts in questionnaire form, Cicero goes on to generalize:

> Men are called both envious and malicious and jealous and fearful and com-
> passionate because of a *proneness* to such disorders, not because they are
> *always* behaving in this fashion. This proneness then of each individual to
> his own peculiar disorder would on the analogy of the body be called sick-
> ness, provided it be understood as proneness to sickness. (p. 356)

States and traits are obviously not unrelated; Allen and Potkay (1981) indeed called the distinction between them arbitrary, and insofar as the distinction is not absolute, they are certainly right. Trait and state anxiety, for instance, show a fairly high degree of correlation (Magnusson, 1979); and, as Howarth (1980a) has shown, some of his state scales were identical with some of his trait scales. All this follows directly from the definition of states and traits; if a trait of anxiety is defined as a *disposition* on the part of the individual to demonstrate anxiety readily in many situations, then a trait inventory applied to particular situations is more likely to disclose anxiety in these situations in this individual than in another who is low on trait anxiety. Nevertheless, the distinction is a useful and important one, and state measures may often be helpful in elucidating specific behaviors in specific situations. Thus, of two people who have equally high scores on trait anxiety, one (who is intelligent and has worked hard) may show less state anxiety in an examination than another (who is dull and has not worked hard). The interaction between the situations and traits makes necessary the addition of states to our conceptual armamentarium.

Of particular interest from some points of view is the degree of *changeability* in states or moods shown by a given person, which in turn may be regarded as a trait. Wilhelm Wundt, as we shall see, elaborated a two-dimensional system of personality description in which he labeled what we would now call extraversion–introversion "changeability-unchangeability," suggesting that extraverts are likely to show much quicker and greater changes in mood (states) than do introverts. As we shall see later on, this hypothesis has indeed been confirmed (see our discussion of happiness further on). This conception is also important to the notion of personality *integration*, as shown in the work of Hartshorne and May to be discussed later. We shall return to a more detailed discussion of some of the issues raised here in subsequent chapters.

TYPE-TRAIT THEORIES AND FACTOR ANALYSIS

The best method for studying the association of individual test variables into traits and the association of traits into types is factor analysis. This is a technique for studying tables of intercorrelations, noting regularities (e.g., sets of high or low correlations) and reducing the total complexity of the sometimes thousands of intercorrelations to the relative simplicity of a few factors, the interaction between which can account for all the observed correlations. It is an empirical fact that when the study is based on a reasonable theory, it is usually possible to carry out such an analysis successfully and emerge with a few meaningful factors that account for most of the variance. Factor analysis is a useful servant but a bad master; it has often been misused in the past, and much of the criticism which has been directed at it arises from this misuse. We will consider some of these criticisms, but first let us briefly describe the nature or logical basis of factor analysis (H. J. Eysenck, 1953).

Consider Table 1, which lists 12 questions. The first six questions pertain to a hypothetical trait of *neuroticism,* emotionality or instability, and the second set of six questions to a hypothetical trait of *extraversion,* as opposed to introversion. Table 2 shows the intercorrelation between these 12 traits, and even a casual glance will show that the six neuroticism questions are intercorrelated positively and quite highly, as are those on extraversion, but that the neuroticism questions are not correlated with the extraversion questions (H. J. Eysenck, 1970c). We can now construct, from the observed patterns of intercorrelations, two factors, *E* and *N,* and calculate the degree to which any given question is correlated with these two factors. The statistical methods of carrying out these calculations are not relevant here; let us merely note that they are based on sound statistical principles and present no great difficulties. These correlations between items and factors are called "factor loadings" and are given in Table 2.

Note first of all that a table of 2×66 correlations can be represented by two columns of 12 loadings each; there is an obvious descriptive advantage in adopting the simpler and shorter method of using factor

TABLE 1. Items in Short Extraversion and Neuroticism Inventory

Questions	Key
1. Do you sometimes feel happy, sometimes depressed, without any apparent reason?	*N*
2. Do you have frequent ups and downs in mood, either with or without apparent cause?	*N*
3. Are you inclined to be moody?	*N*
4. Does your mind often wander while you are trying to concentrate?	*N*
5. Are you frequently "lost in thought" even when supposed to be taking part in a conversation?	*N*
6. Are you sometimes bubbling over with energy and sometimes very sluggish?	*N*
7. Do you prefer action to planning for action?	*E*
8. Are you happiest when you get involved in some project that calls for rapid action?	*E*
9. Do you usually take the initiative in making new friends?	*E*
10. Are you inclined to be quick and sure in your actions?	*E*
11. Would you rate yourself as a lively individual?	*E*
12. Would you be very unhappy if you were prevented from making social contacts?	*E*

TABLE 2. Intercorrelation between Items Given in Table 1 and Factor Loadings on E and N

	Intercorrelations												Factor loadings	
	1	2	3	4	5	6	7	8	9	10	11	12	E	N
1		.65	.48	.38	.29	.50	-.04	.08	-.04	.09	-.07	.01	.01	.75
2	.65		.60	.35	.27	.46	.01	.02	-.10	-.11	-.10	.05	-.06	.74
3	.48	.60		.30	.25	.45	-.04	.02	-.06	-.15	-.15	.08	-.09	.71
4	.38	.35	.30		.50	.31	.03	-.08	-.04	.17	-.04	.06	.02	.58
5	.29	.27	.25	.50		.32	-.04	-.09	-.14	-.14	.17	.02	-.06	.58
6	.50	.46	.45	.31	.32		.02	.12	.04	-.02	.07	.13	.09	.63
7	-.04	.01	-.04	.03	-.04	.02		.40	.12	.17	.20	.16	.48	.00
8	.08	.02	.02	-.08	-.09	.12	.40		.19	.38	.26	.21	.59	.04
9	-.04	-.10	-.06	-.04	-.14	.04	.12	.19		.08	.44	.53	.59	-.06
10	.09	-.11	-.15	.17	-.14	-.02	.17	.38	.08		.42	.13	.49	-.04
11	-.07	-.10	-.15	-.04	.17	.07	.20	.26	.44	.42		.41	.68	-.02
12	.01	.05	.08	.06	.02	.13	.16	.21	.53	.13	.41		.64	.09

loadings. Note secondly that the factor loadings introduce a considerable degree of *order* into the table of intercorrelations; it is much more obvious what is happening in a table of factor loadings than in a table of intercorrelations. (The table of intercorrelations has been presented by giving the intercorrelations between N items first, between E items second, so that the table presents a fairly clear picture. Normally the order of the items would be quite higgledy-piggledy and the correlations much more difficult to reduce to any obvious pattern by simple observation.) Note further that the factors correspond closely to our original hypothesis, and note finally that we now have quantitative estimates of the value of any given item for the measurement of the factors involved. These are powerful advantages that factor analysis bestows, and when the number of items is much greater (as is usually the case) these advantages become much more obvious. It is possible to see a clear-cut pattern in a table of intercorrelations between 12 items, but suppose we had a table of 250 items, producing $2 \times 31{,}125$ correlations in all; it is quite impossible to bring any order into such a table by simple scrutiny, and some statistical method clearly is needed. It is likely that a much larger number of factors would be needed, but even if four or five factors were required there would still be a tremendous simplification and clarification in the factor pattern over the huge table of intercorrelations.

One further advantage of the factor analytic solution is that it can be represented geometrically, in terms of Cartesian coordinates as in Figure 5. Here the factor loadings are plotted on two dimensions, labeled neuroticism and extraversion, and the clustering of the items will be quite obvious. Such pictures help in understanding the patterns that occur in a table of intercorrelations, and even with small numbers the figure will be quite obviously much more easily intelligible than the intercorrelations given in the original table. Much more could be said, of course, about factor analysis, but this is not a treatise on statistical methodology, and what has been said must suffice for the time being.

In the example in Figure 5 we have, for the sake of simplicity, gone directly from items to type factors, E and N. An alternative method would be first to define a series of trait or primary factors, such as those shown in Figure 2 (sociability, liveliness, activeness, assertiveness, sensation seeking, carefreeness, dominance, surgency, adventuresomeness), and then show, by intercorrelating these factors, and factor analyzing their intercorrelations, that a type factor or second-order factor of extraversion can reasonably be postulated. H. J. Eysenck and S. B. G. Eysenck (1969) have shown that identical factors emerge whichever method is adopted. Our brief example contains too few questions to make it possible to extract primary factors first and then base type concepts on their intercorrelations.

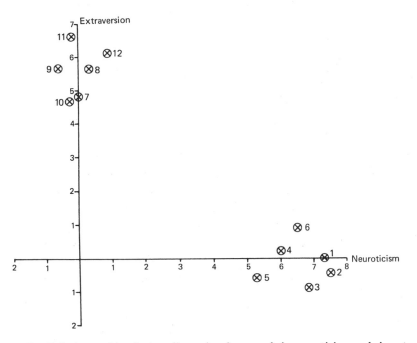

FIGURE 5. Relative position in two-dimensional space of six neuroticism and six extraversion questionnaire items.

Now let us turn to a consideration of some of the criticisms that have been made of factor analysis in the past. In the first place, it is suggested that what factor analysis does is a task of supererogation. Factor analysis, so it is said, can only extract from the data what the investigator originally put into the study. This, of course, is perfectly true and would be equally true of qualitative analysis in chemistry. When we are furnished with a given specimen to analyze and we can only get out of it what was originally put into it, the real point is that we do not know, to begin with, what was in fact put into the specimen and hence must carry out the analysis in order to discover this. In the same way, the investigator may have a *theory* as to what in fact he put into the investigation, but his theory may be (and only too frequently is!) quite erroneous, so that the anlaysis can show him that (a) what he thought he had put into the study did not actually emerge, and (b) the factors that emerge are quite unlike those he thought he had actually included. Thus factor analysis provides us with a check on our hypotheses, and a very necessary one, too. A few examples may make this point clear.

These examples may also serve to illustrate a point, related to the one discussed above, which is often raised as a criticism of trait psychology as a whole. What is suggested is that trait psychology involves a circular argument. We observe that some people are sociable, others not, and to explain this we posit a trait of *sociability*. But of course the trait itself is deduced from the facts it is supposed to elucidate and explain; hence the argument is circular. The point is sometimes made that in this the concept of *trait* is similar to that of *instinct;* instincts, too, were originally posited to explain certain types of behavior, which had in the first place served to produce the concept of the instinct in question! The argument sounds convincing, but it is erroneous in that the original motive for positing the existence of traits was simply the need for a proper *description;* causal analysis comes later and is not implied. It is easy to posit a trait of sociability, but we have to prove that the original observations actually correlate in such a manner as to make such a concept reasonable. This, of course, can be done only by factor analysis, and hence there is a close relationship between criticisms of factor analysis and criticisms of trait theory. Another point in answer to the critic is that we cannot even begin to undertake a causal analysis until we have settled, at least in a preliminary manner, the problem of description. There is no circular argument involved in all this, as the following examples will show.

Let us consider an investigation in which Brengelmann (1952) attempted to investigate Kretschmer's hypothesis of the existence of a schizothymia–cyclothymia dimension. Eight tests were used in this study, all of which had originally been suggested by Kretschmer, on the basis of long-continued experimental work and theoretical conceptualization, to be measures of this particular dimension. These tests were applied to 100 normal subjects and then intercorrelated. Reliabilities of the tests were found to be satisfactory.

According to the critics of factor analysis, the fact that apparently measures of schizothymia–cyclothymia had been put into the analysis should mean that such a factor should also come out of it. However, the analysis showed that out of the 28 correlations only one was significant, and this in a direction contrary to expectation. The study thus conclusively disproved Kretschmer's theory, at least as far as the measurement of the hypothesized dimension by means of these tests is concerned. He clearly had not "put in" what he thought, and the correlational and factor analytic study demonstrated this very clearly. Thus we may believe that we put into the analysis something which simply is not there. This is a very important finding, and no other technique is available to enable us to make such discoveries.

As another example of the fact that factor analysis may disclose an outcome quite different from that predicted, we may consider a study of

Guilford's factor of social shyness. Guilford had advanced this concept as a unitary one and had suggested a large number of questionnaire items to define it. H. J. Eysenck (1956b) investigated the alternative hypothesis that introverted social shyness is different in many ways from neurotic social shyness. He suggested that introverts do not care for people, would rather be alone, but if necessary can effectively take part in social situations, whereas the neurotic is anxious and afraid when confronted with social situations, seeks to avoid them in order to escape from this negative feeling, but frequently wishes that he could be more sociable. In other words, the introvert does not *care* to be with other people; the neurotic is *afraid* of being with other people. To test this hypothesis, the different items in the Guilford S (social shyness) scale were correlated with measures of extraversion–introversion and neuroticism–stability, using a population of 200 men and 200 women.

It was indeed found that some items correlated with N but not with E, whereas others correlated with E but not with N; in fact, the items split up almost completely into two uncorrelated groups related to introverted social shyness or neurotic social shyness, respectively. Some of the questions related to extraversion were the following: Can you usually let yourself go and have a hilariously good time at a lively party? Do you like to mix socially with people? Would you rate yourself as a lively individual? Are you usually a good mixer? Are you inclined to keep in the background on social occasions? (No). Do you usually take the initiative in making new friends? These items are all scored in the extraverted direction, that is, they imply a denial of social shyness.

Typical items related to neurotic shyness are these: Do you often experience periods of loneliness? Are you troubled about being self-conscious? Are you self-conscious in the presence of your superiors? Are you usually well poised in your social context? (No). Are you worried about being shy? Do you often feel ill at ease with other people?

The results appear to leave no doubt that Guilford was mistaken in thinking that the items he selected measured one single trait. Factor analysis demonstrates that instead two traits are being measured, which are essentially independent of each other. Thus again we see that what Guilford thought he had put into the scale was not actually there.

Another example is the work of Eysenck and Furneaux (1945) on suggestibility. Taking a series of experimental tests that had been suggested in the literature as measures of this hypothetical trait, they found that when these tests were correlated they fell into two independent groups, labeled respectively primary and secondary suggestibility. Primary suggestibility is of the ideomotor type and is correlated with hypnotizability; secondary suggestibility is of the gullibility type and is not at all correlated with hypnotizability. They suggest that there is a third type, tertiary suggestibility, that relates to social attitudes and change in these attitudes consequent upon intervention of prestigious figures. Thus this alleged single trait is also found to break up into several independent constituents when subjected to factor analytic study.

Even more impressive evidence comes from the literature on factor analysis itself. Traits believed by factor analysts to be univocal, that is, to be made up of items measuring one single concept, have often been shown in later analyses to break up and to recombine in different ways. The evidence is fully discussed in a later section, where it is shown, for instance, that Cattell's 16PF scale, which allegedly consists of 16 source traits that form separate and univocal measures, has not been found to give rise to such factors in anything like the same form by later investigators. Still worse is the fate of nonfactorial scales, such as the MMPI, as again will be demonstrated later on; each of the MMPI scales is in fact made up of items belonging to many different factors, and none of the scales is unifactorial in any real sense. Hence quite clearly psychologists, even when using fairly sophisticated methods of putting into the factor analysis what they hope to get out of it, are not very successful in doing this. The reader may convince himself of this by looking at a typical personality questionnaire *without consulting the key* and trying to determine just how the items correlate together and what kinds of factors might emerge from such correlations. It will be found that even in very rough outline the task is an impossible one, and when one realizes that factor analysis not only performs the task but also gives exact quantitative results for the nature and relative positions of the factors, one will see that the criticism under discussion is a wholly unreasonable one.

Another criticism, often made of factor analysis, is that the technique does not give a completely objective and universally agreed upon answer to the problem of taxonomy. This objection has a certain amount of truth in it, but it could be made to all taxonomic methods that have been used in botany, zoology, or physics. Taxonomy, because it aims to formulate certain general rules and concepts, is inevitably to some extent subjective; we can classify objects in many different ways, depending on our interest and our purpose. Even if we are agreed on the purpose, we may lack the necessary means of carrying out the requisite research. Thus in zoology we would like to classify animals according to their evolutionary development, but we normally lack the requisite knowledge and hence have to rely on morphology and other indirect and imperfect methods.

Some of the newer methods evolved for the purpose come closer to the ideal. Thus the construction of phylogenetic trees has been made easier by biochemists who use quantitative estimates of variance between species as regards substances such as DNA and cytochrome c. Fitch and Margoliash (1967) have succeeded in constructing such a tree, based on data relating to the single gene that codes for cytochrome c, which was very similar to the "classical" phylogenetic tree. The method is based essentially on the appropriate "mutation distance" between two cyto-chromes, which is defined as the minimal number of component nucleo-tides that would have to be altered in order for the gene for one cyto-chrome to code for the other. This number is considered proportional to the number of mutations that have taken place in the descent from the apex of one cytochrome as compared with another. Thus it is claimed that this new method, which gives a quantitative measure of the event (mutation) that permits the evolution of new species, must give the most accurate phylogenetic trees. In this way it may be possible to overcome the difficulties in the evolutionary method of classification by descent, and it is reassuring that even when it is based on only a single gene the phylogenetic scheme is remarkably like that obtained by classical methods.

These classical methods rely on what Sokal and Sneath (1963) call "polythetic" arrangements. Such arrangements, as they say, "place together organisms that have the greatest number of shared features, and no single feature is essential to group membership or is sufficient to make an organism a member of the group" (p. 3). They credit Adamson (1727-1806) with the introduction of the polythetic type of system into biology. He rejected the *a priori* assumptions of the importance of dif-ferent characters and correctly realized that natural taxa are based on the concept of *affinity*, which is measured by taking all characters into consideration, that is, that the taxa are separated from each by means of correlated features. (For a more detailed discussion of the principles of numerical taxonomy, see H. J. Eysenck & S. B. G. Eysenck, 1969.)

The method of *polythetic* arrangements, which is very much in line with the factor analytic type of classification, is opposed to the so-called *monothetic* type of arrangement. The ruling idea of monothetic groups is that they are formed by rigid and successive logical divisions so that the position of a unique set of features is both sufficient and necessary for membership in the group thus defined. They are called monothetic because their defining set of features is unique. Such systems will always carry the risk of serious misclassification if we wish to make natural phe-netic groups. This is because an organism that happens to be aberrant in the feature used to make the primary division will inevitably be removed to a category far from the required position, even if it is iden-tical with its natural congeners in every other feature. As an example,

consider whales or dolphins; if we make "swimming in water" the monothetic characteristic for defining fish, we will seriously misclassify what
are in essence mammals. Reliance on any single feature, such as physique in Kretschmer's system, can be equally misleading in relation to
psychological divisions.

Analysis by phenetic relationships of the polythetic kind has become
all but universal in biology, and it is obviously of considerable advantage
to find that this system agrees so well, whenever it has been tested, with
Darwinian principles defining relations by descent from a common
ancestor. It is very desirable in psychology, too, that we should have
external criteria, as will be argued later on in this book, whether these
be defined in terms of genetics or of biological functioning. We probably
differ from many factor analysts in calling for such external criteria;
most of the leading factor analysts seem to be quite content to remain
within the rather circular argument of correlational analysis without
wishing to go outside of it. This, to us, seems a serious error.

The inevitable subjectivity of factor analytic or any other taxonomic
method can be avoided only by having external criteria, such as Darwinian evolution in the case of zoology and botany; subjectivity is not necessarily a criticism of a taxonomic method, but it does carry disadvantages which if possible should be avoided.

Other criticisms of the factorial method may be dismissed as largely
misconceptions. Thus it is sometimes suggested that the results of factor
analytic studies, that is, the factors identified, have no real existence.
This is obviously true; they are concepts, and concepts in science never
have any kind of "real" existence. We have already referred to this point
in another context and need not dwell on it here. Reification of factors
would certainly be a grave error, but only singularly naive factor analysts would fall into this trap. Essentially, factors, as psychological concepts, are statements of theoretical import; they require, as we have
already pointed out, our going beyond the circle of correlational analyses
in order to test the adequacy or usefulness of these theories. Thus this
argument again serves only to emphasize the need for the independent,
experimental study of concepts derived from factor analysis. It would
certainly not be adequate, in the majority of cases, to rest content with
the results of factor analysis; they pose a problem and suggest a theory,
but they do not tell us whether the theory is adequate or not.

This is linked with another criticism, namely, that the naming of
factors is subjective. Naming is of course not particularly important; the
essence of the factor consists in the observed relationships and the items,
or tests, involved in a given factor. However, the naming of a factor is
useful in making clear the hypotheses adopted by the writer who

extracted the factor in the first place, and such hypotheses as embodied in the name of the factor should certainly always be tested by reference to outside criteria. Thus, if we find in questionnaire studies a factor labeled "neuroticism," we would be justified in expecting that neurotic patients would have much higher scores on this factor than would non-neurotic, normal persons. Figure 6 shows the distribution of scores on the Maudsley Medical Questionnaire (H. J. Eysenck, 1952b) of 1,000 normal and 1,000 neurotic soldiers. Arbitrarily divided at a score of 9, it is found that only 10.6% of normals are misclassified as neurotics, and 28.6% of neurotics as normals. There clearly is a great difference between the groups on a measure of neuroticism. There are many other means of testing the correctness or otherwise of the label; psychologists should never be content with extracting and labeling a factor and then not going beyond this point. The fact that many of them do precisely this is a reasonable criticism, not of factor analysis as such, but of the inappropriate use of factor analysis.

This, indeed, may be a general comment on most of the criticisms that are made of factor analysis. Like all statistical techniques for analyzing data, factor analysis can be and often is misused. So, of course, are all existing statistical techniques. Even the humble chi-square is often misused, and as regards analysis of variance, what sins are committed in thy name! It would not be untrue to say that many psychologists, not knowing how to make any sense of a fairly arbitrary collection of data brought together for no obvious purpose, and with no reasonable theory in mind, proceed in despair to calculate correlations and carry out a factor analysis in the vague hope that this may lead them to some kind of publishable results. Needless to say, such desperate measures indicate only that a thoroughly bad research has been embarked upon, and it is doubtful whether factor analysis or any other statistical methods could rescue such data from a much deserved oblivion. The results, when they

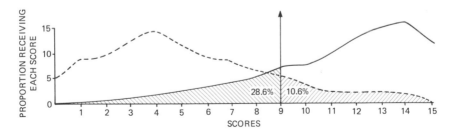

FIGURE 6. Proportion of normals (left) and neurotics (right) receiving each score upon neuroticism questionnaire.

do find their way into the journals, serve only to bring factor analysis into disrepute. But this is a criticism of particular research and particularly research workers; it does not affect the value of the method as such.

It should equally be noted that many research workers misuse factor analysis, even when it has been applied to data and hypotheses that may be perfectly adequate. In relation to personality, Kline and Barrett (1983) have laid down the rules which should govern the collection of data and the method of analysis and rotation adopted; they also note that in the majority of cases these rules have not been followed. If, therefore, the results do not always agree, that should not be laid at the door of factor analysis as such, but rather the inadequacies of the investigators should be held responsible. Factor analysis, as already pointed out, is a good servant but a bad master; it is idle to expect good results from the mechanical application of computer programs that are ill-understood to data not collected with any particular theory in mind. This is the very negation of scientific research, and the fact that journals can still be found that are ready to print such abominations should not mislead the investigator into thinking that such so-called research is adequate for the testing of scientific hypotheses. Factor analysis gives us unique quantitative information with regard to certain problems; it does not aim to answer all problems, and it is a necessary but not a sufficient condition for the development of scientific theories. It has in the past been unreasonably criticized and rejected, and unreasonably praised and adopted as a panacea; both attitudes are inimical to the development of a proper estimate of the limited but real usefulness of the method in the measurement of personality and intelligence.

One particular criticism still requires discussion. Let us go back to Figure 5. It will be seen that the two axes, extraversion and neuroticism, are at right angles; this means that extraversion and neuroticism are uncorrelated with each other. (In the geometrical picture, the cosine of the angle between two lines represents a correlation between them, and of course the cosine of 90° is zero.) Some critics, Carrigan (1960), for example, have doubted whether E and N are truly orthogonal, that is, whether the correlation between them is really zero, so that they can be considered as independent. Carrigan demonstrates that in many studies slightly negative correlations have in fact been found; in more recent studies (e.g., Farley, 1967), correlations have been mostly insignificant and close to zero. What does this prove?

Let us consider Figure 5. Suppose we were to introduce two additional new items, both measures of impulsiveness. The evidence (to be discussed in due course) tells us that impulsiveness correlates with both extraversion and neuroticism; hence these items would appear in the first quadrant, that is, in the presently blank space that constitutes the major part of the figure. We would find that if we performed a factor analysis of the intercorrelatons between the 14 items, E and N would now show a *positive* correlation! If, instead, we had added two items in the fourth quadrant, defining stable extraversion (e.g., items measuring venturesomeness), we would have obtained a *negative* correlation between E and N. Thus we can change the correlation within certain limits from negative through zero to positive simply by adding certain types of items to the item pool. Will the real factor correlations stand up please!

The error in the argument, and indeed in the question itself, is simply that to ask "What is *the* correlation between E and N?" is to reify both E and N. It assumes that there is out there in the cosmos a *real* neuroticism having a unique correlation with extraversion, and that our tests attempt to approximate these *real* factors. But of course, as we have already seen, this is not true. We are dealing with concepts, which are not *real* in this sense but are merely inventions of the human mind. It is not meaingful to ask if two of these concepts are *really* independent or not.

The proper question is: If we administer many reasonable scales representing selections of items measuring the various traits which hypothetically characterise E and N to random samples of the population, do the correlations cluster round zero sufficiently to make the belief tenable that if we define E and N in terms of an orthogonal relation this will prove useful in our research and applied work? We can then proceed to build our tests in such a way that they do in fact give zero correlations. Thus, orthogonality is partly observed in nature, partly introduced into the data by selection of items; there is the usual combination of theory, subjectivity, and objective fact. This may at first sight seem rather high-handed, but it is exactly what the physicist does in his realm. Let us consider this point for a moment.

Let us take an example of what the physicist does in the measurement of heat by means of liquid-in-glass thermometers. What should we put into the thermometer? Clearly water would not be a good liquid to use because it contracts from the ice point (0° Celsius) to the temperature of maximum density (4°C), thus giving an illusory decline in temperature when actually the temperature is increasing! In actual fact the liquids most widely used (mercury and alcohol) are chosen in part because they best fit the kinetic theory of heat, which predicts that the final temperature reading of a fluid obtained by mixing two similar fluids of masses M_1 and M_2 at the initial temperatures T_1 and T_2 should be

$$T = \frac{M_1 T_1 + M_2 T_2}{M_1 + M_2}$$

The linseed oil thermometer was discarded because measurements made with this instrument did not tally with the predictions made by the kinetic theory; mercury and alcohol thermometers do tally. Thus the choice of a measuring instrument is in part based on its agreement with theory, and as we have seen the same is true in psychological measurement (Baker *et al.*, 1975). Psychologists do not always understand the way the "hard" scientist works, and hence they impose restrictions on their own work which would be unacceptable to the physicist, the chemist, or the astronomer.

In searching for the *true* correlation between E and N, the psychologist is chasing a chimera; no meaning attaches to such a quest. To take again an example from the measurement of heat, we may point to the fact that there are many different methods of measuring temperature. There is the mercury-in-glass thermometer, depending on the change in volume of the mercury with increase in heat; the constant-volume gas thermometer, depending on the reactance of the welded junction of two fine wires; the resistance thermometer, depending on the relation between resistance and temperature; the thermocouple, depending on the setting up of currents by a pair of metals with their junctions at different temperatures. Nelkon and Parker (1968) point out that temperature scales differ from one another and "that no one of them is any more 'true' than any other, and that our choice of which to adopt is arbitrary, so it may be decided by convenience" (p. 186). Thus, when a mercury-in-glass thermometer reads 300°C, a platinum-resistance thermometer in the same place and at the same time will read 291°C! There is no meaning attached to the question of which of these two values is correct.

It will be clear from what has been said that some of the questions asked by psychologists of factor analysis are indeed meaningless, and consequently the fact that different answers are returned should not be taken as a criticism of factor analysis, but rather of the failure of those asking these questions to demonstrate a sufficient knowledge of the philosophy of science. It is necessary to understand the element of subjectivity that enters into all scientific measurement and all scientific conceptualizations in order to avoid asking the wrong kind of questions, implying the reification of scientific concepts, and creating difficulties in the understanding of empirical research where none really exist.

One major criticism of modern factor analysis must be that the mathematical methods are far more refined than would be appropriate for the rough-and-ready data that are being analyzed. Revensdorff (1978) has explicated the criticism in detail, showing that the use of very complex methods of statistical analysis is wholly incommensurate with the kind of data available and the hypotheses tested. The often voiced feeling that many psychometrists are more interested in the statistical methodology than in the scientific problem and the psychological results is undoubtedly justified to some extent. Psychologists should always give priority to the nature of the psychological problem involved and the psychological adequacy of the solution; statistical sophistication can never be a substitute for psychological insight.

SITUATIONISM VERSUS TYPE-TRAIT THEORY

What is it essentially that the trait theorists maintain and that critics like Thorndike and Mischel deny? M. W. Eysenck and H. J. Eysenck (1980) have given a thorough review of these criticisms and their own answers to them; here we will deal with these points on a smaller scale. Below is a list of the preconceptions that they perceive as widely accepted by trait psychologists:

1. Individuals differ with respect to their location on important semipermanent personality dispositions known as traits.
2. Personality traits can be identified by means of correlational (factor analytic) studies.
3. Personality traits are importantly determined by hereditary factors.
4. Personality traits are measurable by means of questionnaire data, ratings, objective psychological laboratory tests, and psychophysiological measures.
5. The interactive influence of traits and situations produces transient internal conditions known as *states*.
6. Personality states are measurable by means of questionnaire data, psychophysiological measures, and laboratory tests.
7. Traits and states are intervening variables or mediating variables that are useful in explaining individual differences in behavior to the extent that they are incorporated into an appropriate theoretical framework.

8. The relationship between traits or states and behavior is typically indirect, being affected or *moderated* by the interactions
that exist among traits, states, and other salient factors.

One of the best known of Mischel's criticisms of the state–trait
approach is his assertion that measures of consistency in personality
rarely produce correlations in excess of .30. This criticism is applicable
at most to studies considering specific behavior responses across two dissimilar situations. However, at the empirical level, one inadequacy of
many studies has been the use of very limited and unreliable data sampling. The difference that enlarging the data base can make to correlational measures of consistency was demonstrated clearly by Epstein
(1977). Subjects kept records of their most positive and most negative
emotional experience each day for over three weeks. The mean correlation when either positive or negative experiences were compared on only
two days was less than .20, very much in line with the magnitude of the
correlations discussed by Mischel. However, when the mean for all the
odd days was correlated with the mean for all the even days across subjects, the mean correlation for the pleasant emotions was .88 and only
slightly less for the unpleasant emotions. Similarly, observations made
daily by external judges for four weeks on eight variables related to sociability and impulsivity gave a correlation of only .37 when based on two
one-day samples of behavior, but one of .81 for two 14-day samples. Thus,
if small consistencies are found in empirical work, this is often because
of a small data base, which, if expanded, would give much higher consistencies (Rushton, Brainerd, & Preisley, 1983).

Two further points demand mention. The first one is what Rushton
et al. (1983) called the principle of aggregation. This simply states that
in regarding consistency of conduct we must be careful to look at aggre-
gates of instances rather than at single items. We would not predict educational success or failure on the basis of a single IQ item; such items
only intercorrelate about .2 and are therefore inherently lacking in reliability and validity. However, if we aggregate 100 items, the resulting
estimate of IQ correlates powerfully with scholastic achievement. Much
of the literature deploring the low consistency of conduct relies on single
items of this kind. The work of Hartshorne and May (1928, 1929) and
Hartshorne and Shuttleworth (1930) is a case in point. Because individual tests of honesty, deceit, and so forth correlated only about .2 with
other single tests, the authors argued for lack of consistency in conduct;
however, when scales were constructed using sets of tests, quite high
consistency of conduct was found (Burton, 1963; H. J. Eysenck, 1970c).

The principle of aggregation is considered very thoroughly in the paper by Rushton *et al.*, and no more need here be said concerning this point.

The second point to be made in answer to Mischel and others who emphasize the importance of situations is the obvious fact that we *choose* the situations in which we wish to find ourselves. An interest in books and reading will take us to libraries; an interest in sport will take us to the tennis court or the football field; an interest in music will take us into situations where music is being played. The situation in these cases is not primary; it is a *consequence* of already existing systems of likes and dislikes, values, attitudes, personality traits, and the like, many of them, as we shall see, genetically determined. Life does not *start* with situations as something already given and in which we find ourselves; we have the ability to choose which of many possible situations we wish to enter. This point is of vital importance for a consideration of *situationism;* yet it is never considered by those who deny the consistency of conduct.

Mischel has argued that traits are constructed from global overgeneralizations based on behavior. He has not, apparently, considered the possibility that hereditary factors might be of importance in this context. This is puzzling in view of the fact that the evidence from twin studies consistently indicates the substantial part played by heredity in the dissemination of personality; these studies will be considered later on in this book, but they do suggest that heredity accounts for something like two-thirds of the true variance as far as personality differences in P, E, and N are concerned (Fulker, 1981). It is impossible to combine a belief in inconsistency of human behavior with the facts of genetic determination. Since the evidence indicates that hereditary factors are important in explaining individual differences in personality, and since the trait-state approach is almost the only major theory of personality that acknowledges that fact and incorporates hereditary factors by means of the trait concept, it is incumbent upon theorists of different persuasions to address themselves to this issue.

Mischel has recognized that the task of predicting behavioral responses within a trait-state theory can proceed on the basis of "moderator variables" (Wallach, 1962); the basic notion here is that the influence of any particular trait on behavior will usually be *indirect*, being affected or moderated by a number of other traits, mediating variables, and situational factors. Mischel has criticized this approach, arguing that the more moderators that are required to qualify a trait, the more a trait-based formulation resembles the relatively specific description of a behavior-situation unit. Although it is true that trait-state conceptualizations have become increasingly complex over the last few years, it could very well be argued in view of the complexity of human functioning that this is a necessary and indeed inevitable development. Evidence of some cross-situational specificity of behavior can only be taken as damaging for state–trait theories that assume a direct one-to-one correspondence between internal traits and behavior indices. Since most contemporary state–trait theories postulate the existence of moderator variables and thus claim only an indirect but theoretically predictable relationship between traits and behavior responses, Mischel's evidence loses much of its apparent force.

ˀ   It is of course desirable that those factors emphasized by a theoretical position should account for a sizeable proportion of behavioral variation, but it is also the case that there are various other criteria by which theories can and should be evaluated. One such criterion is a theory's range of applicability. To take but one example: H. J. Eysenck (1967a, 1981) has shown that the personality dimension of extraversion–introversion is related to performance in a theoretically predictable way for the following and other variables: sensory thresholds, pain thresholds, time estimation, sensory deprivation, perceptual defense, vigilance, critical flicker fusion, sleep–wakefulness patterns, visual constancy, figural aftereffects, visual masking, rest pauses in tapping, speech patterns, conditioning, reminiscence, and expressive behavior. Mischel's criticisms suffer from the disadvantage of evaluating the state–trait approach from a rather limited perspective. The later chapters in this book will speak eloquently to this refutation of his criticisms.

Mischel and his followers appear to regard the figure of .30 for an average measure of consistency as meaningful in some way, but it is difficult to see how this can be. Essentially, Mischel is trying to prove a negative, that is, that conduct is not consistent. That clearly is not possible; even if all (n) attempts to discover consistency had been failures, the possibility that the next ($n + 1$) attempt might be successful is not ruled out. To average, as he has done, successful and unsuccessful attempts is meaningless; success clearly depends on having a theory essentially pointing in the right direction, choosing tests that are both reliable and valid, and applying them to an appropriate population under appropriate motivational circumstances. If even one of these (rare and unusual!) preconditions is missing, the failure of the experiment says nothing about consistency of conduct.

This point must be seen against the typical way in which research into personality is conducted. We do not wish to caricature the modal paper in this field, but it would seem that a multiphasic test of personality traits is administered to a population of sophomores and that all the separate scores of this test are then correlated with some criterion or criteria. Usually nothing by way of hypothesis is stated; and even when a hypothesis is allegedly tested, it does not usually specify a particular trait. Although one of the many traits measured by the multiphasic test is probably relevant to the hypothesis, the others most likely are not; yet the correlations with the criterion are usually calculated for all. If a test measures 16 traits (like the 16PF) and if one trait is strongly related to the criterion whereas the other 15 are not, an averaging of all the observed correlations must inevitably give a low mean r; it is this meaningless mean or modal figure that enters into such average figures as those quoted by Mischel.

On another point, it is well known in psychometrics that correlations cannot be interpreted directly without some knowledge of the internal reliability of the scores correlated. Any attempt to estimate the relationship between two variables that relies on unreliable estimates of one or both may grossly underestimate the true correlation, and an attempt should always be made to correct for attenuation. This is practically never done in the studies quoted and averaged by Mischel, although the reliabilities of the variables in question are often known and frequently fall short of what might be regarded as adequate. For this reason, among others, the average correlation of .30 as a meaningful estimate of the true relationships in question must be regarded as underestimating these to an unknown but probably substantial extent. This argument becomes even stronger when we consider that much of the work criticized by Mischel attempts to predict real-life behavior. Such predictions can come to grief for several reasons, among which is the obvious one that there is in fact no consistency of conduct such as is implied in the theory that reasonably high correlations will be found; this is in practice the only one considered by Mischel. It is equally possible, as has already been mentioned, that the wrong theory has been tested, or that the wrong test has been used, or that the tests used were unreliable. It is equally possible, and often demonstrably true, that the criterion may be excessively faulty, that is, either invalid or unreliable. Educational criteria are famous for their lack of reliability (Hartog & Rhodes, 1936); other real-life criteria in industry, psychiatry, and elsewhere often share this fault. Unless we have reason to believe that our criteria are both reliable and valid, the failure of tests to predict these criteria adequately cannot be used as proof of the inconsistency of conduct.

What Mischel, Shweder (1975), D'Andrade (1965), and others have tried to substitute for trait psychology is what is sometimes called *situationism*, the modern title for Thorndike's specificity hypothesis. Thus

Bowers (1973) quotes Mischel as saying that "the utter dependence of behavior on the details of the specific conditions reflects the great subtlety of the discriminations that people continuously make" (p. 308). It is suggested that human conduct is determined entirely ("utter dependence") by the situation in which a person finds himself, although some situationists do not go quite so far as that. Such a position is clearly unacceptable; it fails entirely to account for what is the basis of all trait and type theories, namely *individual differences in behavior in identical situations*. Bowers has provided an excellent critique of "situationism in psychology," and we will not here go into too much detail concerning this general view. Much has been written on it since Bowers's critique, and many experiments have since lent support to his reanalysis of 11 studies evaluating the influence of setting, person, and interaction between the two in analysis of variance studies. He found that in these studies 13% of the variance was attributable to the person, 10% to the situation, and 21% to the person by situation interaction. This finding certainly does not encourage a belief in the overwhelming strength of the situation, and later work (e.g., Sarason, Smith, & Diener, 1975) amplifies this conclusion. The least that can be said, simply from a scrutiny of this particular set of studies, is that personality, whether as a major variable or in interaction with the situation, contributes at least as much, and probably more, to the observed behavior than does situation by itself. In addition, the person variance is essential for explaining interindividual differences in identical situations.

Specific studies on sociability (Gifford, 1981), anxiety (Lazzerini, Cox, & Mackay, 1979), aggression (Olweus, 1979), extraversion (Monson, Hesky, & Chernick, 1982), and many others indicate the great importance of personality traits even when situations are varied (Magnusson, 1981; Argyle, Furnham, & Graham, 1981). These studies deal with cross-sectional consistency; studies on longitudinal consistency will be reviewed later (e.g., Schuerger, Taid, & Tavervelli, 1982). Here let us merely note the paradox that although empirical studies have on the whole decisively disproved Mischel's position, it is still widely and uncritically accepted and referred to as if authoritative in many textbooks. Boring would have invoked the *Zeitgeist* to explain such a paradox, and no better explanation presents itself to us.

We would maintain that not only is situationism contrary to the great majority of empirical findings, but it is internally inconsistent and based on fundamental methodological and theoretical errors. In the first place, the apparently clear-cut distinction between trait and situation is in fact not very meaningful. Usually the name of a given trait also implies the situations in which it can be demonstrated and measured; thus a trait theory implies directly a taxonomy of situations.

Consider traits like sociability, persistence, and personal tempo. We can only measure sociability in certain types of situations, namely those involving the relatively free intercourse between people (or questions relating to such occasions). We cannot measure persistence or personal tempo in these situations, just as little as we can measure sociability in situations that enable us to measure persistence or personal tempo. In other words, trait and situation form the two sides of a coin that cannot be separated from each other, and classical trait theories do not neglect situations; they directly imply them. A typical personality questionnaire relating to, say, sociability will in fact contain reference to a number of situations in which a person can behave in a sociable manner, that is, reference will be made to liking for parties, talking to people, having many friends, and so forth. Similarly a questionnaire on persistence will relate to the individual's tendency to carry on with certain activities in the face of pain, boredom, and so on in many different situations.

Hartshorne and May (1928, 1929) and Hartshorne and Shuttleworth (1930) probably introduced a fundamental error into the discussion when they said that a trait theory of honesty would demand that children who were dishonest in one situation should also be dishonest in another. Such a demand disregards completely the *asymmetric* nature of any scale the items of which are pitched at different difficulty levels. To take a simple example from intelligence testing first, it does not follow that because a child solves an easy problem he should be able to solve a difficult one (although the obverse would indeed by implied by intelligence theory, hence the asymmetry). Acts of dishonesty form a similar scale on which, for instance, cheating at school is an easy item, stealing a more difficult one. One would not necessarily expect that because a child cheats at school he will also steal; there is no such implication in the trait theory in question. When Magnusson (1980) demanded a taxonomy of situations he was clearly conscious of this interrelationship between traits and situations and the need for assigning difficulty levels to different situations. This need has not found much expression in empirical research (Argyle *et al.*, 1981).

Let us now turn to a more detailed discussion of the so called interaction effect. To examine it more closely, let us look at two subfactors that have been suggested to contribute to extraversion, namely, sociability and impulsiveness. S. B. G. Eysenck and H. J. Eysenck (1963) carried out an investigation using a questionnaire with a number of items relating to sociability and a number of items relating to impulsiveness, as well as other extraversion and neuroticism items. The items were intercorrelated and factor analyzed. As predicted, factors of sociability and impulsiveness emerged, but the two were very significantly correlated with each other, a result replicated by Sparrow and Ross (1969) and by Farley (1970). The fact of the intercorrelation between the two gives rise to the concept of extraversion but clearly also implies a certain type of situationism, in the sense that people who are sociable (i.e., react in a certain way in social situations) are not necessarily impulsive (i.e., do not react in an impulsive manner in other types of situations.) The fact that these two quite dissimilar types of situations are nevertheless significantly correlated suggests that the personality trait has a consistency not depending on specific situational variables.

In a similar way, we may look at sociability, as already suggested (H. J. Eysenck, 1956b), and formulate the hypothesis that shyness may be shown in response to different *kinds* of situations, namely those involving anxiety-provoking superiors and those invoking other people in general and the dislike of the shy person to be with them. Any trait theory implies a situational theory, and vice versa; hence, there is an obvious degree of interactionism posited in classical trait theory. The belief of modern writers that they have discovered a compromise between trait theory and situationism in looking for interactions is mistaken; such interaction has always been part of trait theory! We conclude that the attempts of Mischel and others to discount trait theory and substitute situationism are logically inconsistent, historically false, and empirically incorrect. There is a certain degree of consistency about human conduct that extends over many types of situations and which must be taken into account in experimental psychology, social psychology, clinical psychology, educational psychology, industrial psychology, and all other variants of psychology.

Another criticism of the trait hypothesis has been put forward by proponents of "implicit personality theories" (see Wiggins, 1973). Along these lines, Shweder and D'Andrade (1979) have argued that correlations derived from trait ratings are largely representative of the conceptual or semantic associations among rating categories and do not reflect individual differences in personality or behavior at all. This position is supported by findings such as those of Passini and Norman (1966), who found that the factor structure of ratings of unknown strangers was similar to that based on ratings of known individuals. In other studies also, semantic similarity measures have often yielded a factor structure nearly identical to that found when actual responses of subjects or raters were factor analyzed (D'Andrade, 1965; Shweder, 1975; Stricker, Jacob, & Kegan, 1974).

Such a theory implying that laymen possess some intuitive knowledge about trait associations, which would replicate true trait correlations, is contradicted by the empirical evidence indicating correlations between trait assessments either by questionnaire or rating and correlations with external and objectively measurable behaviors and laboratory tests (to be discussed later). A direct test of the hypothesis has been made by Rowe (1982), who used monozygotic twins' responses to personality items involving sociability, emotionality, and impulsiveness. The twins were first entered into a data matrix as individuals, and the resulting trait correlations were factor analyzed. Next, a second matrix was calculated from the MZ twin cross-correlations, that is, trait A in one twin correlated with trait B in his or her twin partner. Rowe argued that the first factor structure should contain "semantic bias" but that the second factor, since the correlations . represent the association between traits independently reported by two individuals, should not.

Contrary to the semantic bias position, the relatively bias-free cross-correlational factor structure was found to replicate the factor structure discovered in the ordinary correlations. Rowe concluded that although there may be some degree of bias inherent in implicit personality structure, it does not appear to distort greatly personality factor structure. Such a conclusion is very much in line with the many findings of correlations between personality factors and external criteria of an objective type. On the whole, the semantic bias hypothesis contributes weakly, if at all, to the observed trait measures and factor structures based on them.

CHAPTER TWO

The Development of a Paradigm

ORIGINS OF PERSONALITY THEORY

It is a well-known saying that psychology has a short history but a long past. It is true that experimental methods have only been used very recently, but various conceptions which still powerfully influence our work certainly date back over 2,000 years. We have already mentioned Cicero's explicit distinction between trait and state anxiety; it might be added that he also had a learning theory of neurosis, well in advance of Watson, Skinner, and Wolpe! In a similar way, two of our major dimensions of personality, namely, extraversion and neuroticism, were anticipated by ancient Greek writers such as Hippocrates, who is credited with the doctrine of the four temperaments and who also advanced an explanatory theory in terms of the so-called humors; this theory was later popularized and extended by the Roman physician Galen.

Galen assigned a definite cause for each of the four outstanding types of individuals, depending on the preponderance of certain bodily humors. The *sanguine* person, always full of enthusiasm, was said to owe his temperament to the strength of the blood; the sadness of the *melancholic* was supposed to be due to the overfunctioning of black bile; the irritability of the *choleric* was attributed to the predominance of the yellow bile in the body; and the *phlegmatic* person's apparent slowness and apathy were traced to the influence of the phlegm.

Absurd as these ideas now sound, they nevertheless embody, if only in embryo, the three main notions which characterize modern work in personality. In the first place, behavior or conduct is to be described in terms of *traits* that characterize given individuals in varying degrees. In the second place, these traits cohere or correlate and define more certain fundamental and all-embracing *types*. In the third place, these types are essentially based on *constitutional*, genetic, or inborn factors, which are to be discovered in the physiological, neurological, and biochemical structure of the individual. There is some evidence in Greek writings of the differentiation so important nowadays between *phenotype* and *genotype*, that is, between behavior as apparent in everyday life and conduct and the genetic basis for behavior. This differentiation, therefore, leads to the important question of the *degree* to which environmental forces determine differences in personality and may thus affect the principles of classification derived essentially from phenotypic investigations.

Admittedly all these points are embryonic and should not be stressed too much; there is always the danger that one tends to read into ancient documents modern ideas not properly contained in them. Nevertheless it can hardly be denied that all of these fundamental ideas are in fact outlined in the early writings of the third and second centuries B.C.

The doctrine of the four temperaments proved a very influential theory of personality for hundreds of years, but there is no point in discussing this in detail. We may perhaps take up the story in 1798 when Immanuel Kant published his famous *Anthropologie*. Kant was not only Europe's foremost philosopher but also an accomplished scientist, and in this book he presents us with what is essentially a textbook of psychology. His chapter on temperament was widely read and accepted in Europe. His description of the four temperaments may therefore serve us as a kind of fundamental and basic theoretical position. This is how he describes the sanguine, the choleric, the melancholic, and the phlegmatic person.

> *The Sanguine Temperament.* The sanguine person is carefree and full of hope; attributes great importance to whatever he may be dealing with at the moment, but may have forgotten all about it the next. He means to keep his promises but fails to do so because he never considered deeply enough beforehand whether he would be able to keep them. He is goodnatured enough to help others, but is a bad debtor and constantly asks for time to pay. He is very sociable, given to pranks, contented, does not take anything very seriously and has many, many friends. He is not vicious, but difficult to convert from his sins; he may repent, but this contrition (which never becomes a feeling of guilt) is soon forgotten. He is easily fatigued and bored by work, but is constantly engaged in mere games—these carry with them constant change, and persistence is not his forte.
>
> *The Melancholic Temperament.* People tending towards melancholia attribute great importance to everything that concerns them. They discover everywhere cause for anxiety, and notice first of all the difficulties in a situation, in contradistinction to the sanguine person. They do not make promises easily, because they insist on keeping their word, and have to consider whether they will be able to do so. All this is so not because of moral considerations, but because interaction with others makes them worried, suspicious and thoughtful; it is for this reason that happiness escapes them.

The Choleric Temperament. He is said to be hot-headed, is quickly roused, but easily calmed down if his opponent gives in; he is annoyed without lasting hatred. Activity is quick, but not persistent. He is busy, but does not like to be in business, precisely because he is not persistent; he prefers to give orders, but does not want to be bothered with carrying them out. He loves open recognition, and wants to be publicly praised. He loves appearances, pomp and formality; he is full of pride and self-love. He is miserly; polite, but with ceremony; he suffers most through the refusal of others to fall in with his pretentions. In one word, the choleric temperament is the least happy, because it is most likely to call forth opposition to itself.

The Phlegmatic Temperament. Phlegma means lack of emotion, not laziness; it implies the tendency to be moved, neither quickly nor easily, but persistently. Such a person warms up slowly, but he retains the warmth longer. He acts on principle, not by instinct; his happy temperament may supply the lack of sagacity and wisdom. He is reasonable in his dealing with other people, and usually gets his way by persisting in objectives while appearing to give way to others. (1798/1912, pp. 114–115)

There are no compound temperaments, for example, a sanguine-choleric; in all, there are only these four temperaments, each of which is simple, and it is impossible to conceive of a human being who combines them in any way.

This view of the four temperaments as quite independent, separate, and unrelated, each presumably inherited through what we now call a Mendelian dominant gene, was clearly not in line with everyday observation, even if a great deal of allowance is made for differences between phenotype and genotype. The modern view originated, oddly enough, with Wundt (1903), who is not normally regarded as a personality theorist. Nevertheless, he was the first (but not the only) psychologist to challenge the categorical type of description of the ancient Greeks and of Kant and to introduce a dimensional one instead. Rohracher (1965) states that already in 1911 Stern was able to describe 15 similar attempts, among them that of Ebbinghaus (who is also not commonly regarded as a personality theorist) and who put forward the two orthogonal factors of optimism and pessimism (extraversion–introversion) and *lebhaftes* versus *verhaltenes Gefühlsleben* (emotional versus nonemotional).

What Wundt suggested was this:

The ancient differentiation into four temperaments ... arose from acute psychological observations of individual differences between people. The fourfold division can be justified if we agree to postulate two principles in the individual reactivity of the affects: one of these refers to the *strength*, the other to the *speed of change* of a person's feelings. Cholerics and melancholics are inclined to strong affects, while sanguinics and phlegmatics are characterised by weak ones. A higher rate of change is found in sanguinics than cholerics, a slow rate in melancholics and phlegmatics.

It is well known that the strong temperaments ... are predestined towards the *Unluststimmungen* (negative affects), while the weak ones show a happier ability to enjoy life ... The two quickly changeable temperaments ... are more susceptible to the impressions of the present; their mobility makes them respond to each new idea. The two slower temperaments, on the other hand, are more concerned with the future. Failing to respond to each chance impression, they take time to pursue their own ideas. (1903, pp. 384)

Using the terms and descriptions of Kant and Wundt in combination, we arrive at a theoretical picture of human personality rather like that given in Figure 7. It will be seen that Wundt has shifted the emphasis from a typology conceived as a *categorical* system which only allocates people to one of four quadrants to a *quantitative*, two-dimensional system in which people can occupy any position and any combination of positions on two major dimensions which he labels "strong emotions" as opposed to "weak emotions," and "changeable" as opposed to "unchangeable." Within this system, which predicts with uncanny accuracy future developments that make it still applicable after 80 years, every person has a position on the two major dimensions, and these type concepts of *emotionality* and *changeability* do not suggest any kind of U-shaped or multimodal distribution; Wundt thought of the distributions as essen-

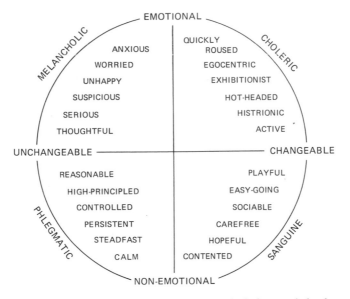

FIGURE 7. Diagrammatic representation of the classical theory of the four temperaments as described by I. Kant and W. Wundt.

tially normal. The typology suggested in this Wundtian scheme owes its origin to unaided but fairly systematic observation.

Psychology has advanced in making the collection of data, either through observation and rating or questionnaire and self-rating, more systematic and in subjecting the results to statistical analysis. In addition, modern psychology has provided theories to explain the observed correlations and has attacked the implicit problem of the determination of observed behavior patterns through genetic and environmental factors. But in spite of all these advances, we have not succeeded in going far beyond the insights of the ancient Greeks, although factors additional to E and N have been discovered and have extended this scheme.

Kant and Wundt were concerned entirely with the descriptive function of typology; two other writers whom we must mention were more interested in the causal factors involved and tried to elaborate a theory which might account for the observed behavior patterns. The first of these was the Austrian psychiatrist Otto Gross (1902, 1909), whose two books (or rather pamphlets) introduce the concepts of *primary* and *secondary function.* These concepts are basically physiological, although the physiology in question is, in part at least, mythological; the concepts refer respectively to the hypothetical activity of the brain cells during the production of any form of mental content and to the hypothetical perseveration of the nervous processes involved in this production. Thus a nervous process that succeeds in rousing an idea in the mind is supposed to perseverate, although not at a conscious level, and to determine the subsequent associations formed by the mind. Perseveration, as conceived by Gross, is similar in nature to the concept of consolidation, which has in recent years acquired excellent experimental support. A review of this work on consolidation and its application to personality differences is given in M. W. Eysenck (1981) and H. J. Eysenck and Frith (1977).

Gross also postulated a correlation between the intensity of any experience and the tendency for that experience to persist secondarily and to determine the subsequent course of mental associations. Most intense and energy-consuming in his view were highly affective and emotional experiences and ideas, and these would therefore be followed by a long secondary function during which the mental content would be influenced and in part determined by the perseverative effects of the primary function. There is of course an obvious similarity between the concept of secondary function and that of refractory period. Gross distinguishes two contrasted types of persons whom he labels "deep-narrow" and "shallow-broad." In the deep-narrow type we find characteristically a primary function that is highly charged with emotion and loaded with affect, causing the expenditure of great nervous energy and requiring a lengthy period of restitution during which the ideas involved in the primary function go on reverberating and perseverating (long secondary function). In the shallow-broad type, on the other hand, a much less intense primary function, necessitating the expenditure of comparatively little energy, is followed by a short period of restitution (short secondary function).

Certain personality characteristics follow from the type hypothesis briefly described above. In the broad-shallow person the short secondary function enables a much greater frequency of primary functions to take place within a given time; this constant readiness for brief actions and reactions suggests a certain superficiality, a distractibility, as well as a prompt reaction to external events; it clearly aligns this type with Wundt's changeable type. In the deep-narrow person the long perseverative secondary function makes the integration of different sets of what Gross calls *themas* (sets of emotions, associations, determining tendencies, complexes, and sentiments centered around one idea which is the object of primary function) more difficult and leads to a sejunctive (dissociated) type of personality. Dissociation leads to damming up of the available libido, to inhibition, and on the behavioral level to absorption in thought and social shyness. It is the basis for Wundt's unchangeable type.

The physiological theories of Gross are of course quite outmoded and bear little relation to reality. However, if we substitute for his primary mental function the concept of the ascending reticular formation and the increase in the alertness or arousal of the cortex produced by this system, then we can see at once that his ideas are by no means as irrelevant to modern theorizing as they may at first seem to be (H. J. Eysenck, 1967a). The functions of the activating reticular system are precisely those stressed by Gross, that is, the arousal of the cortex and the facilitation of future activation of the cortex along the lines laid down by present stimulation of ideation. There is no need to insist on this comparison, of course; as before, we may here be in danger of reading into older writers ideas which they could not possibly have foreseen. However, the resemblances are striking enough to deserve at least passing mention.

The second writer to be mentioned in this context is Jung (1921), the Swiss psychiatrist and one-time follower of Freud. Basing his work on that of many predecessors, Jung saw the main cause of typological differences in the extraverted or introverted tendency of the libido, that is,

...the tendency of the individual's instinctual energies (not only sexual!) to be directed mainly toward the outer world (objects) or toward his own inner mental states (subject):

> When we consider a person's life history we see that sometimes his fate is determined more by the objects which attract his interest while sometimes it is influenced rather by his own inner subjective states.... Quite generally one might characterize the introverted point of view by pointing to the constant subjection of the object and objective reality to the ego and the subjective psychological processes.... According to the extraverted point of view, the subject is considered as inferior to the object; the importance of the subjective aspect is only secondary. (p. 38)

It is somewhat hazardous to try to give descriptions of the behavior of typical extraverts and introverts because Jung is concerned far more with attitudes, values, unconscious mental processes, and so on than with behavior. Furthermore his account is complicated to an almost impossible extent by his insistence that people who are consciously extraverted may be unconsciously introverted, and vice versa, and by the further insistence that these tendencies may find expression according to the four main mental functions. Jung regards extraversion and introversion as the two major attitudes or orientations of personality, but these find expression in the functions of *thinking, feeling, sensing,* and *intuiting.* Thinking and feeling are called *rational* functions because they make use of reason, judgment, abstraction, and generalization. Sensation and intuition are considered to be *irrational* functions. Jung weaves a very complex web stressing the superiority of some functions, the auxiliary role played by others, and so forth. There is little point in recapitulating Jung's complete system as no modern psychologist has adopted it in its entirety and as in any case it seems difficult to apply in any rational manner.

However, making allowances for these complications and the obvious distortions that such neglect must introduce into his system, we may say that from Jung's accounts the extravert emerges as a person who values the outer world both in its material and its immaterial aspects (possessions, riches, power, prestige); he seeks social approval and tends to conform to the mores of his society; he is sociable, makes friends easily, and trusts other people. He shows outward physical activity, whereas the introvert's activity is mainly in the mental and intellectual sphere. He is changeable, likes new things, new people, new impressions. His emotions are easily aroused but never very deeply; he is relatively insensitive, impersonal, experimental, materialistic, and tough-minded. He tends to be free from inhibitions, carefree and ascendant. The traits listed will perhaps suffice to show how closely Jung approximates to the Hippocrates–Galen–Kant–Wundt model and to the extension of it given by Gross. When in the rest of this book the terms *extraversion* and *introversion* are used it should always be borne in mind that they do *not* refer to the conceptions specifically introduced by Jung, but refer rather to the changeable-unchangeable dimension of Wundt, being perhaps more appropriate terms to use than the Wundtian ones, or the rather clumsy nomenclature introduced by Gross. In other words, just as we have been unable to follow Gross in his explanatory physiological conceptions, so we must refuse to take too seriously the technicalities of the Jungian system with its notions of libido, mental functions, and racial unconscious. It may seem unjust that we should take over terms from one author to characterize notions that may be characteristic of others, but it should be remembered that Jung did not in fact originate the terms *extraversion* and *introversion;* they had been in use in Europe for several hundred years before he popularized them, and there is no reason, therefore, why their use should remain sacrosanct. (Browne, 1971, has collected relevant evidence on this point.)

Jung made one important addition to the ancient system of typology by linking his notions of extraversion and introversion with a distinction between the main neurotic disorders as given by Janet (1894, 1903). As is well known, Jung believed that the extravert in case of neurotic breakdown is predisposed to *hysteria*, the introvert to *psychasthenia:* "It appears to me that much the most frequent disorder of the extraverted type is hysteria" (p. 38). On the other hand, speaking of the introvert, he maintains that "his typical neurotic disorder is psychasthenia, a disorder which is characterized on the one hand by marked sensitivity, on the other by great exhaustion and constant tiredness" (p. 39). Nowadays we would probably refer to anxiety state or reactive depression and to phobia or obsessional state rather than use the obsolescent term psychasthenia. H. J. Eysenck (1947) suggested the term *dysthymic* as a more modern equivalent to cover this syndrome of correlated affective disorders.

Jung never formally elaborated this part of his hypothesis, but it can be seen that implicit in his scheme is a second dimension or factor additional to, and independent of, that of extraversion–introversion. This factor we may provisionally call emotionality or instability or neuroticism; it is identified as that particular quality which hysterics and psychasthenics have in common as compared with normal persons. The independence of introversion and neuroticism is especially stressed by Jung: "It is a mistake to believe that introversion is more or less the same as neurosis. As concepts the two have not the slightest connection with each other." Thus if we wish to represent Jung's complete scheme in diagrammatic form then we require two orthogonal axes or dimensions very much as in Figure 8. Hysterics would then occupy the first quadrant, that labeled "choleric," and psychasthenics or dysthymics would occupy the second quadrant, that labeled "melancholic." As given by Jung, this scheme is of course purely hypothetical, but there is some evidence from empirical studies to support it (H. J. Eysenck, 1947).

It is important to note that Freud differed from Jung essentially in identifying introversion with incipient neurosis, a view that has been followed by many American writers and in particular the earlier constructors of personality questionnaires (Collier & Emch, 1938). According to Freud: "An introvert is not yet a neurotic, but he finds himself in a labile condition; he must develop symptoms at the next dislocation of forces if he does not find other outlets for his pent-up libido" (1920). As we have already seen, Jung's position is exactly the opposite; he regards neurosis and extraversion–introversion as essentially independent. Many misun-

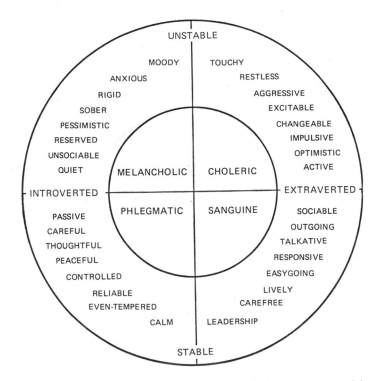

FIGURE 8. Relation between the four temperaments and the modern neuroticism–extraversion dimensional system.

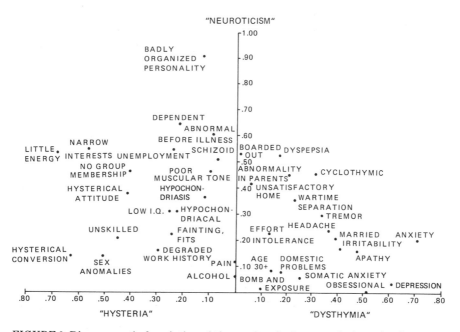

FIGURE 9. Diagrammatic description of the results of a factor analytic study of neurotic soldiers.

derstandings have arisen from this difference in position between Freud and Jung.

A study by H. J. Eysenck (1944, 1947) is relevant to the Jungian view. Seven hundred neurotic soldiers were selected from a total group of 1,000 by excluding all cases of epilepsy, cases involving head injury or previous organic illness, and cases wherein physical illness was an important factor. Ratings were obtained on 39 items, including one test of intelligence, and the intercorrelations between these items were submitted to a factor analysis. The first two factors extracted are shown in Figure 9. It will be seen that we have here a clear verification of Jung's hypothesis dividing neurotics into dysthymics and hysterics, as well as evidence for a general factor of neuroticism. The distribution of these two type factors was found to be reasonably normal, with no indication of bipolarity. (The distributions were plotted for 1,000 male and 1,000 female neurotics by a weighted combination of ratings for the various traits which go to make up these two factors.) Additional evidence for the postulated factor of neuroticism was found by Slater (1943) and Slater and Slater (1944). Howarth (1973) presented a critical reanalysis of Eysenck's original data

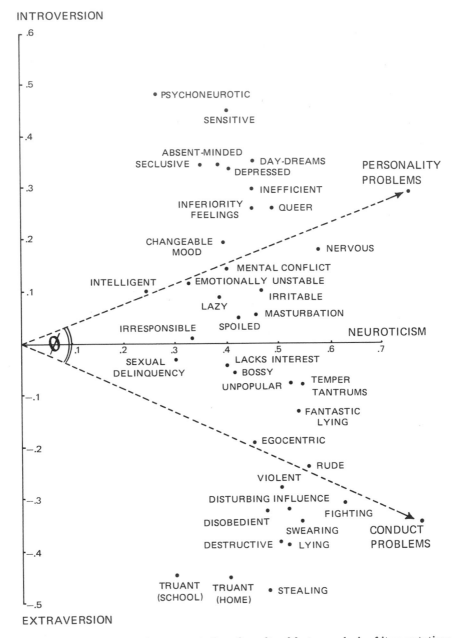

FIGURE 10. Diagrammatic representation of results of factor analysis of item notations for neurotic children.

but concluded that the original solution "appears. . .to be sound in regard to the mathematical aspect" (p. 86).

H. J. Eysenck (1970c) also reports the results of a factor analysis of data collected by Ackerson (1942) for a sample of 2,113 white boys and 1,181 white girls from the files of the Illinois Institute for Juvenile Research. This author selected a number of traits the incidence of which seemed large enough to justify such procedures and calculated tetrachoric intercorrelations between them. Fifty traits were selected and the intercorrelations subjected to a centroid analysis, the results of which are reported in Figure 10. The existence of a general factor of neuroticism is quite marked, as is the division in the second factor between introverted items (sensitive, absent-minded, seclusive, depressed, daydreams, inefficient, queer, inferiority feelings, and nervous) and extraverted items (such as stealing, truancy from home and school, destructive, lying, swearing, disobedient, disturbing influence, violent, rude, and egocentric). Also shown in the figure are two general ratings made by Ackerson, namely, personality problems (dysthymic disorders) and conduct problems. Many other studies supporting Eysenck's interpretation of Jung's theory are discussed by H. J. Eysenck (1970c).

THE BEGINNINGS OF MODERN RESEARCH

The precursor of the large number of factorial rating studies that have been carried out in the last 50 years to test these theories is a *Massenuntersuchung* carried out by two Dutch psychologists, G. Heymans and E. Wiersma (1909). This study differs somewhat from most of those which succeeded it in two ways. In the first place, it was based on a definite hypothesis; in the second, it used statistical methods which, although they resemble factor analysis in their import, were yet rather simpler and more easily understood by the nonmathematical.

Both the hypotheses to be investigated and the method used are clearly brought out in what is essentially a preliminary paper separate from the main work. In this paper Heymans (1908) analyzed biographical material derived from 110 historical persons about whom a great deal was known. These persons were rated on a large number of traits that were considered to be interrelated in such a way as to give rise to three main factors, dimensions, or principles. These three principles are *emotionality*, or emotional instability; *activity*, or general drive; and what we would now call a bipolar factor opposing dominance of the *primary function* to dominance of the *secondary function*, using Gross's terms.

This scheme was later applied by Heymans and Wiersma (1906a,b, 1907, 1908a, 1909) to a rating study in which 3,000 doctors in the Netherlands were asked each to pick one family and rate each member of it by a simple method of underlining or double-underlining a large number of traits. Four hundred doctors responded and sent in material on altogether 2,523 individuals. Most of the papers analyzing this material are concerned with the interpretation of intrafamilial similarities in terms of hereditary hypotheses; as such an interpretation is clearly arbitrary, and as in any case it is not germane to our present topic, nothing will be said here about it. It is in the last paper that we find a detailed analysis and justification for the threefold classificatory system adopted by the authors.

It is clear that if we consider each person to be either above or below the average with respect to each of the three factors, then there are eight possible combinations which might be considered to create separate types. This general scheme links up with the French rather than with the German and Austrian typologists; the eight types show correspondence with the work of such typologists as Ribot (1892), Malapert (1897), Queyrat (1896), and Martiny, (1948). As the last-mentioned points out, it has certain affinities also to Jung's model when the four functions—sensation, intuition, thinking, and feeling—are combined with the extravert–introvert dichotomy.

It seems reasonable to identify Heymans and Wiersma's *emotionality* with emotional instability or neuroticism, and primary and secondary function with extraversion and introversion respectively. To illustrate this correspondence we may set down some of the traits found by Heymans and Wiersma to be characteristic of persons in whom the primary or the secondary function predominated. Those with predominantly primary function are impulsive, give up easily, are always on the move, are jocose, superficial, vain, demonstrative, tend to exaggerate, and are given to public speaking, telling jokes, and laughing a lot. On the other hand, the person with predominant secondary function is quiet, persistent, grave, shut-in, reliable, given to introspective thinking, laughs little, has depressive tendencies, and is not given to indulge in the pleasures of the body.

If our identification of these two factors is correct, we should expect that the emotional person with predominantly primary function (i.e., the nervous type of Heymans and Wiersma) would show the characteristics of the neurotic extravert, which if carried to an extreme would result in the diagnosis of hysteria; whereas the emotional person with predominantly secondary function (the unsentimental type) would resemble the neurotic introvert, or, in extreme manifestations, the person suffering from anxiety and reactive depression (the dysthymic). H. J. Eysenck (1970c) has translated the accounts given by Heymans and Wiersma of these types, and the fit is excellent.

The Heymans and Wiersma study may be considered the first empirical and statistical analysis of personality, and it is interesting that the two major factors discovered by them correspond to E and N. (Activity was found by H. J. Eysenck, 1970, to be correlated with both E and N, more strongly with the former, and is therefore, not an independent dimension. The crudity of the statistical methods used by Heymans and Wiersma did not enable them to sort out these more refined relationships.) Few English-speaking psychologists have taken up their notions; almost the only example is Baehr (1951), who carried out a second-order factor analysis of some factors discovered by Thurstone (1951). Baehr found two second-order factors, the first of which had high positive saturation on Thrustone's sociable, confident, and emotionally stable primaries, and a slight negative loading on active: "The emotionally-toned responses on this factor are generally adjustive. . . . The easy-going and uncomplicated behaviour evident here has caused us to designate this factor *Emotionally Stable*" (p. 43). The second factor has high saturations on Thurstone's impulsive and dominant primaries: "The picture is one of impulsive, carefree and general out-going behaviour responses, all of which are facilitated by spontaneous reaction to stimuli. We designated this factor *Primary Function*" (p. 44). Baehr followed up this reanalysis of Thurstone's data with one of her own, again finding two major factors similar to E and N.

Heymans must also be considered to have been the first to use objective laboratory tests based on a definite theory of personality for the investigation of personality traits. He and Wiersma tried to link tests of perseveration with primary and secondary function, assuming that persons showing strong secondary function (introverts) would show greater perseveration. Many different types of perseveration were already known to the ancients; thus *ideational perseveration* and *emotional perseveration* were described by Aristotle, and *sensory perseveration* was described as follows by Newton:

> If a burning coal be nimbly moved around in a circle with girations continually repeated, the whole circle will appear like fire: the reason of which is, that the sensation of the coal, in the several places of that circle, remains impressed on the sensorium, until the coal returns again to the same place. (p. 13)

Figural aftereffects and flicker fusion may be more modern versions of this type of perseveration, as would be the phi phenomenon.

Heymans and Brugmann (1913) used a fourth type of perseveration, namely *motor perseveration*. They compared speed of writing with speed of writing in reverse, on the assumption that high perseverators would be penalized on the second of these tasks. Wiersma (1906) had used tests of sensory perseveration, and although he did not actually use correlations between tests in his work, this fault was remedied by Heymans and Brugmann, who used six tests, including two of motor perseveration, in addition to tests of sensory perseveration. This is the first table of intercorrelations between objective laboratory tests used as measures of personality, and Heymans and Brugmann also linked up their experimental results with independent ratings of primary and secondary function and found that "we can already go so far as to say that it will be possible to use the after-effects of sensation as a reliable measure of the degree of development of the secondary function" (p. 329).

Spearman (1927) used this work to frame his general law of inertia, which reads: "Cognitive processes always both begin and cease more gradually than their (apparent) causes" (p. 412). He also influenced many of his students to carry on experimental tests of perseveration, thus making a beginning of the work in this field of the so-called London School. A thorough review of all this work is given by H. J. Eysenck (1970c); this would not be the right place to repeat such a review. We can merely state that although there were many suggestive and interesting results, the reliability of the tests used and the accuracy of the personality measurement of extraversion–introversion were never of a sufficiently high order to enable investigators to arrive at agreed results sufficiently clear-cut to make these tests acceptable. Nevertheless they are of historical interest in being the first real attempt to measure personality objectively. From all this work, Heymans emerges as a true pioneer, having been the first to use ratings of personality and objective laboratory measures of personality, the first to use correlations to define factors, and the first to link all this work together with an underlying general theory, namely, that of primary and secondary function. His reward has been a universal disregard by textbook writers and later workers in the field, with the solitary exception of Spearman, whose studies in Leipzig acquainted him with the work of German-speaking psychologists.

In America there was considerable interest in the work of Jung on extraversion–introversion, and many writers made use of his concepts (White, 1916; Wells, 1917; Tansley, 1920, McDougall, 1921; Nicoll, 1921; and many more). Some writers, such as Conklin (1923) and Freyd (1924), made up questionnaires to suit their theoretical convenience; Heidbreder (1926) continued the work of Freyd and Conklin (1927) and developed his "*E–I* interest ratio" on the basis of avocational interests. Other makers of questionnaires were Travis (1925), Neymann and Kohlstedt (1939), and Gilliland and Morgan (1931); a good review of this early work is given by Browne (1971). In addition to questionnaires, rating scales were also devised, such as those of Laird (1925) and Marston (1925). The most successful was that of Heidbreder (1926).

It cannot be said that these early efforts to measure extraversion-introversion and/or neuroticism were very successful. As Vernon (1938) points out after a very thorough review of the early literature:

> The attempts to classify test items or symptoms logically into distinct groups has not, we must admit, been successful. On the one hand, it is found that tests of presumably different traits intercorrelate very highly; on the other hand, different tests of nominally the same trait ... tend to give very poor correlations with one another. It is doubtful, then, whether most of the traits at which the tests are directed are unitary and discrete. (p. 53)

This overlap between hypothetically different traits is most apparent in attempts to measure neuroticism and introversion–extraversion. Vernon quotes the results of 40 experiments showing that the average correlation between different introversion tests and the average correlation between introversion and psychoneurotic tendency tests are practically identical, namely, $+.36 \pm .10$. Tests of inferiority feelings also agree quite closely with tests of introversion.

Somewhat later, Fiedler, Lodge, Jones, and Hutchings (1958) reported, in relation to tests of adjustment:

> This paper has presented intercorrelations among a variety of indices which have been used as measures of personality adjustment. The most important factor which emerges is the general lack of correlation among different indices—even among those which are reliably measurable and which could be expected to correlate with each other. Thus our data yield no evidence justifying the assumption that adjustment in its present state of definition should be considered a unitary trait in clinically unselected populations. (p. 350)

This failure to find agreement is largely due to the failure to use psychometric techniques such as factor analysis in the construction of tests; clearly commonsense and untested theory will not do! Also important was the confusion between Jung's and Freud's conceptions of introversion, as already noted.

While American psychologists indulged in this odd game of "hunt the slipper," the Spearman school in London pursued a rather more systematic path to the discovery of the major dimensions of personality. This work represents a continuation of the Heymans and Wiersma studies and bears the same hallmarks as did theirs. Perhaps the most important of these studies, which are reviewed in detail by H. J. Eysenck (1970c), has been the work of Edward Webb (1915), who was the first to use a method of factor analysis in the nonintellectual field. Although the statistics which he used are far from adequate by modern standards, they are definitely superior to those used by Heymans and Wiersma. From several points of view Webb's research, which was carried on under the guidance of Spearman himself, is methodologically superior to much work that has followed in later years. This judgment is supported by the fact that his table of intercorrelations has been subjected to analysis by more powerful modern methods by a variety of later students. Some of these reanalyses will be described in detail later, as they are essential to a complete understanding of Webb's contribution.

The subjects of his enquiry were two groups of 98 and 96 students respectively, and four groups of schoolboys of an average age of 12, numbering 140 in all. Assessments were made by at least two judges working as independently of each other as possible. These judges were in a position to make observations of their subjects under conditions relatively free from restraint. The subjects were made available for observation under a wide range of environmental conditions—in the lecture room, common room, the social gathering, the playing-field, at home, during holidays. Subjects throughout were unaware that assessments were being made concerning them. Ratings were carried out by fellow students of the subjects in the two experimental student groups and by two class masters in the case of the children. Thirty-nine traits were rated in the case of the students and 25 in the case of the schoolboys, grouped under the headings of "Emotions," "Self-qualities," "Sociability," "Activity," and "Intellect." Tests of intelligence were also administered and estimates of physique and records of examination ability obtained.

Reliability of the ratings varied between .5 and .7 on the average, with one or two occasional values dropping below the former or rising above the latter. Average reliability of ratings retained for the calculations was .55, and product moment correlations were calculated between the averaged ratings. In addition to the raw correlations between all the traits, tables in which the correlations were corrected for attenuation are also presented.

Using Spearman's well-known method of intercolumnar correlations and tetrad differences, Webb proceeded to extract a general factor of intelligence, which was based primarily on the tests administered to the subject but which correlated quite highly also with ratings on such items as quickness of apprehension, profoundness of apprehension, originality of ideas, and power of getting through mental work rapidly.

These correlations range from .5 to .6; there is also a correlation of .67 between intelligence tests and examination ability. These data, which Webb considers as supporting Spearman's theory of *g* (general intelligence), are of no great interest here, although the correlation between ratings and test results may perhaps be considered evidence of validity of the ratings. Webb goes on, however, to show that his correlations cannot be accounted for entirely in terms of this general factor and shows that a second factor, independent of intelligence, can be extracted from the intercorrelations of the data. He is therefore led to put forward the hypothesis "that a second factor, of wide generality, exists; and that this factor is prominent on the 'character' side of mental activity (as distinguished from the purely intellective side)" (p. 33). This factor he considers to be in some close relation to "persistence of motives." He goes on to say, "This conception may be understood to mean *consistency of action resulting from deliberate volition or will*. For convenience, we will in future represent the general factor by the symbol *w*" (p. 34).

The traits which characterize the person possessing a high degree of *w* are: tendency not to abandon tasks from mere changeability, tendency not to abandon task in fact of obstacles, kindness on principle, trustworthiness, conscientiousness, and perseverance in face of obstacles.

This *w*, or will factor, appears in many ways to be the opposite of Heymans and Wiersma's *emotionality*. In part it may perhaps represent a halo effect in the sense that it has often been shown that judges tend to group favorable qualities together because of their general like or dislike of the subject under investigation (Flemming, 1942). We will find ample evidence later, however, that this factor cannot be explained away entirely in terms of errors of rating, and there is no doubt that Webb in his study has made a significant contribution to the development of psychology. In many ways, his study is typical of what has become known as the London school; as Mabille (1951) puts it:

> La caractéristique de l'école anglaise moderne semble bien être d'équilibrer harmonieusement les conceptions théoriques et les points de vue expérimentaux, les nécessités cliniques et les exigences scientifiques de la statistique. (p. 36)

The first reanalysis of Webb's material, and the only one to make a genuine contribution to its understanding, was carried out by Garnett in 1918. His paper, which in its statistical development anticipates much of what was to become important later, such as the geometrical representation of patterns of correlations in terms of scalar products and the rotation of factor axes, used these methods to show that in addition to g and w another factor was contained in Webb's table of intercorrelations. This he called c because he conceived of this factor as being characterized by the trait of *cleverness;* seldom can brilliant mathematical treatment have resulted in a less appropriate naming of the factor discovered!

This will become apparent when we study the traits characteristic of c, in both its positive and its negative aspects. On the positive side we have traits like cheerfulness, aesthetic feeling, sense of humor, desire to excel, desire to impose one's will, desire to be liked by one's associates, impulsive kindness, wideness of influence, and quickness of apprehension; on the negative side we have liability to extreme depression, unsociableness, lack of corporate spirit, tactlessness, little bodily activity, and pure-mindedness. This factor in many ways resembles the primary and secondary function of Heymans and Wiersma or Jung's extraversion-introversion and thus brings into fair agreement the results obtained by all the investigators mentioned so far.

Another reanalysis of Webb's data has been carried out by McCloy (1936), using the multiple-factor technique; Reyburn and Taylor (1939) also reanalysed Webb's data, using only 19 traits from Webb's student sample. This analysis is probably the best that has been carried out on Webb's data and, interestingly enough, it shows the closest similarity to the Heymans and Wiersma data, as well as to a great deal of work reported elsewhere in this book.

Other students of Spearman's continued the Heymans and Wiersma tradition of using objective laboratory tests of personality. The first of these studies were carried out by Oates (1929), Line and Griffin (1935), and others; these studies are reviewed in H. J. Eysenck (1970c). They suggested the possibility of measuring the major dimensions of personality objectively, although by modern standards the data are of course inadequate to prove the point.

From here on the British and the American work begins to coalesce, partly due to the fact that some British workers like R. B. Cattell left the country and went to America; his wide-ranging studies will be referred to later. In America it was largely Guilford who inaugurated the factor analytic study of personality, although Thurstone and others had done some pioneering work previously. All these studies have been reviewed in detail by H. J. Eysenck (1970c), and we will not repeat the summary here. In a later chapter we will attempt to show to what degree all the more recent large-scale work on personality has come together to demonstrate the major importance of the N and E personality dimensions in this field. In this chapter we will now conclude the historical survey by looking at the third major dimension, namely psychoticism.

PSYCHOTICISM AS A DIMENSION OF PERSONALITY

Psychoticism has also frequently appeared in more recent descriptive systems, although usually under another name and often looked at from the positive rather than from the negative end; a detailed discussion of the relationship of psychoticism to such other labels as "superego functioning" and "impulse control" will come later (Chapter 4). Historically, the medieval concept of madness (perhaps allied with genius?) and Wernick's notion of *Einheitspsychose* might be considered somewhat remote ancestors of the modern diathesis–stress type of theory. Essentially, as H. J. Eysenck (1970a) has pointed out, the existence of psychotic disorders faces us with two questions. In the first place, we must decide whether, as Kretschmer (1948) surmised, there is a continuum from normal to psychotic or whether schizophrenia, manic-depressive illness, and other functional disorders are really states which are qualitatively different from normality. If the decision were to be that a continuum existed from normal to psychotic, the second question would be whether this continuum was *collinear* (identical) with that from normality to neurosis, as Freud surmised, with neurotic disorders being intermediate between normality and psychosis (continuum of *regression*) or whether two separate continua are required.

The question of continuity is a difficult one to answer; H. J. Eysenck (1950) suggested the method of criterion analysis for the purpose and later applied it to psychotic illnesses (H. J. Eysenck, 1952a). This method contrasts the *continuum* theory and the *categorical* theory, that is, the view that psychosis is categorically different from normality, and makes deductions from these theories on the kinds of intercorrelations to be found between tests which empirically differentiate between normality and psychosis. The method is a powerful one, and it unambiguously favored the continuum theory.

Other studies (H. J. Eysenck, 1955b, S. B. G. Eysenck, 1956) equally unambiguously demonstrated that neuroticism and psychoticism were independent dimensions. A detailed discussion of the evidence is given elsewhere (H. J. Eysenck, 1970c); it is based on many studies using different statistical and methodological approaches but agreeing on the final verdict of *continuity* between normality and psychosis and *separateness* of neurotic and psychotic types of abnormality. This view is in good agreement with many psychiatric investigations which also attack a third question, namely, that of the *specificity* of psychotic disorders. If schizophrenia and manic-depressive disorder, for instance, were found to be entirely separate disorders, then the use of the term *psychosis* to embrace all the so-called functional disorders would obviously be inappropriate, and equally inappropriate would be any claim for a psychoticism continuum. This problem has been discussed elsewhere by H. J. Eysenck (1972a).

First of all, let us consider certain genetic consequences of the theory of genetic specificity of schizophrenia and contrast them with the consequences expected in terms of the theory of a general factor of psychoticism. Consider, for example, Ödegard's (1963) report on 202 successive patients with psychotic disorders; he found, among the first-degree relatives of his schizophrenics, 45 schizophrenics, which is expected on the specificity theory, but also 40 nonschizophrenic psychoses, which is unexpected and clearly counter to the theory. Already Rüdin (1916) had reported that among the parents of schizophrenic patients there were to be found many manic-depressive patients; and Schulz (1940), who studied 55 couples with affective psychosis, found among their children a *higher* incidence of schizophrenia than that reported in all children of parents one of whom was in fact schizophrenic! Misdiagnosis and differential fertility have been suggested as possible reasons for part of these findings.

Data from twins tend to support schizophrenic specificity (Kringlen, 1967). Planansky points out, "the twin partner of a typical schizophrenic may develop practically any clinical type of schizophrenic psychosis.... No documented case of a non-schizophrenic psychosis has been reported" (Planansky, 1966a, p. 322). Atypical psychoses, as studied in detail by Kraulis (1939), led him to the conclusion that when schizophrenia, manic-depressive illness, and atypical psychoses were concentrated in certain families these psychoses might be considered part of a *single, independent syndrome*. These studies and many others of a similar kind reviewed by Planansky (1972) suggest the existence both of specificity and of generality; neither theory by itself is supported very strongly, and any realistic model requires that one take into account both mechanisms.

The families of schizophrenics show not only an increase in the probability of psychotic, particularly schizophrenic, disorder but also an increased incidence of minor psychosocial defects. Ödegard (1963) reported that relatives of psychotic probands were classified as psychopaths, criminals, or alcoholics in some 10% of all cases. Planansky (1972) comments:

> It might be tempting to interpret the uniformity as an expression of a biological homogeneity, invoking, perhaps, a polygenic hypothesis. However, a concept of single genetic mechanism underlying all endogenous psychoses and their associated disorders would lead to the ancient doctrine of single mental disease. Such concept would render the search for genetic specificity redundant, but would also preclude all discovery. (p. 562)

As we shall see, a proper model based on all the facts need not have such calamitous consequences. Of particular interest in this connection is the close relationship between schizophrenic disorder and psychopathic behavior:

> The psychopathic personalities are the most persistently reported group among close relatives of schizophrenics; and certain forms of these vaguely defined disorders appear to be, not only structurally, but also developmentally, connected with schizophrenic psychosis. (Planansky, 1972, p. 558)

Planansky traces the history of this association from Kahlbaum through Kraepelin to Delay and Schafer and summarizes the empirical literature by saying, "There is an abundance of reports concerning incidence of schizoid psychopathic personality in families of schizophrenic probands" (p. 552). Mostly, these studies have started from the psychotic proband (Essen-Möller, 1946; Planansky, 1966a), but of equal interest are studies starting at the other end, using psychopathic probands (Meggendorfer, 1921; Riedel, 1937; Stumpfl, 1935). These studies have dealt with so-called schizoid psychopathic personalities; however, there is also a considerable agglomeration of nonschizoid psychopathic personalities in relatives of schizophrenics (Rüdin, 1916; Medow, 1914). The most important study in this field is Heston's (1960), in which children of schizophrenic mothers were removed immediately after birth and raised by foster parents. Of 47 children, 9 were diagnosed to be sociopathic personalities with antisocial behavior of an impulsive kind and with long police records. Only 4 of these 47 children developed schizophrenia, demonstrating that the incidence of psychotic and nonpsychotic abnormalities is very high in the progeny of schizophrenics when direct environmental determination is ruled out. (It should perhaps be added that Heston found other behavioral abnormalities in some 20 additional cases.)

Planansky (1972) summarizes his extensive review by saying:

> Thus, the search for the boundaries of schizophrenic phenotype focused on a minimal but still specific single trait, derived either from the clinical picture of the psychosis, or from a theoretical construct of essential psychopathology, may be in vain. Selection of pathognomic functions for the detection of minimal, specific psychopathology is difficult, considering that even in frank, acute psychotic states, disruption of conceptual thinking, considered exclusive for schizophrenia, was also found in patients with non-schizophrenic diagnoses as well as in some neurotics. (p. 570)

This conclusion is strongly supported by experimental studies on overinclusiveness of thinking and retardation (Payne & Hewlett, 1960); retardation tests discriminate between schizophrenics and manic-depressives on the one hand and between neurotics and normals on the other, suggesting an experimental basis for the general factor of psychoticism; overinclusiveness strongly characterized some schizophrenics, but by no means all. As the authors (Payne & Hewlett, 1960) say, "Perhaps the most striking fact in the present study is the heterogeneity of the schizophrenic group." These two findings are supported by many other experimental studies (H. J. Eysenck, 1973a).

It is interesting that neurotic disorders, very much as required by our dimensional hypothesis, are *not* imbricated genetically with schizophrenia to any considerable extent; as Planansky points out, the "ordinary symptom neuroses seem to occur rarely in the parents of schizophrenics (Delay *et al.*, 1957; Alanen, 1966)." Cowie (1961) found no increase in neurotic disorder over the normal in the children of psychotics. These studies, as well as the experimental ones of H. J. Eysenck (1952a,b, 1955b) and S. B. G. Eysenck (1956), leave little doubt about the orthogonality of the neurotic and psychotic dimensions.

If we must conclude from this hurried survey of empirical studies that the evidence favors a general factor of psychoticism, rather than sharply segregated schizophrenic and manic-depressive illnesses, we must add that there is quite strong evidence for the existence of more sharply demarcated subtypes within the schizophrenic diagnosis. Experimental studies such as those of Payne and Hewlett (1960) carry perhaps most conviction, but genetic evidence is also strong (Ödegard, 1963; Kringlen, 1967; Rosenthal, 1963; Slater, 1947, 1953); curiously enough, this does not seem to extend to the level of primary symptoms, for example, thought disorder. Perhaps a combination of the experimental and the genetic method would give more convincing evidence; thought disorder as clinically assessed tends to be an all-or-none affair rather than a quantitatively graded characteristic, and the inclusion of tests of overinclusiveness and other modes of thought disorder might help to objectify the assessment.

Our concept of psychoticism probably has most similarity to that of *unspecific vulnerability* of Weiner and Stromgren (1958). Their data and those of Faergemann (1963) are well in accord with such a notion of a general factor, predisposing persons to psychosis in varying degree and inherited as a polygenic character; this predisposition would extend into the psychopathic and criminal, antisocial field, but not into that of the dysthymic neuroses. However, it seems likely that this modern version of the *Einheitspsychose* theory may have to be supplemented by a theory of specific (major) genes giving rise to special subvarieties of psychotic behavior. The evidence does not make it entirely clear whether such subvarieties would be coextensive with such broad categories as schizophrenia and endogenous depression, or whether we would postulate rather a larger number of more sharply delineated categories normally subsumed within these great groups of psychotic disorders. Our own preference, based on the existing evidence, is in favor of the latter alternative; but clearly in the absence of genetic research specifically directed at the solution of this problem it is impossible to come to any firmly based conclusion. What we think can be said with some confidence is that a general factor of psychoticism, quantitatively varied and strictly independent from the general factor of neuroticism, reconciles many of the reported findings, both from the experimental and from the genetic literature, and is not essentially contradicted by any psychiatric findings yet recorded (H. J. Eysenck & S. B. G. Eysenck, 1976).

It might be retorted that the recently established notions of reactive and process types of schizophrenia (translations of the rather older notions of malignant and benign schizophrenia) seem to contradict these conclusions. This is not so; there appears to be a marked correlation between this particular continuum (for it is difficult to accept the notion of a categorical distinction) and that of extraversion–introversion. Critics of the dimensional system often fail to realize that, in dealing with a particular axis or factor, all other axes or factors must also be taken into consideration. In saying that two people are both six feet tall we do not imply that they are also equal in weight or hirsuteness or drinking habits; similarly, two people equal in psychoticism are not therefore equal in neuroticism, extraversion–introversion, or intelligence. These orthogonal factors may determine, to a very marked extent, the particular *expression* of a person's degree of psychoticism. In any case, as Planansky (1972) points out:

> Since no unequivocal correlations could be established to support a genetic division of schizophrenia in relation to the type of onset and characteristics of the preclinical phase, the 'reactive-process' classification would hardly describe two biologically different groups, although it may be useful as an empirical means for the prediction of the length of stay in hospital. (p. 548)

It will be seen from all the studies summarized so far that there is good evidence for the existence of a continuum from normal, through criminal, psychopathic, alcoholic, and drug addictive behavior generally, to schizoid and finally entirely psychotic states. Such a hypothesis was put forward first by H. J. Eysenck (1952b), has been elaborated in *Psychoticism as a Dimension of Personality* (H. J. Eysenck & S. B. G. Eysenck, 1976), and embodied in the form of a questionnaire (the Eysenck Personality Questionnaire; H. J. Eysenck & S. B. G. Eysenck, 1975).

Fig. 11 illustrates the hypothesis of such a continuum, which is represented as the abscissa. The large normal curve represents the general population, the small normal curve on the extreme right represents schizophrenic patients. The curved line marked P represents the probability of any person at a given place on the abscissa developing schizophrenia. The abscissa thus measures degrees of psychoticism, from low (on the left) to high (on the right).

How can we tell that the P scale actually measures psychoticism? Many people, Davis (1974), Bishop (1977), Block (1977a,b), and Claridge (1983), for instance, have doubted whether the scale actually measures the diathesis related to psychosis and have suggested alternative interpretations, for example, psychopathy or paranoia.

The answer may lie in a particular variant of the method of criterion analysis which has been called the *proportionality criterion* (H. J. Eysenck, 1983c). This may be put in the form of an equation, which states that if the P scale actually measures psychoticism, then any objective test that discriminates between psychotics and normals should also discriminate between high and low P scorers. In other words, psychosis:normality = high P:low P scorers. The work of Claridge may serve as an example. The theory he suggested originally (Claridge, 1967) was based on the idea that psychosis does not involve a simple shift in, say, emotional arousal but represents instead a much more complex dissociation of CNS activity. He suggested that in the schizophrenic phys-

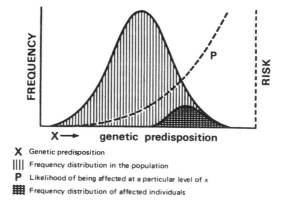

X   Genetic predisposition
|||| Frequency distribution in the population
P   Likelihood of being affected at a particular level of x
▦   Frequency distribution of affected individuals

FIGURE 11. Diagrammatic representation of the diathesis-stress model of psychosis. The abscissa denotes the continuum representing predisposition; the curve marked P denotes probability of breakdown. The large normal curve denotes the distribution of the diathesis in the population, and the small normal curve shows the population of actual psychotic patients.

iological mechanisms that are normally congruent in their activity and thereby maintain integrated CNS function become uncoupled and dissociated. He concentrated on two aspects of central nervous functioning which he considered to be particularly involved in this uncoupling process: *emotional arousal* and a mechanism concerned with the *regulation of sensory input,* including variations in perceptual sensitivity and in the broadening and narrowing of attention. He called this the "phenomenon of reversed covariation" (Claridge, 1981) and the many studies reviewed in this last reference (e.g., Claridge, 1972; Claridge & Chappa, 1973; Claridge & Birchall, 1978; Robinson & Zahn, 1979; Venables, 1963) give ample support for the hypothesis.

What is notable is that this is not just a theory of schizophrenia, but also a theory of psychoticism; normal high P scorers behave like schizophrenics, low P scorers like normals in these various tests, thus supporting very strongly the identification of the factor as one of psychoticism. At the same time Claridge's work supports the drug postulates, (H. J. Eysenck, 1963) which form part of the general P-E-N paradigm. According to these postulates, psychotropic drug actions are collinear with the personality dimensions and act in such a way as to shift temporarily a person's position on these axes in predictable directions. Thus LSD-25 would be classed as a hallucinogen and should in normal subjects produce psychoticizing effects on the Claridge type of test. Claridge (1972) and Claridge & Clark (1982) have provided good evidence for this. Thus Claridge has given a most impressive theoretical account linking the descriptive dimension P with a causal physiological hypothesis.

The work of Claridge just cited is one example; another is the work of Gattaz (1981) and Gattaz and Seitz (1984). This is concerned with a human leukocyte antigen (HLA B-27), which is found significantly more frequently in schizophrenic patients than in healthy controls (Gattaz, Ewald, & Beckmann, 1980; McGuffin, Farmer, & Yonace, 1981). If P does indeed measure a diathesis relevant to psychosis, and particularly schizophrenia, then we would expect that HLA B-27 would be found significantly more frequently in schizophrenics with high P scores as compared with schizophrenics with low P scores, and in normals with high P scores as compared with normals with low P scores. Both these deductions have been verified by Gattaz, and thus constitute strong support for the hypothesis. The work of Gattaz and Claridge does not constitute the only evidence available to strengthen the hypothesis that P is truly dealing with psychoticism rather than with psychopathy or paranoia; this is not the place to go into this question. We merely want to draw attention to the need for some such use of criterion analysis in order to establish the interpretation of factors in an objective manner, and to suggest that the proportionality criterion may be a form of criterion analysis particularly useful in this connection (see also Iacomo & Lykken, 1979).

Of equal interest are deductions from the genetic hypothesis of the psychotic *Erbkreis*. If the *P* scale does indeed measure psychotic diathesis, then we should be able to look at first-degree relatives of schizophrenics compared, say, to a control sample of first-degree relatives of neurotics and predict that the former group should show elevated *P* scores as compared to the latter. One would also expect that measures of Claridge's "phenomenon of reversed covariation" would show similar differences between relatives of psychotic and neurotic probands. Results in support of these hypotheses have been reported by Claridge, Robinson, and Birchall (1983), and it is particularly interesting that the rather unusual pattern of psychophysiological responses "was especially evident in a small subgroup of schizophrenics' relatives whose personality profiles tend to differ in the predictable direction, towards greater psychoticism." Findings such as these powerfully reinforce the interpretation of psychoticism as being truly related to psychosis.

IMPULSIVENESS AND SENSATION SEEKING: A SPECIAL CASE

As shown in this chapter, the development of *E* and *N* preceded the development of *P*, both as concepts and by way of scale construction. It would be idle to pretend that the addition of a new concept (and scale) to already existing concepts (and scales) could be done without producing certain changes. As Rocklin and Revelle (1981) have pointed out, there has been a tendency for some impulsiveness items to be dropped from the *E* scale in the EPI and to appear on the *P* scale of the EPQ. They insist that:

> The distinction between impulsivity and sociability is important. Although psychometric methods will probably never settle the issue of whether extraversion is best thought of as a single construct or a mixture of impulsivity and sociability, experimental methods have shown that the two components of extraversion have entirely different patterns of results in a variety of paradigms. (p. 282)

As we shall see, there is some truth in the assertion that sociability and impulsivity give different correlations with some experimental variables, such as conditioning (H. J. Eysenck & Levey, 1972). Nevertheless, we feel that the statements made by Rocklin and Revelle are exaggerating the effects of whatever change has taken place and that indeed the very terminology carries implications that cannot be supported empirically.

Kline and Barrett (1983), commenting on the correlations between EPI and EPQ N and E factors, as reported by Rocklin and Revelle, consider the assumption by Eysenck and Eysenck (1975) that the E and N of the new test are identical with the previous factors. They conclude:

> Given the reliability of the tests, the assumption of identity is well-founded. This is, perhaps, slightly surprising in view of the lack of impulsivity items in the E scale of the EPQ, but there can be no gainsaying the result. Thus the EPQ factors, we argue, represent Eysenck's factors. (p. 160)

The actual correlations between EPI extraversion and EPQ extraversion was 0.74 and that between EPI neuroticism and EPQ neuroticism was 0.83. Campbell and Reynolds (1982) reported even higher correlations of .80 and .87; these are as high as the reliabilities of the scales.

In actual fact, the problem is of course much more complex than simply a consideration of whether impulsivity goes with E or with P. There are two complicating factors. The first is that impulsivity can itself be subdivided into factors, so that it becomes doubtful whether we can talk about impulsivity as a simple unitary concept. In the second place, the subfactors of impulsivity cannot be said simply to correlate with either P or E; what happens is that they correlate with both, to varying degree. Thus we cannot say that impulsivity has been changed over from E to P altogether; what has happened, rather, is that some former impulsivity items were found to correlate more highly with P than with E and have transferred, while others correlate more highly with E than with P and have been retained. The full story is told in a series of articles by S. B. G. Eysenck and H. J. Eysenck (1977, 1978) for adults, and by S. B. G. Eysenck and H. J. Eysenck (1980) and S. B. G. Eysenck (1981) in another set of papers dealing with children. In addition, papers by Eaves, Martin, and Eysenck (1977) and by H. J. Eysenck (1983a) deal with the genetic and environmental factors determining impulsivity, sensation seeking, and their relation to the major personality dimensions.

In the first of these papers, H. J. Eysenck and S. B. G. Eysenck (1977) carried out factor analyses of sets of items traditionally used to measure impulsiveness. It was found that impulsiveness in the broad sense breaks down into four factors: narrow impulsiveness, risk taking, nonplanning, and liveliness; these factors were found replicable from sample to sample and from males to females. These factors were positively correlated with each other and also with sociability to varying degrees. Broad impulsivity, that is, the sum of all the four factors, correlates quite well with extraversion, but even better with psychoticism. Narrow impulsivity correlates positively with N and P, suggesting that this trait is somewhat pathological. Risk taking shows a clear relationship with extraversion and an almost equally clear one with P. Nonplanning is positively related to P and negatively to N, and there is no obvious relationship with E. Liveliness shows clear correlations with E (positive) and with N (negative); it does not appear to correlate at all with psychoticism.

The correlations between the four impulsiveness factors are all positive for both men and women, but they are not very high, averaging only about .3. Table 3 gives the actual correlations, and it will be seen that to talk about impulsivity as a general factor is decidedly dangerous, as the four subfactors share only about 10% common variance. Clearly some aspects of impulsivity, such as liveliness, should be associated with extraversion; others such as impulsivity in the narrow sense, with psychoticism. Thus it can be seen that the criticism by Rocklin and Revelle (1981) is phrased in too simple a fashion; we cannot just say that impulsivity has gone over from E (in the older questionnaire) to P (in the newer questionnaire). We have to look at each item in turn, consider to which of the four subfactors of impulsivity it belongs, and judge its position in terms of its overall loadings on the three superfactors. When this is done, it will be seen that Kline and Barrett are right in supporting the view that extraversion as a factor has remained essentially unchanged (but see Campbell & Reynolds, in press).

The investigation of the various subfactors of impulsivity raised the question of their relationship to the four factors of *sensation seeking* isolated by Zuckerman (1979b). In pursuing the concept of sensation seeking, Zuckerman discovered in factor analyses of the items used that there were four major factors called by him thrill and adventure seeking (TAS), experience seeking (ES), disinhibition (DIS), and boredom susceptibility (BS). The title "thrill and adventure seeking" is descriptive of the contents of this factor. Experience seeking appears to involve the

TABLE 3. Intercorrelations between Four Impulsiveness Scales:
Males above, Females below Leading Diagonal

	1	2	3	4
1. Narrow impulsiveness		.43	.32	.22
2. Risk taking	.45		.26	.18
3. Nonplanning	.50	.52		.22
4. Liveliness	.33	.22	.24	

Note. From "The Place of Impulsiveness in a Dimensional System of Personality Description" by S. B. G. Eysenck and H. J. Eysenck, *British Journal of Social and Clinical Psychology*, 1977, *16*, 57–68. Copyright 1977 by the *British Journal of Social and Clinical Psychology*. Reprinted by permission.

TABLE 4. Correlations between Four Sensation-seeking Scales for Males (above Leading Diagonal) and Females (below Leading Diagonal); English and American Samples

	English				American			
	TAS	ES	Dis	BS	TAS	ES	Dis	BS
TAS		.27	.25	.10		.27	.15	.06
ES	.42		.32	.21	.39		.24	.26
Dis	.35	.47		.42	29	.40		.37
BS	.20	.29	.48		.18	.37	.40	

Note. From *Sensation Seeking: Beyond the Optimal Level of Arousal* by M. Zuckerman. London: Wiley, 1979. Copyright 1977 by Lawrence Erlbaum. Reprinted by permission.

seeking of arousal through the mind and senses, through nonconforming life-style, loosely called "hippy" in the 1960s, and through spontaneous, unplanned travel. The disinhibition factor appears to describe a more traditional type of sensation-seeking, which seeks release and social disinhibition through drinking, partying, gambling, and sex. And finally boredom susceptibility was clearly defined in males by items reflecting an aversion for repetitive experiences of any kind, or routine work, or dull and boring people, and extreme restlessness under conditions when escape from constancy is impossible. Correlations between these four factors are slightly but not much higher than those between the four impulsiveness factors (see Table 4), and the descriptions given by Zuckerman indicate that there are some relationships conceptually with the impulsivity factors.

Intercorrelations between the eight impulsivity and sensation-seeking scales and P, E, N, and L are given in H. J. Eysenck (1983a). It would clearly be quite wrong to think of impulsiveness as such, or sensation seeking as such, as belonging either to P or to E. What emerges is a complex pattern, in which all the eight scales correlate to some extent with both P and E. Nonplanning, one of the four impulsivity scales, is almost wholly P, whereas liveliness, another one of the impulsivity scales, is almost wholly E. Disinhibition and thrill and adventure seeking are slightly more E than P, whereas two other sensation-seeking scales, boredom susceptibility and experience seeking, are more P than E. The pattern is just a very complex one, and what has happened, obviously, is that these investigations serve to elucidate in considerable detail that octant of the three-dimensional space which relates the positive aspects of P, N, and E. We thus gain a more thorough insight into the nature of a person located in this particular octant. It is possible to construct scales for a more general conception of impulsivity or sensation seeking (called venturesomeness by S. G. B. Eysenck & H. J. Eysenck, 1978) by omitting items and subfactors which show the least relationship with the others, and for certain purposes (such as the relationship between personality and criminality; S. B. G. Eysenck & McGurk, 1980), such scales may be useful. But it should be realized that the traits in question are not unitary in the psychometric sense but are compounds, just as extraversion or psychoticism or neuroticism are compounds.

These new scales of impulsiveness and venturesomeness show interesting correlations with P and E, as does an additional trait, empathy, also measured in this investigation. Table 5 shows the results, for a population of 641 delinquent subjects and 402 normal controls. It will be seen that impulsiveness correlates significantly more highly with P than with E and venturesomeness more highly with E than with P. In addition, impulsiveness correlates positively with N, whereas venturesomeness correlates negatively with N. Impulsiveness and venturesomeness correlate positively for the two populations (.30 for delinquents, .41 for normals). Empathy correlates negatively with psychoticism for delinquents only and positively with neuroticism with both delinquents and normals.

These results do not tell us all we would like to know about the relationships of all these variables with each other; they merely outline very broadly the kind of task that remains to be done in understanding the relationship between the major dimensions of personality, so-called primary factors like impulsivity and venturesomeness and the various subfactors into which these can be split. In doing this, we find some rather unexpected results. One might have expected, for instance, that the four subscales of sensation seeking or venturesomeness would cor-

TABLE 5. Correlation of Impulsiveness and Venturesomeness
with P, E, N, and L

	Impulsiveness		Venturesomeness	
	Delinquents	Controls	Delinquents	Controls
P	.47	.52	.13	.33
E	.12	.39	.38	.46
N	.24	.38	−.16	−.10
L	−.45	−.43	−.28	−.22

Note. From "Impulsiveness and Venturesomeness in a Detention Center Population" by S. B. G. Eysenck and B. J. McGurk, *Psychological Reports*, 1980, 47, 1299–1306. Copyright 1980 by the American Psychological Association. Reprinted by permission.

relate more highly together than they would with any of the impulsivity scales and vice versa. However, this is not true. Risk taking, for instance, one of the impulsivity scales, correlates more highly with thrill and adventure seeking, one of the sensation-seeking scales, than it does with any of the other impulsivity scales. Similarly, experience seeking correlates more highly with nonplanning than it does with the other sensation-seeking scales. These findings thus pose many problems. In the usual discussions among factor analysts the distinction is usually made between primary factors (like impulsivity and sensation seeking) and higher-order factors like P, E, and N. But clearly primary factors are not really *primary* in the sense of not being themselves analyzable into constituent parts that are not independent of each other, but show a certain degree of independence and may correlate more highly with subfactors of other primary factors than with subfactors of their own primary factor. Nor can it be assumed that these subfactors are themselves not divisible. In fact, the situation appears very much like that in subatomic physics. It used to be assumed that the atom itself was indivisible. When Thompson and Rutherford found that the atom itself could be "smashed" and discovered electrons, protons, and other subatomic particles, it was at first assumed that these in turn were elementary building stones of matter. Now we know that this is not so, and literally hundreds of subatomic particles are known. The situation we find in psychology is therefore not unique, and it need not cause unnecessary depression. It does, however, suggest that it is the superfactors, the major dimensions of personality, which are relatively invariant and replicable and that it is the smaller factors which cause a considerable degree of difficulty in identification, naming, and replication. These difficulties should always be borne in mind when analyzing data in this field or trying to integrate reports in the literature into a consistent framework.

It seemed of interest to extend these investigations into the question of the genetic and environmental contribution to impulsivity, sensation seeking, the four subfactors into which each can be divided, and the superfactors P, E, and N (Eaves *et al.*, 1977; H. J. Eysenck, 1983a). The analyses are too technical to be recorded in detail here, but essentially it was found that for the superfactors, just as for primary factors like impulsivity and sensation seeking and for the various subfactors, a relatively simple model invoking nothing but additive genetic variation and specific environmental factors was sufficient to account for the observed data with considerable accuracy. There was no evidence for common environmental factors (between-family), so that environmental variation, here as elsewhere, is within-family rather than between-family.

The contribution of genetic factors, when we are dealing only with the true variance, that is, variance purged of errors of measurement, is very high, reaching almost the same level as the additive genetic variance in intelligence testing. The various scales of impulsivity and sensation seeking, therefore, behave very much as do the more general dimensions of personality; later in Chapter 3 we will discuss results from genetic analyses of personality scales in greater detail and explain to some extent the terms used here.

Given the complexities involved in the analysis of primary factors and subfactors, is this exercise really worthwhile, and what purpose does it serve? There are two distinct answers to this question, although they overlap to some degree. The first, of course, is that three dimensions are not sufficient to describe in detail the complexities presented by personality; the superfactors or dimensions account for the intercorrelations between traits, but these traits themselves establish the finer structure of the total space within which we locate personality. Each trait measurement, apart from error, contains two (or more) components, one of which contributes to the superfactor, the other being specific to the set of items involved. Thus we can say that some of the variance measured by a nonplanning inventory is taken up by P, but that leaves a reasonable proportion which measures something independent of P. Similarly, a reasonable proportion of a liveliness inventory is taken up by extraversion, but still a good deal is left which is peculiar to liveliness as a personality trait. Again, some part of the variance contributed by a disinhibition factor, or a risk-taking factor, is taken by P, some by E, but leaving still a separate measure of a residual of risk taking or disinhibition to be measured.

We may discern several levels in this hierarchy. The top level, which is obviously the most complex, being derived from intercorrelations between many traits, is that of the superfactors P, E, and N; this also is the one that has been easiest to replicate and to connect with physiological and other types of objective laboratory measures. At the bottom of the hierarchy we have traits like risk taking or boredom susceptibility, which are relatively pure and where it would be difficult to subdivide them again into subfactors. Our own advice, from the point of view of research and the establishment of an agreed taxonomy, would be to concentrate on these two levels.

Most research, unfortunately, seems to have concentrated on an intermediary level, such as that taken by sensation seeking or impulsiveness in the broad sense, or any of the many factors which we will discuss in a later chapter. These have neither the advantage of simplicity nor that of replicability; furthermore, they overlap with each other very much in the way that impulsivity and sensation seeking have been found to overlap with each other. The fact that in spite of over 50 years of work with factor analysis there is still little agreement on the major factors in this primary realm suggests that the task is an impossible one.

Given that P, E, and N are forever recurring in any large-scale study of personality (as will be shown in the next chapter), it would appear to be reasonable, in approaching any trait measurement whatsoever, to determine from the beginning the correlations of the hypothetical trait with P, E, and N and thus see to what extent there is any specific variance left for the postulated traits. Such a procedure would also at the same time guarantee that we would know a great deal about the position of that trait in the three-dimensional factor structure described in this book and would thus be able to relate it to other traits for which also their position in this superstructure was known. Only in this way can personality studies be made *additive*, as any scientific endeavor should be; in other words, it would be possible to proceed in a stepwise direction, rather than each investigator's setting out on his own and in a different direction from every other investigator. As we shall see in Chapter 4, the only link which unites the many different personality questionnaires that have been elaborated in recent years is their relationship to P, E, and N; in the absence of such anchors it would be impossible to translate findings from one individual's personality to that of another.

The first reason for investigating the finer structure of the interstices between the major dimensions of personality is thus to gain a better understanding of such subfactors or traits as contribute variance over and above that which is taken up by P, E, and N. The second reason is that when a particular type of behavior is investigated, whether observed in a social setting (like criminality) or in the laboratory (like Pavlovian conditioning), and which is theoretically and empirically correlated with a personality dimension (like E), it is not necessarily true that the correlation between this particular measure and all the different contributory traits to E will be equally high and positive. H. J. Eysenck and Levey (1972), as already mentioned, found that Pavlovian conditioning correlated more with the impulsivity items of the EPI than with the sociability items; this is an important finding, both theoretically and practically, and the fact that antisocial behavior (criminality) also appears to be more highly related to impulsivity than to sociability (H. J. Eysenck, 1977a) suggests possible links between conditioning and antisocial behavior, links which we will explore later on. It is sometimes said, in criticism, that the three-dimensional scheme which forms the subject matter of this book is inadequate to account for the whole of personality;

this is of course true, and we would not doubt such a statement in any way. The real question is how the remaining vacuum is to be filled, and the procedure here suggested appears to be the best way of doing so, while taking into account what has already been firmly established. In proceeding in this manner we should also take into account, of course, such correlations between external criteria and P, E, and N and their subfactors as have been mentioned. Only in this way will we be able to fill in the whole picture in much more detail than would be possible by simply looking at P, E, and N in isolation. This rather extended and detailed discussion of the relationship between impulsiveness, sensation seeking, and P, E, and N was introduced on purpose to illustrate the complexities of the issues involved and to answer some of the criticisms occasionally made of the three-dimensional system as claiming to be all-inclusive. No such claim has ever been made, and it is clearly recognized that an extension of the study to the smaller subfactors which contribute independent variance is an important method of increasing our understanding of personality in its descriptive aspects, as well as giving us important suggestions for the causal analysis of personality and temperament.

THE QUESTION OF VALIDITY

Validity is usually contrasted with reliability in psychological writings, the latter referring to the agreement between two measures of the same trait or ability administered either at the same time (*internal* reliability) or with a gap of several days, months, or years (*repeat* reliability). Repeat reliability soon merges into longitudinal traits, in particular P, E, and N show considerable longitudinal consistency. Internal reliability from some points of view is less important as it can always be raised by increasing the number of items.

Validity, on the other hand, refers to the much more important question of whether a given test actually measures what it is intended to measure. This is easy to establish when we have a criterion against which the test can be evaluated; unfortunately, in most psychologically interesting fields there is no criterion that is universally or even widely accepted. The absence of such a criterion raises many fundamental problems, and hence a few words will be said here about the validity of the P, E, and N scales in the hope of allaying some of the confusion and alerting the reader to the relevance of some of the discussions in this book.

The recommendations for APA test standards regarding construct, trait, and discriminant validity (Campbell, 1960) suggest a nomenclature which we shall follow. The most obvious type of validity is of course what is called *content* validity, that is, validity built into a test through the choice of appropriate items. As Anastasi (1982) points out, for aptitude and personality tests content validation is usually inappropriate and may in fact be misleading. We have already discussed this point in connection with the need to do correlational and factor analytic studies of items; inspection of such items, and *a priori* selection, does not insure that items apparently measuring certain factors do actually do so. Hence content validation and the allied concept of *face validity* are of little interest in connection with personality tests, although, of course, the apparent content of a test may cause us to name factors derived from the intercorrelations of such items in a certain manner. This fact is responsible for the dissatisfaction often felt with such names given to factors; being subjectively derived from impressions given by the overt content of a test, such names cannot confer validity on the test or on the factor. Other methods are required to do that.

Concurrent and *predictive* validation are much more important, being related to some form of criterion. If the test and the criterion are administered in close temporal succession, we deal with concurrent validation; if the criterion situation occurs much later than the testing situation, we speak of predictive validation. Most instances of concurrent and predictive validation deal with cases wherein there is a fairly obvious criterion, such as job performance and satisfaction. In other words, tests are often specifically constructed to measure or predict compound abilities; another example would be the use of intelligence tests to predict scholastic achievement. Uses of tests in education, industry, and other practical areas are socially important but may not be of great psychological interest. Often the criterion is in fact embodied in a *nomological network*, in which case it becomes, just as much as the test itself, part of a general theory. When that occurs we are dealing with what is usually called *construct validity*, to which we will turn in a minute.

Before we do so, however, we may with advantage look at one type of concurrent and predictive validity which might be called *consensual* (McCrae, 1982). By this term we mean simply that we validate questionnaire responses by correlating them with ratings made by external assessors who know the ratee well. This consensual validation can be both concurrent and predictive, and it is important in countering the suggestions of situationalists that such consistency of conduct as is found in questionnaires or ratings is due to artifacts of one kind or another. As McCrae (1982) has pointed out, ratings share with self-reports the use of a human observer who can interpret specific behaviors as evidence of underlying traits, but this observer is objective and free from the particular biases of self-report methods. Self-reports, on the other hand, possess the unique advantage of access to the private thoughts and fantasies of the individuals and are unlikely to be influenced by stereotypes. Since none of the usual artifacts are shared between self-reports and ratings, none of them account for agreement observed between the two sources. As Wiggins (1973) has argued, well-replicated agreement across a two-measurement tradition would thus constitute compelling evidence for agreement on the real dispositions and provide consensual validation of personality traits.

Shrauger and Schoeneman (1979) have recently published a review of agreement between self-reports and ratings from a symbolic interactionist perspective, with the conclusion that evidence in support of agreement between self and others was weak. However, such a term as *weak* has to be seen in the context of the quantitative results; and of the 36 correlational studies in their table, 17 showed clear support and an additional 7 showed at least mixed support for the agreement hypothesis, significant correlations ranging from .14 to .76, and 13 of the studies giving at least one correlation above .50. Edwards and Klockars (1981) responded to this review by conducting another study of the symbolic interactionist position, arguing that only the ratings of a *significant other* should be expected to correspond to self-ratings, as ratings given by persons not knowing the ratee well might be largely irrelevant. They concluded that their results provided consistent evidence of self—other agreement over a wide variety of traits and attributed the agreement to the choice of a knowledgeable and personally significant rater. Many other positive results are quoted by McCrae (1982), who cites a number of very positive comparisons and concludes that:

> One conclusion is indisputable: the agreement between self-reports and ratings is certainly high enough to overturn the notion that personality traits are pure fictions, and that the agreement between two persons on the characteristics of one of them is wholly illusory. (p. 302)

He also provides a study particularly relevant to the theory discussed in this book, as his experiment employed the NEO inventory and the NEO rating form; as explained in another place, the neuroticism–extraversion–openness theory posits three major dimensions of personality that are very similar to those of the *PEN* theory, openness possibly being a personality variable at the opposite extreme of the psychoticism dimension.

The sample consisted of 139 men and 142 women for whom both self-report and spouse rating forms were completed. This study constitutes what Campbell and Fiske (1959) call a "multitrait–multimethod" matrix, there being 18 traits (6 for each of the 3 major dimensions of personality) and 2 methods, namely self-report and rating. Correlations between self-report and spouse rating were .47 for neuroticism, .72 for extraversion, and .64 for openness. McCrae also published correlations between the self-reports for E and N and the EPI E and N scales; these correlations were .56 and .36. Thus there is here ample evidence for the conclusion that consensual validation of personality traits is possible and produces quite high values indicating consistency of behavior.

McCrae and Costa (1983b) added to this proof by publishing further evidence, using the NEO inventory and NEO rating form in which correlations between the 18 traits involved were factor analyzed to see whether identical superfactors would emerge. They did find that the hypothesized structure appeared both within and across self-report and spouse ratings and also found convergent and discriminant validity of the joint factors of the EPI scales. As they point out:

> The results suggest that the effects of method variance can be minimized if well-qualified raters use psychometrically adequate instruments to provide ratings of clearly conceptualised traits. In addition they provide strong evidence for the validity of the three domain model of personality. (p. 245)

A similar conclusion is reached by Amelang and Borkenau (1982), who extracted five factors from a variety of self-rating tests including the 16 PF, the Guilford scales, and the EPI; factor analysis of these showed a very similar factor structure to a set of ratings, with consensual coefficients for factors resembling P, E and N close to .6. We may conclude from all this evidence that as far as concurrent validity is concerned, consensual validation provides good evidence in respect of P, E, and N. As regards predictive validity, it will be seen in the section on longitudinal studies that correlations between ratings at one age and questionnaires at another also show consensual validity.

Additional evidence comes from work on the "Norman Five" major personality components (Norman, 1963, 1969; Norman & Goldberg, 1960). These five factors, replicated several times (e.g., Borgatta, 1964; Conley, 1984; Fiske, 1949; Smith, 1967; Types & Christal, 1961), appear equally in self-ratings and peer-ratings, and cross-correlations between methods have been reported of .35 (neuroticism), .50 (sociability), and .44 (impulse control). For agreeableness correlations are substantially lower, but it may be doubted in any case if this is truly an independent trait; it tends to correlate with extraversion and absence of neuroticism and psychoticism. The fifth factor (intelligence) is well known to be highly consistent and is in any case not part of temperament as here understood; it is dealt with in a separate chapter. It might be argued to what extent sociability equals extraversion and impulse control is the obverse of psychoticism, but certainly the consensual validity of these factors appears to be established.

These results at the same time disprove common objections to the validity of questionnaires and ratings on the basis of response style theory, that is, the notion that acquiescence, indecisiveness, dissimulation (desirability set), and other similar factors have an important influence on questionnaire and rating scores. Special studies of these factors as far as P, E, and N are concerned (H. J. Eysenck & S. B. G. Eysenck 1969; Michaelis & H. J. Eysenck, 1971) have shown that although such factors are not entirely absent, they play a relatively small part in such personality questionnaires as we are discussing here, except under special conditions of motivation, where dissimulation may assume a prominent role. Examples would be situations in which a person is filling in a questionnaire as part of a job interview; under those conditions it has been found that strong distortion occurs. However, under ordinary research testing conditions, questionnaire replies are relatively free from such errors.

We may now turn to the concept of construct validity (Cronbach & Meehl, 1955; Messick, 1975, 1980), which may be defined as the extent to which a given test may be said to measure a theoretical construct or trait. Campbell (1960) suggested within this framework a difference between convergent and discriminant validation. Not only should a test correlate highly with other variables with which it is theoretically connected (convergent validation), but it should also *not* correlate with variables from which it should differ (discriminant validation). A systematic experimental design using both convergent and discriminant validation has been proposed by Campbell and Fiske (1959) under the name of multitrait-multimethod matrix; we have already encountered its use in the study by McCrae. In their study, it will be remembered, they had six traits, each contributing to three major dimensions of personality; they used the methods of self-ratings and ratings by others. Convergent validity is shown when given traits using the one method correlate with given traits using the other method; discriminant validity is shown when a particular trait, using one method, fails to correlate with a different trait, using the other method. Thus the multitrait-multimethod matrix type of analysis can be applied not only to construct validity but also to concurrent and predictive validity, suggesting that these terms refer to somewhat overlapping methodologies.

It may be said that the more complex a given theory, and the more embracing, the greater will be the number of tests of the construct validity type which it enables the research worker to undertake. In this part of the book we have dealt only with the descriptive theory, and as far as that is concerned construct validity is based on correlational and factor analytic studies and the concept of consensual validity explained in this section. We believe that these are essentially weak types of validity; it would be possible to argue that essentially we are dealing here with reliability rather than validity. The fact that a person says that he is sociable or impulsive or persistent and these self-ratings agree with the ratings of someone who knows him well cannot really be said to provide us with much evidence in favor of a theory; the agreement is a necessary but not a sufficient condition for making positive assertions about consistency of behavior, but that is all. Ideally, construct validity should involve a much more abstract type of theory, making possible far more complex and surprising predictions than would be possible on this simple descriptive model.

Such a theory is developed in the second part of this book. This theory posits certain genetically controlled physiological mechanisms, which, in interaction with environmental stimuli, produce a kind of behavior the consistency of which gives rise to the personality dimensions of P, E, and N. These theories enable us to make a variety of predictions in three major domains. In the first place, these theories give rise to direct psychophysiological predictions as to ways in which extraverts and introverts should differ from each other, or stable and unstable persons, or high and low P scorers (Stelmack, 1981). In line with this type of evidence is the set of predictions as to how different people should be affected differentially by drugs (H. J. Eysenck, 1963, 1983c).

The second domain would be psychological experiments on perception, conditioning, learning, vigilance, and so forth in each of which the personality theory has to say something about predicted differences between high and low scorers on any of the three dimensions of personality. Laboratory tests of this kind will form the major portion of the next part of this book.

A third domain is constituted of social behavior which can be predicted from the personality theory through the mediation of the mechanisms studied in the laboratory, such as conditioning, memory retrieval, and the like. Thus, behavior such as criminality or neurosis can be predicted in terms of our personality theory and made part of the construct validity of the concepts in question. Obviously, such social applications of the theory and predictions based upon it are much more hazardous as we have far less control in social life over our subjects than we do in the laboratory; nevertheless these predictions constitute part of the construct validity of the concepts.

These few remarks on validity have been added to the first part of this book partly in order to introduce the second part but also to indicate why we believe that factor analysis and other internal methods of establishing construct validity are in our view relatively weak supports of rather weak theories. It is of course better to have factor analytic support for a personality theory than to construct questionnaires without subjecting them to such careful scrutiny of the underlying theories, but we feel that in order to have a scientifically meaningful concept we must go beyond such simple descriptive models and must take seriously the task of proposing causal mechanisms which can be tested directly in laboratory and social situations relatively independent of the items used to form the questionnaire of the rating scale itself. It is only the use of such independent criteria that can establish proper construct validity for personality concepts such as P, E, and N.

Having thus traced the history of the concepts of psychoticism, extraversion, and neuroticism, we may briefly indicate how they developed into the system of personality description discussed in this book. In the 1940s a position had been reached which is described by MacKinnon (1944) as follows:

> Types are crude pictures of personality. That is why they are so easily drawn, why they invariably overlap, and why such a scheme of interrelationships ... is so easily developed and yet so difficult to prove or disprove, as only that which is precisely stated can be definitely tested. To be sure, many of the relationships assumed to exist among the dichotomous typologies have been investigated both clinically and experimentally, but with little success so far. For the most part these studies have been made by partisan investigators; the details of experimental procedure have not been clear; the bases of selection of subjects, extremely important in studies of this sort, have not been specified; the statistical treatment of results demanded by the very nature of these investigations has been lacking. The problem of the relationships among the various dichotomous typologies remains a problem. (p. 322)

At this time, then, the study of typology (and personality generally) had reached a nadir. The failure to discriminate between introversion and neuroticism, the proliferation of arbitrary questionnaires thrown together on a purely subjective basis and having no firm statistical or experimental foundation, the confusion between the many different typologies (listed in H. J. Eysenck, 1947, but not discussed here in detail because they are only of historical interest), and many other factors had led to a total disbelief in the value of typologies, questionnaire studies, or indeed personality investigations generally. In addition, there was still the suspicion that typology implied bimodal distributions, or categorical separation, and all these problems together made the field a very unattractive one. It was against this background that H. J. Eysenck (1947) published *Dimensions of Personality* attempting to do the following: (1) lay down a firm theoretical basis for personality description, (2) use correlational and factor analytical procedures in evaluating the theory in question, (3) apply these methods to criterion groups, such as hysterics, dysthymics, and normals, which in theory should show differential combinations of the major personality dimensions, (4) express the theoretical conception of traits and types in terms of first- and second-order factors, (5) formulate a detailed model of the two major personality dimensions (E and N) and deduce from it differential behaviors which could be measured experimentally in laboratory investigations, and (6) apply these measures to normal and neurotic populations, clearly specified and large enough to give consistent and meaningful results.

In a continuation of this work (H. J. Eysenck, 1952b), an attempt was made to (7) include psychoticism as a personality variable and (8) construct questionnaires, such as the Maudsley Medical Questionnaire, to measure major personality dimensions. This program was later expanded to include the publication of a series of personality inventories (the Maudsley Personality Inventory, the Eysenck Personality Inventory, and finally the Eysenck Personality Questionnaire). The MPI simply measured E and N; the EPI included an L (lie or dissimulation) scale and had two forms, A and B; the EPQ added a P scale. All these forms were developed using factor analytic methods and present independent dimensions of personality, based on the theoretical and empirical work considered in this chapter and clearly explicated in such a way that new and congruent items could be written for each of these scales.

As a very brief example, consider Table 6, which shows the results of a factor analytic investigation of 10 P items, 10 E items, and 10 N items. Items with factor loading above .4 are in italics, and it will be seen that all 10 P items have loadings only on P, all 10 E items loadings only on E, and all N items loadings only on N. This table should be viewed in conjunction with Figures 1, 2, and 3, in Chapter 3 for an understanding of how the superordinate conception of a *type* finds support in factor analytic studies of items denoting different traits. A look at the items in the table will also give some indication of the nature of the three factors involved.

TABLE 6. Thirty Questionnaire Items Loading on *P*, *E*, and *N*

	P	E	N	
1. Would being in debt worry you?	−.43	.05	−.26	
2. Do you believe that insurance schemes are a good idea?	−.42	−.17	−.03	
3. Do you prefer to go your own way rather than act by the rules?	.47	.08	.06	
4. Do good manners and cleanliness matter much to you?	−.55	−.06	.01	
5. Do you enjoy cooperating with others?	−.46	−.34	.01	P
6. Do you like taking risks for fun?	.50	−.23	−.03	
7. Does it worry you if you know there are mistakes in your work?	−.53	−.04	.25	
8. Do you think people spend too much time safeguarding their future with savings and insurances?	.44	.04	.03	
9. Do you try not to be rude to people?	−.51	.01	.04	
10. Is it better to follow society's rules than go your own way?	−.50	.16	.00	
11. Are you rather lively?	.03	.63	.63	
12. Do you enjoy meeting new people?	−.24	.63	−.14	
13. Do you like going out a lot?	.00	.55	.02	
14. Would you call yourself happy-go-lucky?	.15	.46	−.17	
15. Are you mostly quiet when you are with other people?	−.06	−.44	−.14	E
16. Do you like mixing with people?	−.24	.70	−.12	
17. Do you often make decisions on the spur of the moment?	.13	.44	.28	
18. Do you like plenty of bustle and excitement around you?	−.03	.65	.01	
19. Do you nearly always have a "ready answer" when people talk to you?	−.01	.40	.00	
20. Can you easily adapt to new and unusual situations?	.08	.42	−.30	
21. Does your mood often go up and down?	−.10	.14	.59	
22. Do you feel "just miserable" for no reason?	.04	−.14	.51	
23. Are you an irritable person?	.15	−.11	.48	
24. Do you often feel "fed-up"?	.00	−.08	.67	
25. Are you often troubled about feelings of guilt?	−.10	.13	.58	N
26. Would you call yourself a nervous person?	−.06	−.07	.60	
27. Would you call yourself tense or "highly strung"?	−.07	−.09	.57	
28. Do you often feel that life is very dull?	.06	−.15	.46	
29. Do you often feel lonely?	−.08	.06	.60	
30. Are you easily hurt when people find fault with you or your work?	−.30	.06	.51	

Later books (H. J. Eysenck, 1957, 1967a, 1981) added causal theories to the descriptive ones outlined in the earlier works; these will not be discussed here as they will be dealt with in greater detail later on. Let us merely note that the 1957 book was an attempt, only partially successful, at formulating a causal theory in terms of Hullian concepts. This was given up and a theory in terms of neurophysiological concepts such as cortical arousal and visceral brain activity attempted (H. J. Eysenck, 1967a). This theory gave rise to a large body of laboratory experimental work and psychophysiological studies which will be described in greater detail later on (H. J. Eysenck, 1981), as will alternative theories which have been developed as a consequence of the numerous studies devoted to this topic.

The Universality of *P*, *E*, and *N*

GENETIC FACTORS

Temperament is defined as "the individual character of one's physical constitution permanently affecting the manner of acting, feeling, and thinking" (*Concise Oxford Dictionary*, 1976). This definition corresponds to what in psychology are sometimes called *source traits*, in contradistinction to *surface traits*, that is, traits or combinations of traits which are accidental and variable and have no particular causal background in biological factors and genetic determinants. If *P*, *E*, and *N* are to be regarded as source traits, it would seem that certain testable predictions could be made, and in this chapter we will report on studies that have been done to test these predictions. The first and most obvious one is that if *P*, *E*, and *N* are part of one's physical constitution *permanently* affecting one's behavior, then surely one would expect genetic factors to play an important part in the causation of individual differences along these dimensions. Thus in our first section we will consider the evidence concerning the *heritability of personality*.

If *P*, *E*, and *N* are grounded in constitutional factors of biological importance, then it would seem likely that they could also be observed not only in human conduct but also in the behavior of animals, particularly mammals. Although not much work has been done in this field, our second section will deal with the personality of animals, insofar as this has been studied scientifically and by way of specially devised experiments and observational procedures.

If *P*, *E*, and *N* are not just patterns of behavior to be observed in Western society (particularly Europe and the United States of America), then one would expect that cross-cultural studies would demonstrate the existence of similar factors in other nations and other cultures. Our third section deals with this particular question of cross-cultural iso-morphism.

Fourth, it would seem to follow that longitudinal studies should be capable of explicating the prediction that *P*, *E*, and *N* would remain rel-atively steady over the years, so that early measurement would predict later positions in the dimensions.

The belief is widespread that personality factors of various kinds are largely if not entirely the product of environmental influences, so that genetic factors play little part in causing differences between people. This belief derives largely from a study by Newman, Freeman, and Hol-zinger (1937), in which 100 pairs of twins, 50 identical and 50 fraternal, were compared with respect to a number of physical measurements and mental and educational tests. The authors' conclusion was that:

> The physical characteristics are least affected by the environment, that intelligence is affected more; educational achievement still more, and per-sonality or temperament, if our test can be relied upon, the most. This find-ing is significant, regardless of the absolute amount of the environmental influence. (p. 315)

As far as personality is concerned, this conclusion rests on very uncer-tain foundations, and a detailed critique has been given by H. J. Eysenck (1967a). Among these criticisms are the following: The tests used were of very doubtful reliability and validity. The tests were standardized on adults but used on twins who were children with an average age of about 13 years and some as young as 8 years or even younger. A third criticism, to which we will come back presently, is the use of statistical methods that are not adequate for the analysis of genetic data. Last but not least, the conclusion is not really in accord with the data.

Thus, the authors report on a personality inventory of neuroticism, for instance, a test for which detailed statistics are presented, and it appears that for identical twins the intraclass correlation is .562, for fra-ternal twins it is .371, and for identical twins brought up in separation it is .583. Here we would seem to have quite strong evidence for the importance of heredity, seeing that identical twins are distinctly supe-rior in point of intraclass correlation to fraternals and that identical twins brought up in separation are if anything more alike than are iden-tical twins brought up together! The authors comment that the test in question "appears to show no very definite trend in correlations, possibly because of the nature of the trait, and also because of the unreliability of the measure." It is not quite clear to us why there is this denial of a definite trend; it seems fairly clear that identical twins, whether brought up in separation or together, are more alike than are fraternal twins, and as we shall see, later modern work has amply justified such a conclusion.

Using the intraclass correlation for MZ twins brought up apart, and correcting for the unreliability of the inventory used, we obtain an estimated heritability figure of approximately 70%! Why Newman *et al.* considered this to indicate "no very definite trend" is certainly mysterious, as is the acceptance of this unsupported dictum by subsequent writers. A similar conclusion follows from correcting the MZ twins apart figure for attenuation. Recent work suggests that these estimates are not far from the truth.

The alleged lack of heritability of personality widely claimed as a consequence of the Newman *et al.* study was contradicted by two studies reported by H. J. Eysenck and Prell (1951) and H. J. Eysenck (1956a); in both a large contribution of heredity factors was found for N and E. These were the first genetic studies to use factors obtained by factor analysis, rather than simple scores, to establish heritabilities; future research was to show that indeed personality factors were strongly influenced by heredity.

The main type of evidence considered by Newman *et al.* is differences in intraclass correlation between monozygotic (identical—MZ) and dizygotic (fraternal—DZ) twins, or other statistics corresponding to this. These statistics are sufficient to establish whether or not MZ twins are significantly more alike than DZ twins, but they do not enable us to analyze the available data in a manner which would appear meaningful to geneticists. More modern methods (Mather & Jinks, 1971, 1977) enable us to go well beyond the simple calculation of intraclass correlation and the calculation of heritability estimates, which are meaningful only as long as we are willing to accept a number of underlying assumptions which the method itself is incapable of proving.

In what follows we shall consider some of the results of the application of these modern methods, but these are too technical to be discussed here, and a verbal description of the methods and the purpose of using them must suffice. H. J. Eysenck (1979) gives a detailed description of modern methods as applied to the analysis of the inheritance of intelligence, and Fulker (1981) does the same for the analysis of personality traits. In this discussion it will also be essential to consider certain widely held beliefs about the nature of heritability that are in fact erroneous and misleading.

It would not be accurate to say, for instance, that the behavioral geneticist is exclusively concerned with *heritability*, that is, that proportion of the total variance in the phenotype which is due to the genotype. What the behavioral geneticist is concerned with is to take the total variance contributed by the phenotype, that is, the particular trait measured, whether IQ, *P, E,* or *N,* and partition that variance into various contributory causes which can be said in some sense to determine the phenotype. One part of this variance is of course the contribution made by additive genetic gene loci; the proportion this represents of the total phenotypic variance is the so-called *narrow heritability.* In addition, there are nonadditive factors such as assortative mating (the tendency of like to marry like), dominance of gene action, and epistasis (nonadditive genetic variance due to interaction between different gene loci); these, when added to the additive genetic variance, give us the *broad heritability.*

On the environmental side we have the important distinction between within-family and between-family environmental factors, that is, those factors which distinguish one family from another, such as socioeconomic status and education of the parents (between-family environmental factors), and those environmental factors which differentially affect children within the same family (within-family environmental factors).

Finally, we have the interactions between genetic and nongenetic factors, of which two in particular are distinguished. First is the *statistical interaction,* which means that different genotypes may respond differently to the same environmental effect. For instance, if a particular change in the environment causes some genotypes to gain 20 points of IQ, some 10 points, and some none, while some may show a loss, the environmental change *interacts* with genotypes to produce different phenotypic effects in different genotypes.

A different type of interaction is covered by the term *covariance* of genotype and environment, which arises when genotypic values and environmental values are correlated in the population. An example would be children with genotypes for high intelligence reared in homes with superior environmental advantages for intellectual development.

Another contributor to phenotypic scores on tests, which should be carefully considered, is *error variance,* the failure of the measuring instrument to be totally reliable. By virtue of the methods of calculation used, this is usually added to the environmental variance, but this gives an erroneous impression of the true state of affairs by reducing the calculated heritability of the trait in question, and the results of such calculations should always be corrected for attenuation. Because this is usually not done, most published estimates of heritability are in fact *underestimates* of the true heritability, an error more or less serious depending on the internal reliability of the test used.

From what has been said so far, we discover the first important conclusion: Heritability is a concept that may be defined in different ways, and hence the fact that different investigators obtain different heritabilities may simply mean that they have used different concepts and different methods of calculation, rather than that they disagree on the true state of affairs. Thus the same data may give rise to heritabilities of .6, .7, or .8. The first of these (.6) might be the narrow heritability, uncorrected for attenuation; the second (.7) might be the broad heritability, uncorrected for attenuation; and the third (.8) might be the broad heritability corrected for attenuation. In evaluating data it is always important to know what type of heritability is in fact being considered; otherwise, it is easy to come to the conclusion that estimates differ so widely that none can be regarded as accurate. This is not necessarily so. In particular, in relation to personality and temperament, where reliabilities tend to be lower than in the field of intelligence, the correction for attenuation may raise the estimates of heritability by a considerable amount.

What the behavioral geneticist is trying to do, therefore, is not so much to obtain an estimate of heritability, but rather to investigate the architecture of the genetic and environmental influences which determine the phenotype, in our case, P, E, and N. We thus have the curious situation that the geneticist, allegedly only interested in genetic factors, is in fact concerned with *all* the determinants that contribute to individual differences in the phenotype—genetic, environmental, and interactional factors. This is, of course, inevitable, as a genetic analysis is an analysis of variance and cannot afford to leave out of account sources of variance which are relevant to the phenotype. Environmentalists habitually leave out of account genetic factors and hence look only at a small part of the evidence; this is not usual in scientific work. Usually studies in personality give results which can be interpreted in terms of both environmental and genetic influences, and these can be sorted out only by reference to proper research designs such as those pioneered by behavioral geneticists. Environmentalists rule out genetic factors on *a priori* grounds and interpret their results in environmentalist terms; this is inadmissible scientifically, and evidence should always be demanded before such an interpretation is accepted.

On the other hand, it is also important not to overinterpret the meaning of heritability, particularly when that heritability is strongly marked, that is, accounts for something like 70% or 80% of the total variance. Heritability is always a *population statistic;* in other words, it applies to a particular population, at a particular time, such as native-born English people in England in 1982 or native-born Americans in the United States in 1935. The heritabilities thus discovered do not necessarily apply to populations in other countries, or to populations in the same country at other times. This is an important restriction that must always be borne in mind.

Another consequence of the fact that heritabilities are population statistics is the deduction that they *do not apply to individuals;* like all estimates of variance, it is impossible to apply such statistics to singular cases. The fact that intelligence accounts for 80% of the total variance in British or American populations at the present time cannot be interpreted to mean that for any particular individual in those countries heredity accounts for 80% of whatever IQ they may show; such a statement would be meaningless.

These limitations on the use of heritability estimates also highlight a misuse of the concept often made by critics. Thus textbooks often cite a statement by D. Hebb that it is as meaningless to try and attribute differential importance to heredity and environment in the genesis of differences in intelligence as it would be to say which was more important in determining the size of a field, width or length. It would indeed be impossible to deal with a single field in this manner, but the example is irrelevant; as we have said above, estimates such as heritability are population parameters and do not apply to individual cases! If we had a thousand fields, all differing in length and width and in area, then it would be perfectly meaningful to say that length was more important than width, or vice versa, and to give a numerical estimate of the relative importance of these two factors.

Two errors are frequently made in the interpretation of genetic data and should be corrected here. Consider the statement that 80% of the variance in IQ is determined by genetic factors. This is often erroneously interpreted as implying a greater degree of genetic determination than is actually present, and it is also often interpreted to mean that nothing can be done to alter a person's position on the intellectual dimension, that is, that intelligence, being "fixed," cannot be touched by environment. Let us consider the first point. The 80%/20% ratio of genetic to environmental factors referred to proportional *variance*, which is a concept not too well understood by the general public or even by some psychologists. The variance is the square of the standard deviation, and therefore, if we want to talk in terms of standard deviations we should take the square root; thus, looking at the ratio 80/20, we should take the square root of $8/2 = 4$, which is 2! Thus in ordinary terms environment is only half as important as heredity, but this of course is still quite a lot.

If we assume a heritability of .8 in a perfectly reliable test of IQ with a standard deviation of 16, then a change in environment of 4 standard deviations would make a difference of about 28 points of IQ; this would certainly be well worth having! Our real difficulty, of course, is that we do not know quite how to bring about such a change. We should remember that of the environmental contribution only two-thirds are between-family (which can possibly be controlled and changed); we know nothing about the variables in the one-third of the environmental variance controlled by within-family factors and presumably can do little to alter these.

The notion that any quality that is largely determined by hereditary forces is fixed in the sense that it can never be changed is very widespread but quite false. To take a somewhat farfetched example, consider the size, shape, and consistency of the female bosom. In normal European and North American populations this is determined largely by genetic factors. However, in recent years we have learned to produce pronounced changes in all three measures by means of silicone injections, plastic surgery, hormonal administration, and the like. It is not too fanciful to imagine that in 50 years' time the proportion of the total variance contributed by genetic factors in, say, California may be very much less than it is now and may even be nil! The crucial point is to find new environmental influences that affect the phenotype. Where this is feasible and can be done, the proportion of variance contributed by heredity may decrease very markedly.

A good example to illustrate the point is phenylketonuria. This is a disorder which affects about one child in 40,000. It causes mental defect, and it has been found that about one in every hundred patients in hospitals for severely mentally handicapped children suffers from phenylketonuria. This disorder is known to be inherited and is in fact due to a single recessive gene. The great majority of children suffering from it have a level of mental performance which is usually found in children half their age. These children can be distinguished from other mentally handicapped or from normal children by testing their urine, which yields a green-colored reaction with a solution of ferric chloride due to the presence of derivatives of phenylalanine. Here we have a perfect example of a disorder produced entirely by hereditary causes, where the cause is simple and well understood and where the presence of the disorder can be determined with accuracy.

Is there reason to believe that the low IQ of the children so affected is fixed and that therapeutic nihilism is called for? The answer is that we must go on to demonstrate *in what way* the gene actually produces the mental defect. It has been shown that children affected by phenylketonuria are unable to convert phenylalanine into tyrosine; they can break it down only to a limited extent. It is not clear why this should produce mental deficiency, but it seems probable that some of the incomplete breakdown products of phenylalanine are poisonous to the nervous system. Phenylalanine, fortunately, is not an essential part of the diet provided that tyrosine is present in the diet. It is possible to maintain these children on a diet which is almost free of phenylalanine, thus eliminating the danger of poisoning to the nervous system. It has been found that when this method of treatment is begun in the first few months of life there is a very good chance that the child may grow up without the mental handicap he would have otherwise encountered. In other words, by understanding the precise way in which heredity works and by understanding precisely what it does to the organism, we can arrange a rational method of therapy which will make use of the forces of nature, rather than trying to counteract them.

One of the reasons why we have been relatively unsuccessful in carrying such a program into execution in the field of intelligence and personality is that we have in the past tried to deny the importance of genetic factors in the production of individual differences in temperament and intelligence and hence have not looked for the precise way in which these effects are produced. Had we done so, we might by now have much better methods of influencing and improving both. In part it is the intention of this book to try to go beyond the simple assertion that differences in temperament and intelligence are produced by genetic causes to a marked extent and to suggest theories regarding the biological variables underlying this determination.

It will have become clear that in addition to heritability we want to know about other important factors, such as whether the genes contributing to the given trait are dominant or recessive, whether assortative mating has taken place, whether there is any evidence for epistasis, whether environmental influences are within-family or between-family, whether there is any interaction between environment and heredity, and if so of what kind. Modern methods of analysis to answer questions of this kind exist, but they do require very large numbers of twin pairs for the analysis, and it is only in recent years that large-scale samples of MZ and DZ twins have been employed for the purpose. In addition to twins of these two types, of course, we have studies of MZ twins brought up in separation, correlational studies of different degrees of consanguinity within families, studies of adopted children, and many other methods of estimating the various portions of the total phenotypic variance. As stated above, no attempt will be made here to explicate the complex statistical reasoning involved, the assumptions made, or the qualifications that are necessary before any estimates are accepted. The reader must be referred to the sources quoted above for elucidation of these points. Here we will deal only with the major findings of recent research.

Let us first of all look at some studies of MZ twins brought up in separation. The first satisfactory study of this kind was reported by

Shields (1962); in this study 42 pairs of separated twins were assessed using an early form of H. J. Eysenck's MPI, measuring N and E. The mean age of separation of these pairs was 1.4 years and the age at reunion 11.0 years. Shields also had a control group of twins who had been brought up together. His major results are as follows: For extraversion, MZ twins brought up together had an intraclass correlation of .42, MZ twins brought up in separation one of .61, and DZ twins one of −.17. For neuroticism these three correlations were .38, .53, and .11. Several results are immediately obvious. MZ twins resemble each other much more closely than do DZ twins; MZ twins brought up in separation are if anything more alike than MZ twins brought up together. The data are somewhat similar to those of Newman et al. already quoted but even more decisive in suggesting genetic factors in determining the phenotype. Note, however, that DZ twins show very poor (and even negative) intraclass correlations; this does not accord with genetic theory, which would expect DZ correlations to be roughly half those for MZ twins.

Lykken (1982), who studied 30 MZ pairs brought up in isolation, found similar results. His twins were separated at the age of 0.3 years and reunited at the age of 23.9 years. He also finds a tendency for DZ twins to show very low intercorrelations and explains this in terms of what he calls "emergenesis," a kind of epistasis. As he says: "When the MZ twins' are very similar but the DZ twins' correlations are near zero, that is when there is reason to suspect that the trait is emergenic" (p. 365).

It is sometimes said, in criticism of such research, that MZ twins are treated more alike by their parents, teachers, and others than are DZ twins and that this accounts for the greater similarity of MZ twins. Loehlin and Nichols (1976) specially investigated the early experiences of MZ and DZ twins with particular reference to their being treated alike or differently. They found that indeed MZ twins had been treated more alike than DZ twins but that the method of treatment did not correlate to any significant extent with personality or ability. They conclude:

> It is clear that the greater similarity of our identical twins' experience in terms of dress, playing together, and so forth cannot plausibly account for more than a very small fraction of their greater observed similarity on the personality and ability variables of our study. (p. 52)

This objection, therefore, cannot be regarded as a serious obstacle to accepting the results from twin studies.

There are three major sources for data on personality studies of twins in recent years, and these contain adequate summaries of earlier studies. The first is Buss and Plomin's *A Temperament Theory of Per-*

sonality Development (1975). It also contains extensive data on longitudinal studies of personality development, showing how even with neonates it is possible to make predictions of later personality traits. The number of twins in this study is rather small and the statistical treatment somewhat old-fashioned; as the authors recognize, more advanced methods would be inappropriate for samples numbering only in the low hundreds.

The second source is the study of 850 sets of twins by Loehlin and Nichols (1976), *Heredity, Environment, and Personality;* not only is this work on a much larger scale, but also the statistical analyses are much more in line with modern theory. Many of the results to be quoted stem from their work.

The third set of data comes from the work of the Maudsley school and is summarized by Fulker (1981). Good summaries of the methods used are given by Eaves *et al.* (1977, 1978). Details of the studies themselves will be found in Eaves (1973), Eaves & H. J. Eysenck (1975, 1976a,b, 1977, 1980), Eaves, Martin, and H. J. Eysenck (1977), and Martin and H. J. Eysenck (1976). A full discussion of twin methodology in particular is given by Eaves (1978), and a thorough review of the literature on the genetics of intelligence is given by Fulker and H. J. Eysenck (1979).

The most recent and by a large margin the most comprehensive twin study has been reported by Floderus-Myrhed, Pedersen, and Rasmusson (1980) using 12,898 twin pairs. Estimates of heritability were .50 and .58 for *N* (males and females separately) and .59 and .66 for *E*. No correction was made for attenuation; since a short form of the EPI was used, such a correction would have been substantial, giving estimates of heritability between .7 and .8. This Swedish sample was perhaps more random than any other so far studied, and this investigation gives perhaps the most acceptable estimate of the heritability of *E* and *N* available at the moment.

In the following few paragraphs we will summarize the major results from these various studies and many others not quoted directly; there are literally hundreds of studies now which on the whole give a very congruent picture of the genetics of human personality. We will also discuss here the major results from the study of the genetics of human intelligence, partly because there are interesting similarities and differences between the temperament and intelligence fields and also because later (Chapter 5) we will be describing the physiological measurement of intelligence and basing some of the arguments there on findings reported here.

To begin with, then, it is found that for practically all traits and dimensions of personality there is a considerable degree of genetic determination of individual differences (Loehlin & Nichols, 1976). The amount of heritability varies according to whether we refer to the narrow or the broad heritability and whether we correct for attenuation or not. Broadly speaking, twin studies suggest a narrow heritability for temperamental traits of around 50%, which, when corrected for attenuation, suggests heritabilities between 60% and 70%. The exact figures differ from study to study, as they should; there are differences in the selection of twins, their ages, the tests used, the reliabilities of the tests, the countries in which the studies were carried out, and so on. All of these might be expected to exert an influence on the data. No serious worker in this field denies that genetic factors account for at least something like half of the variance, and equally none would deny the importance of environmental variables.

As for intelligence, the narrow heritability is probably something like 60%, the broad heritability something like 70%, and the broad heritability corrected for attenuation something like 80% (Fulker & Eysenck, 1979). Again, some writers give rather lower figures, but the differences are never terribly large, and again no one would doubt that there are important genetic factors in the determination of individual differences in intelligence, just as no one would doubt the importance of environmental factors or the fact that on the whole heritability for intelligence is somewhat higher than for temperament, but perhaps not so very much.

The genetic component in the temperament field is made up almost exclusively of additive genetic variance; there is little or no evidence for dominance or assortative mating. Like does not marry like more frequently as far as personality traits are concerned; correlations tend to run around zero, although when they depart from it to a minor degree the departures are usually in the positive direction, but seldom higher than .2. Dominance also appears to be absent, suggesting that evolution does not prefer one extreme of P, E, or N to the other but places its bets near the middle range. In the virtual absence of nonadditive genetic variance, broad heritability = narrow heritability. For intelligence, of course, the nonadditive part of the genetic variance is considerable, with assortative mating and dominance being prominent. Thus, although both temperament and intelligence show a strong degree of hereditary determination, the architecture of this hereditary determination is quite different for each of these two aspects of personality.

Similar differences can be found on the environmental side. In the case of intelligence, about two-thirds of the environmental determinants are found to be of the between-family kind and one-third of the within-family kind. For temperament, on the other hand, practically all the environmental variation appears to be of the within-family kind, with little or none due to between-family variation. On this surprising finding there is a good deal of agreement (Loehlin & Nichols, 1976; Fulker, 1981), and the consequences for the study of temperament can be quite far-reaching. Thus Freudian and most other psychiatric theories tend to implicate factors such as the personality structure of the parents (e.g., the "ice-box mother") in the causation of schizophrenia, or the "double-bind" structure of the environment by the parents of schizophrenics; yet all these are of the between-family kind, in the sense that the causative factors are related to aspects of a given family that would not be found in other families. The evidence suggests, therefore, that all these theories must be false, an important conclusion, reached along a line that is not very frequently adopted in studies of personality and mental abnormality.

A great deal more could of course be said about the genetics of temperament and intelligence, but this is not a textbook of behavioral genetics, and our main point in discussing the issue is to relate it to our general belief that P, E, and N are source variables and hence that differences in the position of people along these dimensions would be strongly determined by genetic causes. For a more detailed treatment of methods and results, readers must refer to the works cited; for the present we need only conclude that the evidence strongly supports the hypothesis.

PERSONALITY IN ANIMALS

To some people, the notion of applying the concept of personality to animals, particularly to lower mammals such as the rat, may seem preposterous. Yet Pavlov noted marked differences in the behavior of his dogs, particularly in relation to conditioning experiments, and found these differences to be persistent over long periods of time. Similarly, Scott and Fuller (1965) not only found marked differences in aggressive and other types of behavior in dogs but demonstrated a strong genetic basis for these differences, which were clearly related to strain. In primates, Stevenson-Hinde, Stillwell-Barnes, and Zunz (1980), using behaviorally defined adjectives and a 7-point scale, rated all individuals over a year old in a colony of rhesus monkeys every November for four years. Factor analysis of these data resulted in three major factors, labeled confidence, excitability, and sociability. At all ages, confidence scores were stable from year to year, and excitable and sociable scores were stable once adulthood was reached. Clearly the social behavior of at least the higher mammals (and, as we shall see, the lower ones as well) shows evidence of large individual differences in behavior, giving rise to traits which are to some degree genetically determined and are persistent over time (see also Locke *et al.*, 1964).

In terms of Darwinian evolutionary principles, it seems unlikely that humans should have developed genetically based behavior patterns corresponding to our three major dimensions of P, E, and N and based on certain physiological structures and hormonal secretions independently of animals lower in the evolutionary scale. Indeed, one might say that these three dimensions correspond to the major ways in which one organism can logically respond to another in a social situation. The three ways of interaction are: (1) one organism shows suspicion, hostility, and aggression toward the other (P); (2) one organism shows anxiety, fear, and withdrawal toward the other (N); (3) one organism interacts pleasurably and peacefully with the other (E). Animals of course cannot answer questionnaires to facilitate our research, but ratings of their behavior can be obtained, and these should give rise to dimensions similar to those observed in humans.

A study explicitly devoted to this purpose has been reported by Chamove, Eysenck, and Harlow (1972). In these studies, 168 *Macaca mulatta* rhesus monkeys were separated from their mothers at birth and reared in individual mesh cages. They were given daily peer experience starting at between 15 and 19 days of age. All subjects were assigned to a group composed of four age-mates, and all social experience, both pairings and group sessions, involved these group members. The behavior of the animals in social situations was observed and recorded, in particular social exploration, social play, nonsocial play, nonsocial fear, appropriate withdrawal, inappropriate withdrawal, hostile contact, nonhostile contact, social cling, and noncontact hostility. All the interobserver reliability coefficients were very high, hardly ever falling below .9.

Observations were intercorrelated and factor analyzed; three major factors emerged which were interpreted as affiliative, hostile, and fearful: "These factors were almost entirely independent and resembled the extraversion, psychoticism, and emotionality factors frequently found in humans" (Chamove *et al.*, p. 496). It was thus seen that in rhesus monkeys, at least, factors similar to those obtainable in humans can be found. This finding is similar to results reported by Van Hooff (1971), who reported a component and cluster analysis of 53 behaviors recorded in a stable group of 25 chimpanzees; 69% of the variance was accounted for by components termed affiliative or social positive (E), aggressive (P), and submissive (N); in addition, there was a play component and small factors related to grooming, excitement, and display.

Turning now to the large-scale work that has been done on rats, we will concentrate particularly on the studies begun in 1954 on the Maudsley reactive and nonreactive strains (H. J. Eysenck & Broadhurst, 1965); this work used a restandardized open-field test of the kind pioneered by Hall (1938), and work done on it has been summarized by Broadhurst (1975). In these studies, rats are exposed in a circular, confined space to bright lights and loud white noise, and their defecation, urination, and ambulation measured. The major measure of emotionality is the defecation score, and genetic studies were done on specially bred reactive and nonreactive strains. The table summarizing past research (Broadhurst, 1975) contains 280 items, and many more have been added since then; obviously it would be impossible to summarize all this work here, or the numerous publications on emotionality and other "personality" traits in mice (e.g., Royce, Poley, & Gendall, 1973a,b).

The major finding of the studies of the Maudsley strains is that the trait of emotionality, as measured by the open-field test, is strongly heritable and is related to hormonal secretions and the size of the adrenal glands. Emotionality appears to be closely related to neuroticism, and the open-field test seems to be a good way of measuring this trait in rats.

The question immediately arises as to whether we can really regard defecation in the open-field test as a measure of such a very general trait as emotionality or neuroticism; could it not be that this is a very specific test simply measuring predisposition to defecate? Savage and H. J. Eysenck (1964) tried to argue the point along the following lines: There are certain situations which, by common consent, produce emotional arousal. Thus in humans the threat of a strong shock leads to greater generalization of a voluntary response than does threat of a weak shock (Rosenbaum, 1953). Hence, on the hypothesis that threat of strong shock produces greater emotion than a threat of weak shock, we may say that generalization of a voluntary response is a consequence of emotion such as that produced in the experiment. If, now, a questionnaire measure of anxiety is a true measure of emotionality, then anxious subjects should show greater generalization to identical stimuli (threat of shock) than should nonanxious subjects (Rosenbaum, 1956). This was actually found to be the case, and hence we may equate anxiety (as measured by a questionnaire) with emotion as produced by the experimental design. One experiment and one such generalization would of course not suffice to prove the identity of the concepts, but if a number of such experiments could be found, each different from the other, then one might have greater faith in the results.

Consider now an experiment by Savage and H. J. Eysenck (1964), using the same kind of argument. In the experiments the constitutional differences in emotionality between the Maudsley reactive and nonreactive strains were measured by their response in a simple approach-avoidance conflict situation (Miller, 1944). The animals first learned to approach a goal box for food reward by running down an alleyway. Fifty approach trials were given before avoidance, which consisted of one trial, was introduced. Electric shock was used in the avoidance trial. The effect of the avoidance training on subsequent approach responses was measured. In this situation the gradient of approach was relatively strong compared to the avoidance gradient and the point of intersection of these two such that all animals continued to enter the goal box for the food reward. The high emotionality of the reactive strain in the situation should lead to larger conflict scores, that is, their avoidance gradient should be steeper and lead to slower approach responses. This was exactly what was observed.

This and other experiments reported in the chapter by Savage and H. J. Eysenck (1964)

> support the hypothesis that emotionality as measured in the open field situation is not specific to that situation, nor to the defecation response, but has predictable properties which generalise to other situations and responses....We may say that the concept of "emotionality" has been defined and made amenable to measurement by aligning with the concept of "emotion" in general and experimental psychology. In all the emotional situations investigated, the reactive strain showed behaviour significantly different from that of the non-reactive strain. The reactive animals were more susceptible to fear, anxiety, frustration and conflict. The evidence strongly supports the view of the differential-inherited basis for emotional behaviour. (p. 312)

Further evidence will be found in the Broadhurst (1975) paper.

Attempts to measure extraversion–introversion in the rat by Broadhurst (1973) and Weldon (1967) were largely unsuccessful, due probably to the differences in emotionality in the two strains used, which would interfere with the measurement of another personality dimension. There are obvious difficulties in the way of measuring extraversion–introversion in rats because the major behavioral component of that trait in humans, sociability, is not easily observed in rats.

However, the solution to the problem may be found by going to the underlying psychophysiological variable which, according to Eysenck's (1967a) hypothesis, is responsible for differences in extraversion–introversion, namely cortical arousal. Low cortical arousal, characteristic of extraverts, might lead to exploratory behavior, which might be regarded as producing a rise in cortical arousal, thus acting as positive reinforcement for animals whose lower arousal was productive of boredom and other negative emotional states. Broadhurst (1960), Hayes (1960), and Whimbey and Denenberg (1967) had concluded on the basis of their observations that ambulation in the open field was a type of exploratory behavior which would make it a suitable measure of extraversion-introversion. Unfortunately Broadhurst (1960), Denenberg (1969), and Mikulka, Kandall, Constantine, and Posterfield (1973) found a negative correlation between ambulation and defecation, which made Broadhurst conclude that ambulation could be considered a somewhat inferior index of emotional reactivity. Russell (1973a) and others had not found such a correlation, and Russell (1973b), in reviewing this subject, concluded that new stimuli would provoke both fear and exploration and that possibly exploratory behavior could be interfered with by fear. In the type of situation used by Broadhurst, frightening the animals by very high intensities of light and of noise, fear would be the major emotion shown. However, if the intensity of light and sound stimuli were to be lowered, fear reactions might be lessened or eliminated, and exploratory behavior as a measure of extraversion-introversion might become independently assessable.

A full series of investigations into this question was carried out by a group of psychologists at the Barcelona *Universitas Autonoma*, and a brief survey of the work is given by Tobena, Garcia-Sevilla, and Garau (1978). This group used the same open field as had been used by Broadhurst, but they eliminated the 78 dB white noise used by him. Eight independent investigations showed that there was no correlation between ambulation and defecation, demonstrating that independence of the two measures had been achieved. Genetic studies showed that ambulation as measured in this particular situation had a hereditary component. Other tests of exploratory behavior, such as rearing and Boissier's test (number of holes entered by the rat in the course of exploratory behavior), correlated significantly with ambulation (Goma, 1977). Various conditioning experiments, using training with positive reinforcement, extinction, and discriminated extinction, showed that the majority of experiments favored the hypothesis, being in line with the results of similar experiments with human extraverts and introverts (Garau, 1976; Sevilla, 1974; Sevilla & Garau, 1978). In shuttlebox avoidance conditioning (Tobena, 1977) and in tests relating to aversive thresholds (Duran, 1978), results were only partially favorable to the hypothesis and difficult to interpret. Results of experiments on the effect of drugs (Sevilla, 1974; Garau, 1976) were mostly in favor of hypotheses deduced from the general theory and from work with humans (H. J. Eysenck, 1963).

These studies, supplemented by others of Pallares (1978), Garau *et al.* (1980), and Garcia and Garau (1978), suggest that ambulation in the rat under conditions that do not produce fear responses may be a useful measure of extraversion–introversion. Unfortunately most of this work has been published in Spanish and hence has not been taken up by English-speaking psychologists. A survey of much additional material, mostly favorable to the hypothesis that open-field ambulation can indeed by used as an analogue to extraversion, is given in an important paper by Sevilla (in press). Being published in English, it is perhaps the best introduction to this important line of research.

Similar work on the third dimension, psychoticism, has unfortunately not been reported, although there is a large literature on aggressiveness in rats and other mammals. There is little in the way of factor analytic studies of intercorrelations between different tests and little attempt to relate these differences in a systematic way to personality theories. This would seem to be a promising area, but at the moment it would not be possible to consider the available evidence as strong support for the hypothesis that aggressiveness in animals resembles psychoticism in humans (Beilharz & Beilharz, 1975; Blanchard *et al.*, 1975; Blanchard *et al.*, 1977; Lagerspetz & Lagerspetz, 1971; Vale, Ray, & Vale, 1972).

On the whole, evidence from the animal field is sketchy as far as similarity of personality patterns to humans is concerned, but as far as it goes it tends to be confirmatory rather than critical. It is unfortunate that animal psychologists have not on the whole been very interested in individual differences and have not adopted the methods of research used by psychometrists in the human field. One hopes that in future the promising beginnings here recorded will be followed up and will lead to more definitive results.

CROSS-CULTURAL STUDIES

If the three super factors *P*, *E*, and *N* are as fundamental and important as we have suggested, then one would expect that they should also be universal, in the sense of not being restricted to those cultures where they were first isolated but becoming apparent in many different cultures. This leads us to the problem of cross-cultural studies of personality (H. J. Eysenck & S. B. G. Eysenck, 1983) and an analysis of the investigations which have been done to establish the comparability, or otherwise, of personality patterns in different countries.

We may begin by stating that there are three problems rather than one in comparing personalities from one culture to another, and the failure to realize this has led to many difficulties and complications. The first problem is the simple descriptive or structural one: Do the same dimensions suffice adequately to describe certain areas of personality in the two cultures being compared? This question is absolutely fundamental, and a positive answer is required before any further steps can be taken; yet it is usually aborted and the assumption made that the same dimensions, traits, or factors which account for the major part of the variance in one population will suffice to do so in a second population also. As we shall see, this hypothesis can be shown to be erroneous in many instances, although it appears to be correct in others; clearly, empirical investigations are needed before we can proceed with the investigation of our second problem.

The second problem, given that identical (or very similar) dimensions, traits, or factors are found to account for a major part of the variance in the two cultures, relates to the problem of *measurement* in the two cultures. Even though the *factors* in the two cultures may be identical, it can and does occur that individual *items* in the scales show different factor loadings. When this happens, clearly a different *weight matrix* (or scoring key) has to be constructed for measurement in the second, as compared with the first culture, choosing appropriate factor loadings to determine the nature of the weight matrix. Thus in this case we are concerned with constructing a suitable measurement instrument for Culture 2; we cannot simply take the original weight matrix in order to do so, and although changes may be minimal, they should nevertheless be made in line with the above discussion.

This leads us to the third problem, namely, the actual cross-cultural comparison between the two cultures or nations involved. Clearly, if different weight matrices are needed in order to score the test in the two cultures, the test scores are not strictly comparable. This is true whether we use the same weight matrix or different weight matrices; if we use the same weight matrix, then scores will be based on items having different loadings and therefore will not be strictly comparable. If the weight matrices differ, then clearly again no direct comparison can be made. What is required is a single *reduced-weight matrix*, including only those items having identical (or nearly identical) loadings for the two sets of factors. Proper comparison should be based only on such reduced-weight matrices.

Our first problem, as mentioned above, is the question of the *comparative dimensionality* in the two different populations. Psychometrically, dimensions of personality, factors, or traits are defined in terms of factor analytic investigations which identify groups of items sharing common variance and setting them off against other groups of items not sharing this common variance. It cannot be assumed that the same items will be found to share common variance when different cultures are being studied, and the assumption that they do so must be empirically verified. A few examples may suffice to indicate the importance of this prescription.

Materanz and Hampel (1978) conducted factor analyses of interitem correlations for the FPI (*Freiburger Persönlichkeitsinventar*, a German personality questionnaire that contains measures for a number of traits as well as the major type factors extraversion–introversion and neuroticism–stability). Questionnaires were applied to German and Spanish probands and separate factor analyses carried out. These analyses demonstrated that while there was considerable invariance for extraversion (E) and neuroticism (N), the other traits of the FPI gave quite different results in the two countries. It is clear that it would be unacceptable simply to administer the Spanish version of the FPI to groups of Spanish probands and score them according to the original manual. This procedure might be admissible for E and N, but it would be completely meaningless as far as the other traits in the questionnaire are concerned. Even for E and N, as we shall see, there may be difficulties in spite of the apparent identity of factors in the two matrices.

A second example comes from the extensive work that has been done, nationally and internationally, on the 16PF scale of Cattell. He himself has of course always been fully aware of the necessity of comparing factor structures across cultures before using the test and has indeed suggested a rigorous and original method of carrying out such comparisons (Cattell, 1970). Even in his own hands (Cattell, Schmidt, & Pavlik, 1973) cross-cultural comparisons were more often found to be incongruent than congruent, and the large number of people who have tried to match factors in other countries (even ones closely similar to the original American culture, such as England, Germany, and New Zealand) show on the whole a far-reaching failure to obtain congruence (e.g., Adcock, 1974; Adcock & Adcock, 1977, 1978; Amelang & Borkenau, 1982; Comrey & Duffy, 1968; H. J. Eysenck, 1972; H. J. Eysenck & S. B. G. Eysenck, 1969; Greif, 1970; Howarth & Browne, 1971; Levonian, 1961; Schneewind, 1977; Sells, Demaree, & Will, 1968, 1970; Timm, 1968). These many results indicate that Cattell's factors are not replicable in other countries (and often not in the United States of America either), that items scored for one factor in his manual may have much higher loadings on other factors in other countries, that unitary factors in his analysis emerge truncated or separated into two or three, or associated with other factors in other analyses, and so on. Clearly the many studies which have simply translated his scale and used the original weight matrices may give meaningless results in other countries.

The method used by us has been suggested by Kaiser, Hunka, and Bianchini (1969). It is essentially concerned with the relative positions of the factors extracted in *n*-dimensional space, enabling us to interpret similarity between sets of factors derived for identical items from different populations in terms of indices of factor comparison which range from 0 (no similarity at all) to 1.00 (perfect agreement). We have used these indices in many studies, adopting the somewhat arbitrary criterion of .95 for *similarity* and .98 for essential *identity* of factors between populations. Obviously indices of factor comparisons below .95 are still indicative of similarity provided they exceed .80, but for the purpose of making sure that factors in different cultures are strictly comparable we will use the term *similar* only for indices of .95 or above and the term *identical* only for indices of .98 and above. Here too, of course, actual identity would demand indices of 1.00, but we have followed the above definition of these two terms.

In our work we have used carefully translated versions of the Eysenck Personality Questionnaire (EPQ; H. J. Eysenck & S. B. G. Eysenck, 1975). Translated questionnaires were then applied to samples of 500 males and 500 females, sometimes more, occasionally somewhat less, making up a reasonable sample of the population of that country. We have found (H. J. Eysenck & S. B. G. Eysenck, 1975) that social status variables are not very relevant to personality; this is fortunate, as it makes the selection of a reasonable sample very much easier. Age and sex are relevant and hence require to be controlled. The concept of a "reasonable sample" as opposed to a random or quota sample is discussed elsewhere (H. J. Eysenck, 1975). Here we may ask to what extent a reasonable sample would give similar or identical results to those obtained from a proper quota sample, when both are taken in the same country. Such a comparison is reported by H. J. Eysenck (1979); identity was obtained on all comparisons.

Apart from the indices of factor comparison, our published data usually give the alpha coefficient reliability of the scales in the two countries, intercorrelations between factors in the two countries, and, when available, information on scores of specially selected additional groups such as criminals, psychotics, and neurotics (H. J. Eysenck & S. B. G. Eysenck, 1983). As the EPQ comes in two forms (adult and junior), we have worked both with adults and with children. Detailed comparisons for many different countries are given below in Table 7 (for adults) and Table 8 (for children), giving in each case the reference to the authors of the study and the particular country in which the work was carried out.

These tables speak for themselves. It will be seen that the vast majority of indices indicate the essential *similarity* of factors in the dif-

TABLE 7. Indices of Factor Comparisons for Adult Samples of Males and Females, Compared with British Samples

EPQ	Country	M N	F N	Males				Females			
				P	E	N	L	P	E	N	L
Adult 1	Greece	639	662	.941	.992	.983	.977	.892	.999	.961	.999
Adult 2	France	428	383	.987	.998	.992	.993	.983	.996	.996	.996
Adult 3	Nigeria	329	101	.980	.990	.990	.980	.660	.910	.920	.930
Adult 4	Australia	336	318	.933	.997	.994	.993	.995	.996	.994	.988
Adult 5	Japan	719	599	.946	.990	.978	.981	.994	.994	.997	.992
Adult 6	Yugoslavia	491	480	.994	.990	.987	.997	.967	.970	.999	.982
Adult 7	Bangladesh	544	531	.998	.984	.998	.980	.989	.991	.996	.991
Adult 8	Brazil	636	760	.998	.992	.997	.999	.992	.981	.996	.990
Adult 9	India	509	472	.981	.986	.985	.997	.968	.992	.991	.964
Adult 10	Sicily	376	409	.982	.995	.998	.994	.934	.978	.997	.992
Adult 11	Spain	435	595	.972	.998	.990	.980	.966	.998	.994	.994
Adult 12	Hungary	548	414	.997	.995	.998	.981	.936	.999	.961	.991
Adult 13	Hong Kong	270	462	.962	.995	.997	.950	.993	.995	.998	.996
Adult 14	Singapore	493	501	.993	.995	.999	.987	.964	.997	.996	1.000
Adult 15	Bulgaria	520	518	.993	.999	.998	.985	.979	.992	.995	.986
Adult 16	Egypt	641	689	.996	.990	.994	.990	.984	.996	.985	.986
Adult 17	Germany	745	591	.963	.991	.995	.999	.997	.999	.998	.997
Adult 18	Iceland	577	567	.981	.994	.998	.995	.992	.999	.999	.998
Adult 19	Iran	347	277	.983	.981	.990	.992	.938	.999	.975	.996
Adult 20	Israel	688	362	.985	.980	.997	.985	.983	.995	.993	.978
Adult 21	Puerto Rico	536	558	.956	.993	.993	.966	.996	.999	.977	.992
Adult 22	Sri Lanka	506	521	.979	.993	.998	.984	.995	.977	.996	.995
Adult 23	Uganda	921	555	.941	.993	.988	.993	.955	.995	.998	.999
Adult 24	Lebanon	634	605	.859	.967	.889	.980	.991	.996	.994	.989

Factor Comparisons

TABLE 8. Indices of Factor Comparisons for Junior Samples of Boys and Girls, Compared with British Samples

EPQ	Country	M N	F N	Males				Females			
				P	E	N	L	P	E	N	L
Junior 1	Hungary	1150	1035	.913	.996	.930	.997	.735	.949	.923	.994
Junior 2	Spain	976	1002	.967	.995	.996	.993	.974	.984	.986	.990
Junior 3	Japan	261	228	.855	.990	.982	.875	.955	.987	.989	.893
Junior 4	New Zealand	644	672	.999	.999	.996	.994	.983	.993	.998	.997
Junior 5	Hong Kong	698	629	.981	.985	.988	.993	.975	.975	.988	.959
Junior 6	Singapore	520	508	.962	.998	.990	.991	.917	.974	.970	.965
Junior 7	Canada	546	512	.976	.997	.999	.997	.981	.984	.993	.998
Junior 8	Denmark	536	575	.966	.964	.994	.997	.952	.987	.998	.993
Junior 9	Greece	1117	1199	.982	.991	.964	.967	.987	.991	.949	.974
Junior 10	Yugoslavia	601	638	.931	.987	.988	.996	.955	.997	.963	.998

Factor comparisons

ferent countries, and an astonishingly high number speak for *identity*. This is true both for males and females, in the adult samples, and for boys and girls, in the children's sample. Not wishing to paint the lily, we will refrain from commenting at too great length on the results, except to say that they are strongly in support of the view that *essentially the same dimensions of personality emerge from factor analytic studies of identical questionnaires in a large number of different countries*, embracing not only European cultural groups but also many quite different types of nations. This of course was to be expected in view of the strong genetic components underlying these major dimensions of personality (Fulker, 1981). We have given a more detailed discussion of the results of our work elsewhere (H. J. Eysenck & S. B. G. Eysenck, 1983) and will not repeat the major points made there. Altogether, we feel that we have succeeded in demonstrating that sufficient identity obtains as far as the structure of personality is concerned in these different populations to allow us to proceed to a consideration of the other two points.

The occasional low values pose a problem. In some cases, such as the low values of the females in Nigeria, the cause may be the small number of cases (101) involved; it has been our experience that numbers substantially below 500 in a given group lead to unstable factor loadings and hence to poor indices of factor comparison. The same may apply to the adult Japanese groups but really cannot be the reason for the low indices of factor comparisons for the Hungarian females on P. Such occasional deviations are difficult to explain, but they do not detract from the generally high level of the indices found in these tables.

It is interesting to note that alternative methods of measurement for E and N have given similar results when we used different nations as variables. Several sets of data are discussed by H. J. Eysenck and S. B. G. Eysenck (1983); we will just refer to the very original and innovative work of Richard Lynn (Lynn, 1971, 1981; Lynn & Hampson, 1975, 1977). His approach consists of taking demographic phenomena such as national rates of suicide, alcoholism, accidents, and so forth and treating them as manifestations of the underlying trait of neuroticism and introversion in the population. The various indices are intercorrelated, a factor analysis is performed, and finally the major factors are interpreted in terms of what is known about the correlations between personality and the various indices used. When quantitative results on E and N are compared, using Eysenck's questionnaire data and Lynn's demographic data, a correlation is found across countries of .70 for neuroticism and .84 for extraversion. No similar data are available, unfortunately, for P.

The question must arise as to the causes of the observed differences between different countries. There are three major types of theories put forward in this connection. The first of these would relate personality differences to such aspects of the environment as the climate or the type of country involved; Lynn (1971) has favored such a hypothesis and gives some factual evidence in support. The second type of hypothesis would favor accidental features in the history of a given population, such as winning or losing a war or remaining neutral; Lynn and Hampson (1977) give some evidence to support this view. It seems likely that both the first and second hypotheses play prominent parts in the differences between scores on personality tests for different nations.

However, there is also a third possibility, namely, that there are genetic differences between populations and that these are linked in some way to the observed differences in personality. It is of course difficult to find a methodology that would enable us to sort out the genetic from the environmental factors, but one such approach has been suggested by H. J. Eysenck (1977c).

His argument starts from the findings of Angst and Maurer-Groeli (1974) in Switzerland that there are significant differences in the frequency of blood groups found among European introverts and extraverts and between highly emotional and unemotional persons. Introversion was found significantly more frequently among persons having AB blood group. Emotionality was significantly more frequent in persons having blood group B. If we can interpret these findings as evidence for some pleiotropic mechanism linking blood groups and personality, then it will become possible to put forward testable hypotheses relating to national and racial comparisons in the personality field. In particular Eysenck predicted that Japanese as compared with English would have a significantly higher proportion of persons with the AB blood group and also that they would have a significantly smaller number of persons with blood group A than with blood group B; he used the ratio A/B to test this hypothesis. The reason for contrasting the Japanese and the English was that the Japanese had exceptionally high scores on introversion and neuroticism and that detailed figures for blood groups for both the Japanese and the English could be found in Mourant, Kopec, and Domaniewska-Soblezak (1976).

For the percentage of persons with blood group AB, the proportions were 3.01 in England and 9.98 in Japan, a very sizeable difference in the predicted direction. There was some variation in different samples and in different parts of the countries in question; the extreme values of large samples were 1.63 and 4.11 in England and 6.63 and 12.88 in Japan; there was no overlap.

The ratio A/B was, as predicted, larger in England than in Japan; the mean values were 4.54 and 1.64, respectively. Variability within each country was large, ranging from 2.0 to 7.33 in England and from 1.32 to 1.95 in Japan; again there was no overlap. There seemed to be little doubt that with such very large samples (the total British sample amounted to 616,106 persons, the total Japanese sample to 421,151 persons) the predicted differences are significant statistically. These data bear out the hypothesis that there are genetic factors predisposing the Japanese to be more introverted and more neurotic than the British.

In a recently published paper, Jogawar (1983) studied 590 young Indian subjects with respect to blood group and neuroticism, using the Cattell 16PF. He found that persons with blood group B were more affected by feelings (Factor C), were more apprehensive (Factor O), were less self-sufficient (Factor Q2), and were more tense (Factor Q4) than persons with other blood group antigens. These results are all in line with the Swiss data in linking blood group B with emotional instability (Jogawar did not test for extraversion–blood group relation).

We may now look at countries high and low on neuroticism, respectively; these values are taken from the table given by Lynn (1981), who translated results from different questionnaires into standard scores, with the mean of 50. Results are shown in Table 9, and it is found that the mean percentage of blood group antigen B for the countries with high neuroticism scores is indeed higher than that for countries low on neuroticism. Mean average B blood group percentage is 16.73 for the high neuroticism countries, whereas the other countries have an average of 13.94, which goes down to 11.11 when we exclude India, where the per-

TABLE 9. Percentage of Blood Group Antigen B for Countries High (+) and Low (−) on Neuroticism

+ Countries	B%	Standard score	− Countries	B%	Standard score
Egypt	24.28	63.0	Australia	8.54	50.6
France	8.59	54.1	Canada	10.17	50.7
West Germany	12.34	51.8	India	36.64	48.6
Greece	13.38	54.5	Italy	11.11	50.5
Iran	23.79	55.2	Sweden	10.12	41.7
Japan	21.04	53.8	Turkey	15.73	44.6
Poland	19.04	55.1	U.K.	8.52	50.0
South Africa	11.38	52.2	U.S.A.	9.90	50.0
			Yugoslavia	14.76	49.2
			Mean	13.94	
Mean	16.73		(without India) =	11.11	

TABLE 10. Percentage of Blood Group Antigen AB for Countries High (+)
and Low (−) on Extraversion

+ Countries	AB%	Standard score	− Countries	AB%	Standard score
Australia	3.16	51.5	Egypt	7.57	48.5
Canada	4.34	53.8	France	3.44	48.1
Greece	4.93	52.5	West Germany	5.23	49.1
India	7.74	50.8	Iran	6.73	48.0
Poland	7.95	51.9	Japan	8.86	46.6
South Africa	4.18	51.7	Turkey	7.23	49.4
Sweden	4.77	50.8	Yugoslavia	7.02	47.6
U.K.	3.04	50.0			
U.S.A.	3.75	56.6			
Italy	4.05	50.5			
Mean	4.79		Mean	6.68	

centage of blood group B is very much higher than in any other country
whatever, thus unduly weighting the average.

Table 10 shows the relationship between extraversion and AB blood
group percentage, again for the data from Eysenck questionnaires tab-
ulated by Lynn (1981). It will be seen that the extraverted countries have
an average AB proportion of 4.79 and the introverted countries have one
of 6.68, again a difference in the predicted direction. The difference is
only 2% in absolute terms, but introverts gave a 50% higher score than
extraverts, using the percentages as they stand.

These and other data are discussed in H. J. Eysenck (1982a), and on
the whole they appear to support at least to some extent the possibility
that differences in personality between countries and cultures may be
genetically determined to some extent. The paper points out many necessary qualifications and difficulties, the major one of which appears to
be the fact that the personality data and the blood group data were not
obtained from the same people. Much further work remains to be done
in relation to the study of genetic markers for personality and the comparison of different countries and cultures in this respect.

LONGITUDINAL STUDIES OF PERSONALITY

Conclusions based on studies of cross-sectional consistency demand
support from studies of longitudinal consistency. These are hardly ever
mentioned by Mischel and his supporters, but they form an important

part of the evidence, and we shall now turn to a brief discussion of this evidence.

Actually there are two questions that should be carefully distinguished here. As Hindley and Giuganino (1982) point out:

> One concerns the extent to which the behavioural characteristics assessed can be regarded as similar in nature at different ages: the issue of what Emmerick (1964, 1967, 1968) and Baltes and Nesselroade (1973) term continuity versus discontinuity of variables. The other concerns the extent to which individuals maintain their relative status across ages on the variables in question. (p. 127)

As they go on to say, it would make most sense in studying stability or change in subjects' behavior if we employed variables displaying a high degree of continuity from age to age, so that the subjects were repeatedly assessed against a similar yardstick.

We will concentrate on the second question, devoting only a few lines to a consideration of the first. Smith (1974) conducted an investigation the purpose of which was to produce a downward extension of the Junior Eysenck Personality Inventory suitable for use with children of 5 and 6 years of age. The questionnaire was derived from the JEPQ (Junior Eysenck Personality Questionnaire), a number of the questions being modified in order to make them suitable for use with the age groups concerned. Children aged 5 and 6 constituted the sample, and a factor analytic examination of the data suggested that it was possible to group questions included in this infant personality questionnaire in terms of those questions measuring neuroticism, those measuring extraversion–introversion, and those suggesting that the child was "faking" his responses. In other words, it was possible even at this very young age to replicate the factors of N, E, and L (a dissimulation or lie scale).

At a slightly older level, the fact that H. J. Eysenck and S. B. G. Eysenck (1965, 1976) in their work on the EPI and the EPQ have found it possible to produce junior scales containing factors similar to those of the adult scales suggests again that similar trait groups can be found at these various ages. In a similar way Cattell has claimed that the same factors he discovered in his adult populations could also be found in school children (Cattell, Eber, & Tatsouka, 1970).

Of particular interest is the extensive work of Hindley and Guiganino (1982) already mentioned. A longitudinal sample of 97 subjects was rated at 3, 7, 11, and 15 years on variables selected as indices of extraversion–introversion and neuroticism. Principal factor analysis and Varimax rotation were performed at each age, and continuity of patterning of the variables across the four ages was determined before and after rotation of the items to maximal contiguity by Kaiser's method. Two principal factors, clearly identifiable as E and N, displayed very high continuity. This careful and detailed study is an example of how such research should be conducted, and also contains historical references; it demonstrates fairly conclusively that identical factors are to be found at these various age levels.

Coming now to the second type of longitudinal consistency (Schuerger *et al.*, 1982), we may begin with a study by Conley (1984) who looked at the longitudinal consistency of the personality dimensions of neuroticism and social introversion-extraversion over a 45-year period. The data were drawn from the Kelly longitudinal study, a panel involving an original group of 300 males and 300 females. The personality inventories used in this study were developed during the earlier decades of this century, and some of their scales have problems concerned with high intercorrelations and or low internal consistencies. Factor analysis was used to derive revised scales that are adequate in both internal consistency and discriminant validity. Scales related to neuroticism and social introversion–extraversion had rather substantial correlations across time periods and across personality inventories. For these two traits, there is a moderate level of personality consistency across the entire adult life span. Correlations ranged between .3 and .6, which, considering that different and rather old-fashioned inventories were used, is a surprisingly positive outcome over such long periods of time. As Conley points out:

> Mischel (1968) argued that .3 represented an *upper* limit on the cross-situational correlations of personality traits. The present data suggest that .3 may represent the *lower* limit on the longitudinal consistency on the traits of neuroticism and social introversion–extraversion. (p. 22)

These results are not out of line with other longitudinal studies (Moss & Susman, 1980). Thus Leon, Gillenn, Gillenn, and Ganze (1979) reported on the MMPI profiles of 71 men studied from middle to old age, finding the correlations on the MMPI scale across a 30-year period to average above .4. The social introversion scale showed the greatest consistency, a correlation of .74 across three decades! Mussen, Eichorn, Hanzik, and Bieher (1980) examined personality ratings in a group of 50 women over a 40-year period. Two personality dimensions, one related to neuroticism and the other to social introversion-extraversion, had consistency coefficients of approximately .3. Schuerger *et al.* (1982) summarized results of retests of personality questionnaires currently in use and concluded that self-reported traits of anxiety and extraversion had four-year to ten-year consistency coefficients of approximately .6. These data are thus all in good agreement.

Similar results have recently been reported in a study by Giuganino & Hindley (1982) to which reference has already been made. Having

shown high stability of the orthogonal factors of neuroticism and extraversion, they used the subjects' scores on these factors to derive stability
coefficients over the ages of 3, 7, 11, and 15 years. A modest general level
of correlations was obtained, but his was found to be higher for measures
of extraversion and neuroticism than for narrower factors. Correction
for rater unreliability and possible protocol unreliabilities raised the coefficients, although these remained lower than those usually found for IQ
data. Corrected longer-term comparisons gave correlations mostly in the
.5 to .6 range, with about a third exceeding .6. Results at this end of the
age scale thus also give grounds for asserting a satisfactory degree of
consistency, particularly when the great difficulties attending such rating studies are taken into account.

A very large and invaluable body of evidence has been collected by
Costa and McCrae, much of which is summarized in a recent survey
(Costa, McCrae, & Arenberg, 1983). They conclude, after a detailed survey of the evidence: "Within the scope of this chapter we have shown
that traits in the domain of neuroticism, extraversion, and openness to
experience are indeed the 'enduring dispositions' posited by trait theorists." (As will be shown later, the concept of "openness to experience"
posited by Costa and McCrae is possibly the obverse of psychoticism.)
Thus they find that the three major superfactors posited in our theory
show considerable consistency during adulthood, complementing the
work of Hindley and his associates and replicating that of Conley. The
actual studies reported in this series (e.g., Costa & McCrae, 1980b; Costa,
McCrae, & Arenberg, 1960; Costa, McCrae, & Norris, 1981; McCrae, Costa,
& Grenlevy, 1980) are too detailed to be discussed at length.

To discuss just some typical examples, consider the work of Costa et
al. (1980) on the Guilford–Zimmerman Temperament Survey, where subjects were tested at 6- and 12-year intervals. Results showed uncorrected
stability coefficients ranging from .59 to .87; no consistent evidence of
lower stability in younger subjects was found, and neurotic and extraverted traits appeared comparably stable when corrected for
unreliability.

Among other studies that ought to be mentioned is the classic work
on longitudinal consistency of personality undertaken at the Berkeley
Institute of Human Development (Eichorn, 1973; Eichorn et al., 1981).
Block and Haan (1971) reported time periods running from senior high
school to early adulthood for approximately 80 males and 80 females,
using a verion of Q-Sort to summarize impressions from a great supply
of different types of data. Moderate degrees of longitudinal consistency
are found for clusters of traits which resemble neuroticism, extraversion, and the impulse control aspect of psychoticism.

Bronson (1966, 1967), using other portions of the Berkeley studies, analyzed ratings of 45 boys and 40 girls from ages 5 to 16, reporting some degree of consistency for variables obviously related to extraversion and neuroticism. The former seemed more consistent over time than did the latter. When ratings for roughly three-quarters of the sample were again undertaken at age 30, reasonably high correlations were again found of between .40 and .55 from early childhood to age 30, as far as extraversion was concerned; for neuroticism the correlations were smaller.

In another study of the Berkeley material, Mussen *et al.* (1980) had 53 of the mothers of the subjects taking part in the original series of studies rated at ages 30 and 70 by independent judges. Two major factors of neuroticism and sociability (extraversion) gave consistency correlations of .34 and .24, respectively.

Apart from the Berkeley studies, the British National Child Development Study by Ghodsian, Fogelman, and Lambert (1980) deserves mention. Emotional adjustment (neuroticism) was rated by parents and teachers at ages 7, 11, and 16, with a longitudinal consistency over nine years ranging between .3 and .4. In a similar study in Sweden, Backteman and Magnusson (1981) had independent ratings made at ages 10 and 13 of approximately 1,000 youngsters. For neuroticism the longitudinal consistency was over .5; for sociability (extraversion) it was over .4.

On smaller samples consistency of temperament in children was found by Adkins *et al.* (1943), Wiggins and Winder (1961), and Kohn and Rosman (1973). These studies were mostly concerned with factors resembling *E* and *N* and the impulse control part of psychoticism. Palmore (1981) reports on a longitudinal study of the aged.

Other recent studies on a smaller scale could be mentioned, such as those of Schuerger *et al.* (1982), Gabrys (1980), Demangeon (1977), and Backteman (1978), but these would not change the major conclusions to be drawn. Thomas and Chess (1977) have found consistency even from the neonate level through childhood to adolescence and early adulthood; they also give further references to a large psychiatric literature documenting consistency. Stability considerably in excess of that postulated by Mischel is found in longitudinal studies regardless of the age at which the studies are conducted, and although stability declines as the number of years intervening between first and second measurement increases, stability remains high even after very lengthy periods (Conley, 1984).

A possible exception to this rule is found in recovery from neurosis (e.g., Giel *et al.*, 1978; Hallam, 1976), where usually *N* scores decline and *E* scores increase when the first application of the measure is at the beginning of the neurosis and the second one is after recovery. Drug withdrawal does not seem to produce any great change (Ward & Hemseley, 1982), but recovery from mental disorder as such is usually accompanied by the changes mentioned above. This should probably be seen as a return to preneurotic levels, thus isolating a neurotic illness as one kind of event which disturbs the stability of personality measures over time.

A rather different method of establishing consistency of conduct is to relate personality measures at one time with neurotic or criminal conduct at a later time. Burt (1965) has published a report on the follow-up of children originally studied over 30 years previously. Teachers rated 763 children, of whom 15% and 18% respectively later became habitual criminals or neurotics, for N and for E. H. J. Eysenck (1960, 1964a,b, 1977a) had argued that criminals should be high on N and high on E, neurotics high on N and low on E. Of those who later became habitual offenders, 63% had been rated as high on N, 54% had been rated as high on E, but only 3% as high on introversion. Of those who later became neurotics, 59% had been rated as high on N, 44% had been rated as high on introversion, but only 1% as high on E. Thus we see that even the probably rather unreliable ratings made by teachers of 10-year-old schoolboys can predict with surprising accuracy the later adult behavior of these children. Similar results have been reported by Michael (1956), demonstrating again a surprising degree of consistency of behavior from one type of situation to quite a different one.

Yet a different way of approaching this problem has been offered by Young, Eaves, and H. J. Eysenck (1980). This approach should be read in conjunction with the section of genetic factors in the determination of temperament in this book, because it is concerned with the modification of the effects of genes and environment during development, which, as the authors point out, is critical both for our understanding of behavioral changes in general and for the practical application of psychometric tests to the long-term prediction of behavioral patterns. Using designs based on the classical twin study, Wilson (1972) had shown how the profiles of twins' cognitive development are apparently under genetic control. Similar longitudinal studies do not exist in the temperament field, although Dworkin *et al.* (1976, 1977), employing a retrospective longitudinal approach, have produced evidence for significant change from adolescence to adulthood. Arguing from their results, they conclude that there is involvement of genetic factors not only in the expression of traits at both the ages at which their 42 twin pairs were studied but also in the *process of change* in expression over time. However, the small size of the subsample participating at both ages obviously casts doubt on the importance of the finding that for many skills there was an absence of significant genetic variance. Eaves and H. J. Eysenck (1976b) similarly suggested that long-term changes in neurotic behavior may be under genetic control, basing their assertion upon the presence of a significant correlation between DZ twins' intrapair differences and age, matched by the lack of such a correlation for MZ twins.

Young *et al.*, using the balanced pedigree method, studied the juvenile families of twins and singletons together with their parents, as well as a large sample of twins in the adult range. Their major finding was that the covariance of genetic effects as expressed in juveniles and adults was reasonably high for extraversion and neuroticism, especially so in the latter case, for which rate there was a surprising degree of intergenerational consistency. However, in the case of psychoticism the covariance of parents and offspring was low, irrespective of its basis, genetic or environmental. The results suggested that the prediction of adult temperament in childhood may be quite successful along the dimension of neuroticism–stability, somewhat less successful along the dimension of extraversion–introversion, and for the psychoticism dimension prediction would not be expected to be very successful. The failure may reflect the weaker theoretical basis of psychoticism and the poor discriminating power in the normal range of variation of the scales employed in this investigation. On the whole, as far as extraversion and neuroticism are concerned, the results support Rachman's (1969) summary of the available evidence that the personality dimensions of *N* and *E* can be measured in children as young as 7 or 8 years and that the factor structures so defined show a high correlation with those of much older children.

Alternative Systems of Personality Description

In 1973, Royce published a lengthy review of all the available factor analytic studies of personality, paying particular attention to the amount of agreement between different investigators. He concluded that the available evidence strongly supported the existence of 11 first-order factors and 3 second-order factors, which he identifies as anxiety, introversion-extraversion, and superego. The similarity to the *P-E-N* system was apparent. More recently, Royce and Powell (1983) outlined a revision of their system, which still describes three major dimensions of personality called by them emotional independence, introversion–extraversion, and emotional stability; again the similarity to the *P-E-N* system is clear. Figure 12 shows the final structure that Royce and Powell recognized as emerging from the very large body of evidence they surveyed. It will be seen that these three factors are essentially the ones postulated in our paradigm, except that the authors use the term *emotional (in)stability* for *neuroticism,* and the term *emotional independence* to characterize the third factor; otherwise, there is clearly considerable agreement between the two schemes.

For many readers used to the apparent disagreement between different workers, it may seem odd that any firm conclusion can be reached on the basis of studies apparently so very divergent. The main reasons for the appearance of disagreement is of course that different writers work at different levels of the hierarchical scheme; if one writer emphasizes the 3 second-order factors and another the 11 first-order factors, there might appear to be complete divergence of opinion, whereas in actual fact the two ways of looking at the hierarchy would be completely compatible with each other. Thus, as we shall see, the fact that Cattell

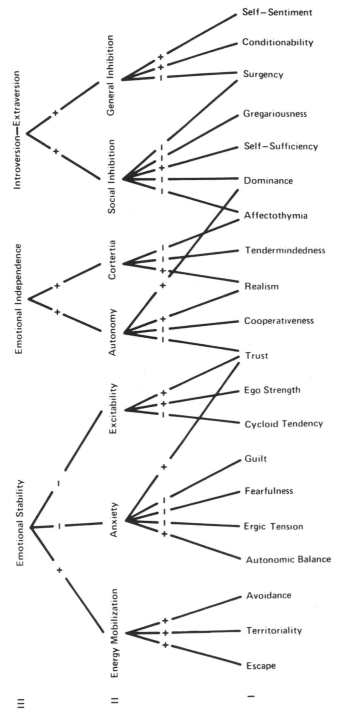

FIGURE 12. The hierarchical structure of the affective system. (From *Theory of Personality and Individual Differences: Factors, Systems, and Processes*, Englewood Cliffs, N.J.: Prentice-Hall, 1983. Copyright 1983 by Prentice-Hall, Inc. Adapted by permission.)

has emphasized the existence of 16 personality factors (Cattell, 1973), while H. J. Eysenck and S. B. G. Eysenck (1976) emphasize 3 superfactors has led many people to believe that there is an essential contradiction between these two views. This is not so; the same superfactors recognized by Eysenck emerge from Cattell's work (H. J. Eysenck & S. B. G. Eysenck, 1969).

Other investigators, such as the authors of the MMPI or the CPI, emerge with large numbers of traits or types which are intuitive rather than based on factor analysis. These at first sight appear incompatible with the three-dimensional scheme adopted in this book, but, as we shall see, when subjected to statistical analysis they also tend to yield super-factors essentially identical with those posited here. Studies which have appeared since Royce's summary have strengthened the view expressed by him, and I think it would be difficult now to deny that much of the work reported over the past 50 years is compatible with the view that major aspects of personality can be described in terms of three super-factors and a number of lower-order factors, with the former emerging from practically every large-scale study reported in the literature, whereas the latter tend to be somewhat less easily replicable and differ in nature to a greater or lesser extent from one study to another.

In this chapter we propose to go through a number of the major systematic approaches to personality description and see to what extent the data bear out this general view. We will find that not all studies give rise to three superfactors, some only giving evidence for the existence of extraversion and neuroticism. Psychoticism clearly is a less readily dis-cernible factor, but the number of replications found is sufficient to establish it as an essential part of any thoroughgoing descriptive schema of personality. Our survey will be confined to the major models of per-sonality description, where correlational and factor analytic data are available to come to a reasonable conclusion; the number of empirical studies in the literature is so large that it would be impossible to aim at completeness in such a survey. Readers searching for a more complete discussion are referred to Morris (1979), who has made a thorough sur-vey of the literature on extraversion and introversion; otherwise, the original studies themselves will have to be consulted. It is our view that such a complete review will not in any way alter the final conclusions arrived at here.

It should be noted that the precise definition of the three superfac-tors will differ slightly from study to study, depending on the items and first-order factors included in the study. This means that some degree of subjectivity has to be exercised in coming to conclusions about the nature of second-order factors, and the reader is invited to look at the details in each case and arrive at an independent conclusion if he is dissatisfied with our interpretation.

Before we turn to the psychometric literature, it may be of interest to point out that not only did psychiatric writers like Jung and Kretschmer have personality theories incorporating concepts similar to P, E, and N, but Freud (1931/1976) also suggested a three-dimensional system of typology which has surprising similarities to the one dealt with in this book. He argues:

> As libido is predominantly allocated to the provinces of the mental apparatus, we can distinguish three main libidinal types. To give names to these types is not particularly easy; following the lines of our depth-psychology, I should like to call them the *erotic*, the *narcissistic*, and the *obsessional* types. (p. 217)

The erotic type, Freud considers, is characterized by having its main interest centered on love, that is, loving, but above all being loved. "From the social and cultural standpoint this type represents the elementary instinctual demands of the id, to which the other psychical agencies have become compliant" (p. 218). His detailed description suggests that this erotic type is very similar to the extravert, and the outstanding sexual interest of the extravert (H. J. Eysenck, 1976c) well illustrates this agreement.

As regards the obsessional type:

> It is distinguished by the predominance of the super-ego, which is separated from the ego under great tension. People of this type are dominated by fear of their conscience, instead of fear of losing love. They exhibit, as it were, an internal instead of an external dependence. (p. 218)

This would be the type corresponding to our notion of neuroticism.

The narcissistic type has his main interest directed to self-preservation; he is independent and not open to intimidation. His ego has a large amount of aggressiveness at its disposal, and in his erotic life loving is preferred to being loved. "They are especially suited ... to give a fresh stimulus to cultural development or to damage the established state of affairs" (p. 218).

To make particularly clear the similarity of Freud's notions to those outlined here, note this penultimate paragraph:

> It seems easy to infer that when people of the erotic type fall ill they will develop hysteria, just as those of the obsessional type will develop obsessional neurosis. ... People of the narcissistic type who are exposed to a frustration from the external world, though otherwise independent, are peculiarly disposed to psychosis; and they also present essential preconditions for criminality. (p. 220)

These relations to neurosis and psychosis clearly agree with the very nomenclature of *neuroticism* and *psychoticism*; note also the relation of the latter to criminality in Freud's speculation, which, as we shall see, finds ample support in the actual data (H. J. Eysenck, 1977a).

Empirical work has been more concerned with the conceptions of the superego, ego, and id (Cattell, 1957; Kline, 1981; Pawlik & Cattell, 1964) and also with notions of the importance of developmental factors in personality organization, such as the concepts of oral optimism and oral pessimism (Abraham, 1952; Freud, 1938; Glover, 1924); these have been studied by Goldman-Eisler (1948, 1950, 1951), showing good agreement with the hypothesis that oral optimism is closely related to extraversion, oral pessimism to introversion.

Kline (1978) and Kline and Storey (1977) have published work on Freudian theories of oral and anal character and find evidence for three factors which they call oral optimism, oral pessimism, and anal character. Howarth (1980b) related these concepts to well-established psychometric scales and suggested that the Kline conceptualizations of Freud's hypotheses are essentially similar to Eysenck's three dimensions. Noting that Kline's superego scale correlated negatively with Eysenck's psychoticism scale, he states that: "If this were the case then all three of Kline's scales would be accounted for largely in terms of Eysenck's 'super factors'" (p. 1041). For a thorough summary of the empirical work of Freudian theories of personality, Kline (1981) should be consulted.

CATTELL AND THE 16 PERSONALITY FACTOR QUESTIONNAIRE

R. B. Cattell is well known as perhaps the leading factor analyst among researchers into personality, and his long-continued search for the major personality factors spans a range too immense to be summarized here. Suffice it to say that on the basis of many different studies he arrived at the conclusion that there were 16 major personality factors, to many of which he gave newly coined names and to all of which he refers in terms of a letter system ranging from A to O and then from Q1 to Q4. He has described the major results of his work on the 16PF in book form (Cattell *et al.*, 1970), and a brief listing of the major primary factors of the system is given in Table 11. Scales are scored in terms of standard scores (Stens), and the table contains a description of the two opposite poles for each factor.

Cattell's scheme is clearly more extensive than Eysenck's; it includes intelligence (Factor B) and social attitudes (Factor Q1). Although there are relations between temperament on the one hand and intelligence and social attitudes on the other, as we shall see, it is nevertheless doubtful

TABLE 11. Cattell's 16PF Scales and Their Interpretations

Factor	Low sten score description (1-3)	High sten score description (8-10)
A	Reserved, detached, critical, aloof, stiff Sizothymia	Outgoing, warmhearted, easygoing, participating Affectothymia
B	Dull Low Intelligence (Crystallized, power measure)	Bright High intelligence (Crystallized, power measure)
C	Affected by feelings, emotionally less stable, easily upset, changeable Lower ego strength	Emotionally stable, mature, faces reality, calm Higher ego strength
E	Humble, mild, easily led, docile, accommodating Submissiveness	Assertive, aggressive, competitive, stubborn Dominance
F	Sober, taciturn, serious Desurgency	Happy-go-lucky, gay, enthusiastic Surgency
G	Expedient, disregards rules Weaker superego strength	Conscientious, persistent, moralistic, staid Stronger superego strength
H	Shy, timid, threat-sensitive Threctia	Venturesome, uninhibited, socially bold Parmia
I	Tough-minded, self-reliant, realistic Harria	Tender-minded, sensitive, clinging, overprotected Premsia
L	Trusting, accepting conditions Alaxia	Suspicious, hard to fool Protension
M	Practical, "down-to-earth" concerns Praxernia	Imaginative, bohemian, absent-minded Autia
N	Forthright, unpretentious, genuine but socially clumsy Artlessness	Astute, polished, socially aware Shrewdness
O	Self-assured, placid, secure, complacent, serene Untroubled adequacy	Apprehensive, self-reproaching, insecure, worrying, troubled Guilt proneness
Q_1	Conservative, respecting traditional ideas Conservatism of temperament	Experimenting, liberal, free-thinking Radicalism
Q_2	Group-dependent, a "joiner," and sound follower Group adherence	Self-sufficient, resourceful, prefers own decisions Self-sufficiency

(Continued)

TABLE 11. (*continued*)

Factor	Low sten score description (1–3)	High sten score description (8–10)
Q_3	Undisciplined self-conflict, lax, follows own urges, careless of social rules Low self-sentiment integration	Controlled, exacting, will power, socially precise, compulsive, following self-image High strength of self-sentiment
Q_4	Relaxed, tranquil, torpid, unfrustrated, composed Low ergic tension	Tense, frustrated, driven, overwrought High ergic tension

whether it is very meaningful to include these as primary personality traits. They appear to be different in nature, and it would seem better to measure and discuss them independently of personality traits as such.

Cattell's personality factors are not independent but are correlated (sometimes highly so); as a consequence, it is possible to extract second-order factors. The two major second-order factors, confusingly labeled by Cattell QI and QII, are clearly recognizable as extraversion and neuroticism; Cattell labels them exvia-versus-invia and adjustment-versus-anxiety. QVIII looks very much like psychoticism, opposing weak superego strength, low ego strength, lack of discipline and low self-sentiment integration with strong superego and ego, controlled behavior, socially precise, high will power, and conscientiousness; it is identified by Cattell as "the real superego factor." In addition there are other, rather minor superfactors, such as one of intelligence, and some which, as Cattell points out, are "not yet good enough to allow identification or extensive use" (p. 17).

Many correlatonal studies have supported the view that Cattell's anxiety and exvia factors correspond closely to Eysenck's neuroticism and extraversion ones. Thus Hundleby and Connor (1968) find a correlation of .60 between N and anxiety, and of .73 between extraversion and exvia. H. J. Eysenck & S. B. G. Eysenck (1969) used items from the Cattell Scale selected by Cattell himself as being the best measures of his exvia and anxiety scales and found them to have high loadings on extraversion and neuroticism factors derived from items in the Eysenck and Guilford Questionnaires. There are no similar studies regarding congruence between psychoticism and QVIII; identification at the moment has to rest on similarity of description and items having high loadings on these factors. However, Cattell and Scheier (1961) are quite clear that

> psychoticism is a direction of abnormality distinct from neuroticism and anxiety. As a rule neurotic-contributory factors are not psychotic-contributory, that is, the neurotic-contributory factors discriminate between neu-

rotics and normals, and between neurotics and psychotics, but they do not discriminate between psychotics and normals. (p. 119)

We would appear to be justified in stating that at the superfactor level the results of the Eysenck and Cattell analyses reveal considerable agreement.

How about Cattell's primary factors? Let us consider first the replicability of the Cattell factors, an obviously crucial issue. If the 16 personality factors only emerge in analyses done by Cattell and his students, then clearly they are of very limited interest. If other psychometrists either in the United States or in other countries fail to obtain results similar to Cattell's, then the meaning and very existence of the 16PF must be very much in doubt.

The evidence suggests that although many investigators have tried to replicate Cattell's results, none outside his own circle has ever succeeded. Levonian (1961), Sells *et al.* (1968), Becker (1961), and Comrey and Duffy (1968) in the United States, Greif (1970) and Timm (1968) in Germany, and H. J. Eysenck and S. B. G. Eysenck (1969) in England have all tried in vain to derive Cattell-like factors from the intercorrelations between Cattell's items. The results will not be discussed in detail; they are pretty conclusive as far as lack of replicability of Cattell's 16 personality factors is concerned. We will look briefly at two more recent and quite extensive studies devoted explicitly to an examination of this point.

The first study is by Howarth and Browne (1971), carried out in Canada. The authors administered the 187-item adult 16PF to 567 subjects and item-factored the results. They obtained ten interpretable factors that bore little relation to the 16 Cattell factors, and they concluded "that the 16PF does not measure the factors which it purports to measure at the primary level" (p. 117).

Howarth and Browne go on to criticize the original method of factoring adopted by Cattell, using the (subjective) packaging of items into parcels which are then correlated with other parcels (rather than using item intercorrelations), and the targeting of the resulting factors to factors obtained by rating. They conclude unequivocally "that Cattell's questionnaire factor system has been developed on the basis of inadequate investigation of the primary factors" (p. 138).

For the sake of completeness, we should also here mention the work of Howarth (1972a) in relation to the analysis of objective personality factors. Howarth has published a large-scale attempt at the replication of a series of Cattell's factors, using objective personality tests. Using 50 variables and 569 subjects, he concluded from his research that:

At neither level (primary or secondary) were the factors other than partly similar, and then only in certain cases to Cattell's factors. . . . Contrary to prior expectation what actually emerged from this group of carefully selected Cattell markers were factors which more closely resembled those of Eysenck, especially at the second order level. (p. 451)

Duckitt and Broll (1982) carried out a factor analysis of the 16PF scales and obtained six higher-order factors, of which three (anxiety, extraversion, and inhibition) are similar to Cattell's anxiety, exvia, and superego control factors and thus are similar to the *N*, *E*, and *P* dimensions. The other factors relate to radicalism ("critical independence"), tendermindedness ("sensitivity"), and shrewdness. The main purpose of this study was to relate personality to the effects of life stress, and Duckitt and Broll found that the major personality factor involved was extraversion; as life stress increases, introverts report substantially elevated levels of anxiety and strain; there is no such increase for extraverts. This, as the authors point out, is very much in line with Eysenck's (1957) prediction.

Of some additional interest is the work of Amelang and Borkenau (1982), partly because it comes from a different language area, partly because the design and the analysis of the study are exceptionally good, but mainly because the authors used Cattell, Guilford, Eysenck, and other variables in a very large-scale study attempting to see to what extent Eysenck's hypothesis was justified that factors similar to *E* and *N* would emerge from any comprehensive study of temperamental variables and that the typical Cattell, Guilford, and others' primary factors would not be likely for the most part to emerge from such an analysis.

The first factor to emerge from this combined analysis of the many different scales was clearly one of neuroticism, and the second factor one of introversion–extraversion, as the authors readily agree. The 16PF scales have high loadings on these two factors, in the expected direction.

Howarth (1976) has gone back to Cattell's original analyses and suggests that these already presaged the faults that the system was later to demonstrate so clearly. Cattell, of course, has reacted to the various criticisms made by H. J. Eysenck (1971a) and in his reply (Cattell, 1972) presented the results of his own item factor analysis of the 16PF. As he himself admitted, the factor pattern does not contain very convincing loadings for the hypothesized primaries (p. 176). Somewhat more positive evidence is put forward in studies by Bolton (1977), Burdsall and Bolton (1979), and Burdsall and Vaughn (1974); and indeed there are now so many studies that one could write an entire book on Cattell's 16PF alone. The fact remains that the great majority of authors have failed to find any convincing degree of confirmation of the Cattell primaries and that even those who have succeeded to some extent still fail consistently to replicate all the scales; even Cattell himself, as pointed out above, did not succeed very well in doing so. The results seem fairly conclusive to us; the reader who is not convinced should go to the original literature to satisfy himself as to the correct conclusions to be drawn.

We must now turn to the work of Barrett and Kline (1982a), which constitutes probably the most competent factor analytic examination of Cattell's model. The analysis is based on data from 491 subjects who completed form A of Cattell's 16PF questionnaire. The data were item analyzed and then factored using both principal component and image analyses. "However, even though five different factor solutions were rotated to a maximum simple structure, the 16 factors did not emerge as expected" (p. 259). Results suggested that between seven and nine factors could be found in the analyses, and these bore little relation to the Cattell factors. It should be noted that Barrett and Kline made a determined effort to try many different methods of rotation, used many different criteria to determine the number of factors extracted, and attempted in every way to find a solution that would agree with Cattell's theory; they clearly failed, and "from a consideration of all the results presented so far, it was concluded that Cattell's 16 factors were not represented in this sample data using form A of the 16PF questionnaire" (p. 266). In another analysis of the same data, concentrating on second-order factors, Barrett and Kline (1980a) found four of these second-order factors which could clearly be identified with P, E, and N, with a possible fourth factor suggestive of L. Thus it would appear that even when an analysis of Cattell questionnaire data is carried out by former students and collaborators of Cattell (both Howarth and Kline fall into this category), the results nevertheless fail to disclose any resemblance of the resulting factors to the original 16 personality factors of Cattell, and on the whole the results much more resemble the Eysenck than the Cattell factors.

We must now turn to the second point at issue, namely, the contribution made to prediction and explanation by having many primary factors, as compared with a few second-order factors. Cattell is quite specific in his claims here:

> The primary factors give one most information, and we would advocate higher strata contributors only as supplementary concepts.... It is a mistake, generally, to work at the secondary level only, for one certainly loses a lot of valuable information present initially at the primary level. (Cattell *et al.*, 1970, p. 111–112)

Eysenck's position is equally clear; he would maintain that second-order factors are far more meaningful psychologically (H. J. Eysenck, 1967a) and that little if any information is lost by disregarding the primaries in such personality studies as those reported by Cattell (H. J. Eysenck, 1971a, 1972b).

Such an argument is of course amenable to factual settlement, and H. J. Eysenck (1972b) has attempted to obtain some evidence by looking at the intercorrelations between Cattell's five anxiety scales and the intercorrelations between his five exvia scales, correcting these in each case for attenuation, that is, giving the correlations that would be obtained if the scales were perfectly reliable. Correlations were given separately for men and women, so that we have 20 correlations for anxiety, and 20 correlations for exvia. Of the 20 coefficients for anxiety, 18 are above .8! This clearly shows that all the different scales are essentially measuring the same underlying trait, namely neuroticism, and can only contribute separate and specific variance to a very small extent, if at all. Intercorrelations between the exvia scales are somewhat lower, but still high enough to make it unlikely that they contribute much beyond the general extraversion–introversion variance.

A study very relevant to this point has been reported by Saville and Blinkhorn (1976, 1981). They used very large groups of over 2,000 randomly selected individuals aged between 15 and 70 years and 1,148 university undergraduates drawn from almost all the British universities. They carried out a very detailed factorial study of the Cattell scales, as well as of the EPI. Two of their conclusions are very relevant to our point. The first is:

> Although Cattell has argued that neuroticism is different from anxiety in that it is a pathological disorder of which anxiety is just one symptom, from these analyses Cattell's neuroticism derived criterion could simply be described as an amalgam of anxiety and introversion (p. 163)

As regards extraversion:

> It would appear that Cattell has successfully split down extraversion into more than one component factor. When the EPI E was extracted from the 16PF, factors A (outgoing), E (assertive), H (venturesome), and Q2 (group-dependence) still appeared as relatively strong factors. (p. 163)

This result is the outcome of a rather interesting method of analysis in which the correlations between the 16PF and the two EPI scales were analyzed in such a way that first of all, all the variance due to E and N was removed in order to see what was left. As the authors say, "the results of this procedure fit pre-existing theory almost too neatly; factor B as one would expect, was the first, followed by factor I and factor Q1, that is to say intelligence, tough/tender-mindedness, conservatism/radicalism" (p. 116). These factors, of course, also appear in the general work of Eysenck, but are usually kept apart from the strict personality (or temperament) field, belonging respectively to the field of intelligence (H. J. Eysenck, 1979) and social attitudes (H. J. Eysenck, 1954; H. J. Eysenck

& Wilson, 1978.) Thus agreement between Cattell's factors and Eysenck's factors extends well beyond the strict personality area to these other areas as well. Something is still left, particularly correlations relating to factor G, and this is presumably an indication of the existence of a psychoticism factor in the Cattell data.

This discussion has been rather more thorough than that devoted to other scales, primarily because the Cattell 16PF is not only the most widely used questionnaire for normal subjects in the United States but is also claimed by the author to be the most thoroughly researched and the one best supported by factor analytic studies. It is reassuring to find considerable agreement between the major Eysenck and Cattell factors and also to find that what is distinctly Cattellian about the 16PF, namely, the stress on the large number of primaries, fails to find support in the empirical literature; no one has been able to replicate the 16 factors, whereas replication of the superfactors has been relatively easy and indeed customary.

These conclusions are similar to those arrived at by earlier writers who have compared the Cattell and Eysenck systems (e.g., Adcock, 1965; Howarth, 1972b). They also go in detail into the question of different methods of factor analysis, rotation and so forth adopted by the two authors, but it would take us too far here to go into such technical detail. The major point to be made is that the similarities in the outcome at the second-order level greatly outweigh any differences and that although the 16PF system appears more inclusive, because of the addition of measures of intelligence, social attitudes, and the like, this is only so because in the Eysenck system these are treated independently and as additional major variables outside the P, E, and N system. For a much more technical review of this whole matter, Kline and Barrett (1983) should be consulted.

THE GUILFORD-ZIMMERMAN FACTORS

J. P. Guilford might rightly be considered the originator of the sustained factor analytic approach to temperament, not in the sense that no one else carried out factor analyses in the personality domain before him, but in that he was the first to do so on a reasonable scale and to continue his work for a sufficiently lengthy period to arrive at conclusions worthy of being taken seriously. The Guilford and Zimmerman (1976) *Temperament Survey Handbook* embodies, as the title suggests, "twenty-five years of research and application" and summarizes in 457 pages a wealth of detail that obviously cannot be incorporated into our brief discussion here. We shall merely take up points of difference with Guilford, which have been aired in previous discussion (Guilford, 1975, 1977; H. J. Eysenck, 1977b). These points will deal mainly with the factorial replicability of factors and with the particular higher-order factors to be extracted from the intercorrelations between the primaries.

Table 12 gives a brief description of the major 11 factors now measured by the Guilford and Zimmerman scales, each referred to by a single letter; the table also indicates some of the traits which contribute to the measurement of each of these factors. Note that each factor is firmly based on factor analytic work of a high order and that the factors themselves are intercorrelated.

Guilford and Zimmerman report results from some 23 analyses involving correlations between scores on their traits, and it is clear that these reflect neuroticism (E, O, F, P) and the two aspects of extraversion, namely, sociability (A, S, G) and impulsiveness—or rather, its opposite, restraint (formerly called rhathymia; R). There is an absence of items or traits which could be characterized as components of P in the scales (with a possible exception of F), and this may be the cause of the absence of P as a factor emerging clearly from the intercorrelations between the scales.

Guilford and Zimmerman also report on factor analysis of items sets; these lead Guilford *et al.* (1976) to postulate a hierarchical model representing their factor traits. This model is shown in Figure 13. E here denotes emotional stability, and Pa (paranoid disposition) is combined

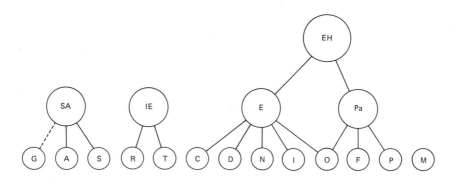

FIGURE 13. A proposed hierarchical model for some of Guilford's (1977) factors of personality. The model shows four second-order factors, namely, social activity, introversion–extraversion, emotional stability, and paranoid disposition; also a third-order factor, emotional health. (From "Will the Real Factor of Extraversion-Introversion Please Stand Up: A Reply to Eysenck" by J. P. Guilford, *Psychological Bulletin*, 1977, *84*, 412–416. Copyright 1977 by the American Psychological Association. Adapted by permission.)

TABLE 12. Guilford's Major Factors and Their Interpretations

Positive qualities	Negative qualities
G—General activity	
Rapid pace of activities	vs Slow and deliberate pace
Energy, vitality	vs Fatigability
Keeping in motion	vs Pausing for rest
Production, efficiency	vs Low production, inefficiency
Liking for speed	vs Liking for slow pace
Hurrying	vs Taking time
Quickness of action	vs Slowness of action
Enthusiasm, liveliness	
R—Restraint (opposite of former trait of rhathymia)	
Serious-mindedness	Happy-go-lucky, carefree
Deliberateness	vs Impulsive
	Excitement-loving
Persistent effort	
Self-control	
A—Ascendance	
Self-defense	vs Submissiveness
Leadership habits	vs Habits of following
Speaking with individuals	vs Hesitation to speaking
Speaking in public	vs Hesitation to speaking
Persuading others	
Being conspicuous	vs Avoiding conspicuousness
Bluffing	
S—Sociability (formerly called "social extraversion," opposed to "social introversion" or shyness)	
Having many friends and acquaintances	vs Few friends and acquaintances
Entering into conversations	vs Refraining from conversations
Liking social activities	vs Disliking social activities
Seeking social contacts	vs Avoiding social contacts
	Shyness
Seeking limelight	vs Avoiding limelight
E—Emotional stability (opposite to a combination of the former traits of _C_, cycloid disposition, and _D_, depressive tendencies)	
Evenness of moods, interests, energy, etc.	vs Fluctuations of moods, interests, energy, etc.
Optimism, cheerfulness	vs Pessimism, gloominess
	Perseveration of ideas and moods
	Daydreaming
Composure	vs Excitability

(Continued)

TABLE 12. (*continued*)

Positive qualities	Negative qualities
Feeling in good health	vs Feeling in ill health Feelings of guilt, loneliness, or worry
O—Objectivity	
Being "thickskinned"	vs Hypersensitiveness Egoism, self-centeredness Suspiciousness, fancying of hostility Having ideas of reference Getting into trouble
F—Friendliness (former trait of agreeableness, *Ag*)	
Toleration of hostile action	vs Belligerence, readiness to fight Hostility, resentment Desire to dominate
Acceptance of domination	vs Resistance to domination
Respect for others	vs Contempt for others
T—Thoughtfulness (formerly called "thinking introversion")	
Reflectiveness, meditativeness Observing of behavior in others Interested in thinking Philosophically inclined Observing of self	vs Interested in overt activity
Mental poise	vs Mental disconcertedness
P—Personal relations (formerly cooperativeness, Co)	
Tolerance of people	vs Hypercritical attitude toward people, faultfinding habits
Faith in social institutions	vs Critical attitude toward social institutions Suspiciousness of others Self-pity
M—Masculinity	
Interest in masculine activities and vocations	vs Interest in feminine activities and vocations
Not easily disgusted	vs Easily disgusted
Hardboiled	vs Sympathetic
Resistant to fear	vs Fearful Romantic interests
Inhibition of emotional expressions	vs Emotional expressiveness
Little interest in clothes and styles	vs Much interest in clothes and styles Dislike of vermin

with it to make up a third-order factor entitled *EH* (emotional health). *IE* is extraversion–introversion, and *Sa* is again social activity.

It will be clear, and Guilford (1975) is explicit on this point, that his introversion–extraversion factor differs from the Eysenck interpretation in that for Guilford extraversion is a mixture of rhathymia (*R*) and lack of thoughtfulness (*T*). *A*, *S*, and *G* form a separate activity factor. For Eysenck *R*, *A*, *S*, and *G* go together to define extraversion–introversion. The matter is discussed in detail by H. J. Eysenck (1977b), but fortunately we have since then the work of Amelang and Borkenau (1982), already described in the previous section. Using scales from Cattell, Guilford, Eysenck, and other sources, they find, as already pointed out, two very strong factors of extraversion and neuroticism. The highest loading on the extraversion factor is obtained by the EPI extraversion scale (.82); Guilford's rhathymia scale, as Eysenck and Guilford both agree, has a good loading of this factor (.56), but thoughtfulness, counter to Guilford's view and in agreement with Eysenck's, has no loading on this factor. Guilford's factor *S* (sociability) has a high loading on this extraversion factor (.75), although in the figure quoted from Guilford he posits that it belongs to a separate higher-order factor. Amelang and Borkenau (1982) conclude, "The highest intercorrelations, clearly and consistently, are between the scales S and R, which speaks for the interpretation of Eysenck, and goes counter to that of Guilford" (p. 124). Campbell and Reynolds (in press) present another detailed comparison between Guilford's and Eysenck's interpretation.

Our own interpretation would then differ somewhat from that of Guilford *et al.* As stated before, we would consider the two second-order factors in his system, *IE* and *Sa*, as the two components of Eysenck's extraversion, namely sociability and (some aspects of) impulsivity; and we are not impressed with the joining of *E* (which seems the obverse of Eysenck's neuroticism) and *Pa* (which in truncated form seems to be Eysenck's psychoticism) in a combined trait of emotional health; the evidence for this third-order factor is distinctly weak. These are relatively minor points; clearly there is considerable similarity between the higher-order factors of Guilford and the *P*, *E*, and *N* factors, and minor differences are to be expected in view of the different universe of items and the many different types of population to whom the questionnaires have been administered. H. J. Eysenck and S. B. G. Eysenck (1969) have reported a joint factorial study of the Guilford, Cattell, and Eysenck items defining the major factors of the three theorists, allowing Cattell and Guilford respectively to choose those items which they considered as the most prominent markers for their various factors. Cattell's work is represented by 15 scales comprising 99 items, and Guilford's work is represented by 13 scales containing 109 items. In addition, the Eysenck Personality Inventory, measuring N and E, and the Lie scale were included. These items were then administered to 600 male and 600 female English subjects, ranging in age from 18 to 40 years; items were then scored for the scales which they represented in the systems of the three authors. The actual primary factor scales which were scored in these three systems are given in Table 13, and the loadings for the two major factors that emerge, E and N, are given in Figure 14 for the male subjects and in Figure 15 for the female subjects.

Looking at the extraversion and neuroticism factors, we find considerable agreement between the two samples. It is perhaps simplest to look at the four quadrants. The choleric quadrant is characterized by the following scales: rhathymia, impulsiveness, protension, dominance, surgency, jocularity, sociability, and liveliness. In the melancholic or dysthymic quadrant, items prominent are guilt-proneness, depression, ergic tension, mood swings, irritability, cycloid disposition, sleeplessness, ner-

TABLE 13. Scales Factor-Analyzed by H. J. Eysenck and S. B. G. Eysenck
(1969)

Eysenck Personality Inventory		22 Autia	M
		23 Sophistication	N
1 Mood swings	M	24 Guilt proneness	O
2 Sociability	Soc	25 Liberalism	Q 1
3 Jocularity	J	26 Self-sufficiency	Q 2
4 Impulsiveness	Imp	27 Self-sentiment control	Q 3
5 Sleeplessness	Sl	28 Ergic tension	Q 4
6 Inferiority feelings	Inf	29 Acquiescence	Ac 2
7 Liveliness	L		
8 Nervousness	N	Guilford Personality Inventory	
9 Irritability	Irr		
10 Sensitivity	Sens	30 Ascendency	A
11 Lie Scale A	Lie A	31 Agreeableness	Ag
12 Lie Scale B	Lie B	32 Cycloid	C
13 Acquiescence	Ac 1	33 Cooperativeness	Co
		34 Depression	D
Cattell Personality Inventory		35 Activity	G
		36 Lack of inferiority	I
14 Cyclothymic	A	37 Masculinity	M
15 Ego strength	C	38 Lack of nervousness	N
16 Dominance	E	39 Objectivity	O
17 Surgency	F	40 Rhathymia	R
18 Superego strength	G	41 Social shyness	S
19 Parmia	H	42 Introspectiveness	T
20 Premsia	I	43 Acquiescence	Ac 3
21 Protension	L		

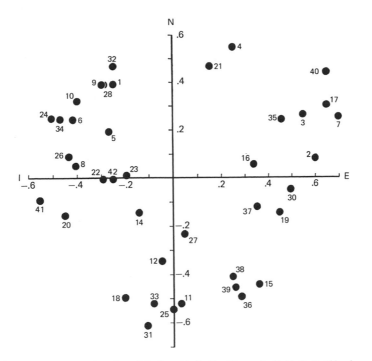

FIGURE 14. Two-dimensional model of traits in the Eysenck–Cattell–Guilford combined study; males. (From *Personality Structure and Measurement* by H. J. Eysenck and S. B. G. Eysenck, London: Routledge & Kegan Paul, 1969. Copyright 1969 by Curtis Brown Ltd. Reprinted by permission.)

vousness, and sensitivity. There are relatively few scales in the phlegmatic quadrant; we may note that the two sexes have in common the cyclothymia scale, superego strength, and agreeableness. As regards the sanguine quadrant, the two sexes have the following scales represented: ego strength, lack of inferiority, objectivity, lack of nervousness, and liberalism. Activity has a loading on extraversion for both sexes but is positively loaded on neuroticism for the men, negatively for the women. These results are all very much in line with expectation; there are minor sex differences which are discussed in detail in H. J. Eysenck and S. B. G. Eysenck (1969).

The book goes in much greater detail into the factor analysis of the individual groups of items which give rise to the various scales; there will be found the justification for the various primary factors in the Eysenck Personality Inventory. Here we are particularly concerned with the factors that emerge in the Guilford Personality Inventory as a check to see

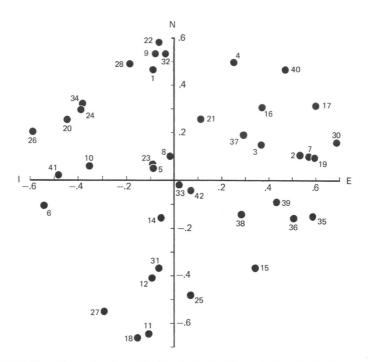

FIGURE 15. Two-dimensional model of traits in the Eysenck–Cattell–Guilford combined study; females. (From *Personality Structure and Measurement* by H. J. Eysenck and S. B. G. Eysenck, London, Routledge & Kegan Paul, 1969. Copyright 1969 by Curtis Brown Ltd. Reprinted by permission.)

whether the Guilford factors are replicable in other samples. The factors that emerged from the Eysenck and Eysenck analysis were sociability, moodiness, carefreeness, sleeplessness, dominance, optimism, compassion, inferiority, nervousness, ascendancy, introspectiveness, and activity. These factors coincided in many aspects with those posited by Guilford, a finding quite different from that which emerged from the Eysenck and Eysenck analysis of the Cattell items, which failed entirely to substantiate and replicate the Cattell factor analysis. However, this agreement between the Guilford and the Eysenck analyses of the Guilford scale items was not perfect. For instance, it was found on occasion that two of Guilford's factors coalesce to give a perfectly interpretable single factor; thus moodiness emerges as a combination of cycloid and depressed and nervousness as a mixture of femininity and (Guilford's) nervousness. Even where several Guilford factors enter the picture, the factors involved are usually relatively easy to understand and interpret.

"We would conclude that the primary factors which emerge from the analysis of Guilford's items are psychologically superior to those which emerge from the analysis of Cattell's items" (p. 235).

It would take us too far to go into the details of all these analyses; the major points have already been made, namely, that (a) Guilford's factors are largely if not entirely replicable at the primary level; and (b) Guilford and Cattell's factors give rise, at the second-order level, to extraversion and neuroticism factors very similar to those which emerge from the items of the Eysenck scales. It is unfortunate that at the time these analyses were done and the data collected for the analysis, no P scale was in existence; hence the analysis is concerned only with two factors. It would be interesting to repeat the work, introducing the P scale and relevant items from the Cattell and Guilford scales, and to attempt to find a third factor representing psychoticism in the joint item pool.

Much early work (not here considered) on the intercorrelations of the Guilford scales is reported in H. J. Eysenck (1970c); the results indicate that practically always the two superfactors of E and N emerge from the intercorrelations between Guilford's primaries.

THE NEO MODEL OF PERSONALITY

The NEO model of personality was introduced by Costa and McCrae (1978, 1980b) and McCrae and Costa (1980) as a combined model for primary and secondary factor structure. It is a conceptual classification of personality traits based on factor analyses of a number of standard and new self-report personality measures, as shown in Figure 16, which constitutes a schematic representation of the 18-facet neuroticism–extraversion–openness model.

Two of these superfactors, N and E, are of course replications of well-known factors; openness, however, is introduced as a new concept, conceived as a broad but continuous dimension of adult personality, characterized by openness to experience in the areas of fantasy, aesthetics, feelings, actions, ideas, and values. In each of these, the more open individual has broader interests, a greater need for variety, and toleration for, if not active pursuit of, the unfamiliar. McCrae and Costa (1980) give some references to earlier writers whose views influenced their adoption of this construct.

In their NEO scale, eight items are used to measure each of six facets, or specific traits, within each of the three broad domains or dimensions. Anxiety, hostility, depression, self-consciousness, impulsiveness, and vulnerability represent the domain of neuroticism. Warmth, gregar-

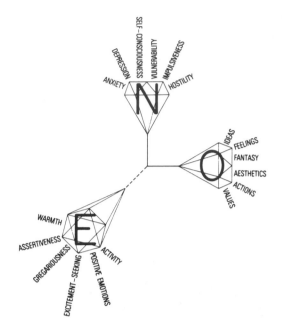

FIGURE 16. Model of the Costa and McCrae three-dimensional system. (From "Still Stable after All These Years: Personality as a Key to Some Issues in Adulthood and Old Age" by P. T. Costa and R. R. McCrae. In P. B. Baltes and O. G. Brim (Eds.), *Life Span Development and Behavior*, Vol. 3, New York: Academic Press, 1980. Copyright 1980 by Academic Press, Inc. Adapted by permission.)

iousness, assertiveness, activity, excitement-seeking, and positive emotions represent the domain of extraversion. Finally, openness to experience is measured in the areas of fantasy, aesthetics, feelings, actions, ideas, and values. Overall domain scores are obtained by summing the scores of the facets in each domain.

The factorial structure of these scales is very much as demanded by the theory (e.g., McCrae & Costa, in press). Those scales hypothetically related to each of the three superfactors actually have their highest loadings on these factors in every case, and appreciable loadings on other factors are quite rare. The nature of the three superfactors, as determined by the content of the questionnaires, suggests that openness represents possibly the opposite end of a continuum to psychoticism; for N and E there are available high correlations with Cattell and Eysenck scales to demonstrate that the concepts are very similar, if not identical.

McCrae and Costa (in press) would seem to deny the likelihood that openness and psychoticism are opposite ends of the same dimension. As they say:

> Individuals high in psychoticism are emotionally isolated or hostile, like odd and unusual things, and are tough-minded in attitudes. Criminals and psychotics are among the groups that score high on this dimension. The tough-minded and emotional isolation are reminiscent of the authoritarian attitudes and lack of empathy found in closed individuals, suggesting a negative correlation between openness and psychoticism. On the other hand, the liking for odd and unusual things and the unconventionality would suggest a positive correlation with openness. However, all these resemblances are superficial. Fundamentally, psychoticism seems to reflect the strength of the bond between the individual and society, other people, and other living things. Unconventionality, suspicion and cruelty may all result from an underlying alienation or pathology. By contrast, openness has to do with the person's preferred mode of dealing with novel experience—a very different content domain. It is unlikely that there is much correlation between the two constructs, although of course an empirical test of the relationship is needed to resolve the question.

Clearly the need for empirical evidence to resolve the difference in opinion cannot be gainsaid, and fortunately work is in progress in the Costa and McCrae laboratories to settle this issue once and for all.

The work of Costa and McCrae has been extended in an interesting direction, again linking it up with E and N through the concept of happiness. Bradburn (1969), in his study of mental well-being, had reported that when positive and negative affects are independently measured the items form two independent clusters. Contrary to natural expectation, Bradburn's positive affect scale (PAS) and negative affect scale (NAS) were not negatively correlated with each other, but were independent and virtually uncorrelated. Others (Andrews & Withey, 1976; Costa & McCrae, 1977; Lowenthal, Thurner, & Chiriboga, 1975) succeeded in replicating the Bradburn findings, and in all we may regard it as established that *happiness* consists essentially of two relatively independent components, one relating to the presence of positive affect, the other to absence of negative affect (McDowell & Praught, 1982).

Costa and McCrae (1980a) formulated the hypothesis that temperamental traits of emotionality, fearfulness, hostility, and impulsivity will be associated with lower levels of happiness and especially with high negative affect; and the temperamental traits of sociability and activity will be associated with higher levels of happiness and with positive affect. Using a relatively random sample of the population, they applied the Bradburn scales, which yielded scores for PAS, NAS, and the difference of these scores (ABS). Other scales used were the Beck Hopelessness Scale, Knutson's Personal Security Inventory, and the Emotional, Activity, Sociability, and Impulsivity Scales from the Buss and Plomin Temperament Survey. Correlations were in the expected direction, the Happiness Scale (ABS) correlating negatively with general emotionality, fear, anger, and poor inhibition of impulse, and positively with sociability, tempo, and vigor.

Costa and McCrae argued that the list of traits provided by this study "begins to take the shape of two established dimensions of personality: extraversion (E) and neuroticism (N)" (p. 673). The authors showed that the Buss-Plomin general emotionality, fear, anger, and poor inhibition of impulse scales defined an N factor, whereas sociability, tempo, and vigor formed part of an E factor. The simplest test of the model, as they say, is direct correlation of measures of E and N with happiness measures.

Again using the Bradburn scales, Costa and McCrae also used the Cattell 16PF and the EPI scales. Having in a previous study (Costa & McCrae, 1976) shown an anxiety or N cluster and an E cluster in the Cattell scales, they went on to identify the 15 items in the Cattell scale that best predicted full N and E scores and then used these 30 items in the present investigation. The new Cattell and EPI scales correlated .65 for E and .68 for N.

The questionnaires were applied four times each, three months apart, and on all occasions the results clearly gave correlations between the negative affect scale and N and the positive affect scale and E both for the Cattell and the Eysenck scales. The affect balance scale (ABS) showed negative correlations with N and positive correlations with E on all occasions and for both sets of personality inventories. The repeated testing and the similarity of results suggests that it is unlikely that temporary states of happiness would substantially alter personality and thus produce the correlations observed. It is much more likely that the causal arrow points the other way, from personality to happiness.

A third study was explicitly devoted to testing this point. As Costa and McCrae point out, the most direct test of the hypothesis would be given by an examination of the predictive relations between personality measures and levels of subjective well-being obtained ten years later. Using the Cattell scales, administered on an earlier occasion, and the Bradburn scales, applied ten years later, they obtained data for 234 men. N cluster scores were significantly related to NAS ($r = .39$) but not to PAS ($r = -.08$). E cluster scores were not related to NAS ($r = .03$) but were related to PAS ($r = .23$). "Knowing an individual's standing on these two personality dimensions allows the prediction of how happy the person will be 10 years later" (p. 675). These data would seem to rule out effectively the alternative explanation that associations between happiness and personality result solely from the mediating effects of temporary moods or states.

Alker and Gawin (1978) have proposed that happiness or well-being is higher among more psychologically mature individuals and that happiness is qualitatively different for individuals of different levels of maturity. McCrae and Costa (1983b) tested this hypothesis using the Loevinger Washington University Sentence Completion Test (Loevinger, 1976) as a measure of development maturity, together with the Bradburn Affect Balance Scale and two additional measures of subjective well-being on a sample of 240 adult males. They failed to discover any association of maturity with well-being and demonstrated that the personality dispositions of neuroticism and extraversion showed significant relations to happiness regardless of maturity level of subjects.

Correlations in agreement with these findings were also obtained for the Knutson Personality Security Measure and the Beck Hopelessness scale (Knutson, 1952; Beck, Laude, & Bohnert, 1974).

Zevon and Tellegen (1982) carried out a mood study in relation to happiness. In a variety of factor analyses by Tellegen and his associates, using subjects' self-ratings on a set of 60 adjectival descriptors of mood or emotion, they found two broad dimensions of self-reported current mood, namely, positive affect and negative affect. The authors carried out a detailed idiographic-nomothetic analysis for a small group of subjects, all of which encouraged their belief that "Positive and Negative Affect are best characterised as *descriptively bi-polar* but *affectively uni-polar* dimensions" (p. 112). (They contrast this conclusion with an alternative one suggested by, among others, Mehrabian and Russell, 1974, who have maintained that affective space is more correctly mapped as a bipolar, good versus bad feeling dimension and a high versus low arousal dimension.) A consequence of all these findings for an important concept in the mood and personality literature concerns the *variability of mood states*. If indeed extraverts tend to show variation between positive affect and neutrality and high N scorers to show changes from negative affect to neutrality, then we can make predictions for various combinations of N and E, such as those shown in Figure 17.

In the choleric quadrant, which combines high N and high E, individuals would show both high positive and high negative affect, that is, great variability. The opposite to this would be found in the phlegmatic quadrant, where stable introverts would show neither high positive nor

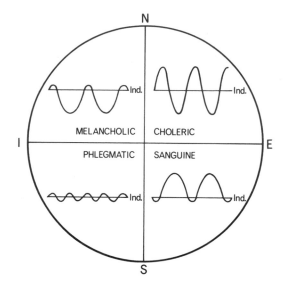

FIGURE 17. Relationship between mood variability and personality.

high negative affect but would go through life with relatively minor excursion above and below the line indicating indifference (IND). Worst off would be individuals in the melancholic quadrant, showing deviations from indifference almost entirely along the negative affect direction; and happiest will be the sanguine person, showing deviations almost entirely in the positive affect direction! It is notable that the very terms used (melancholic, phlegmatic, sanguine, and choleric) seem to carry these implications, with the melancholic showing negative affective states and the sanguine positive affective states, while the phlegmatic shows little by way of positive or negative affect. Cholerics, again, are usually represented as showing strong mood swings, although the motivation for these mood swings may be minimal. Thus here, too, the ancient Greeks appear to have shown an insight into human behavior which modern research must acknowledge.

Moods are usually regarded as personality states rather than as traits, but as we have shown previously the two notions are not entirely independent, and a person's trait measures can predict his likely state in many different situations. When averaging over many situations, as the "happiness" inventories do, we would thus expect a fair correlation between states and traits of a similar nature. This is indeed what Costa and McCrae have found and what is the major content of our figure.

THE MINNESOTA MULTIPHASIC PERSONALITY INVENTORY

We must now turn to the scales that are not factor analytically derived, and the best known of these is, of course, the Minnesota Multiphasic Personality Inventory. The 550-item MMPI is scored in terms of eight clinical scales and the nonclinical social introversion (Si) scale. The clinical scales are supposed to be measuring hypochondriasis (Hy), depression (D), psychasthenia (Pt), hysteria (Hs), hypomania (Ma), psychopathic deviate (Pd), paranoia (Pa), and schizophrenia (Sc). The scales were not derived through factor analysis but by contrasting different populations of psychiatrically ill patients with each other and selecting those items for each scale which best discriminated a particular group. The scales overlap in part, that is, the same item may appear on different scales; this, as we shall see, presents us with an important opportunity for testing certain hypotheses regarding the relationship between the MMPI and the PEN system.

Comrey (1957a, b, c, 1958a, b, c, d, e, f) and Comrey and Margraff (1958) have carried out factor analyses of the different scales in the MMPI and have revealed the presence of many distinct factors in the items of the MMPI. Many of these factors appeared in more than one analysis giving different collections of items from the same data, and the major factors that emerge from these analyses bore little relation to the MMPI scale names, the major ones being shyness, poor physical health, sensitivity, poor concentration, cynicism, sex concern, education, paranoia, and poor reality contact (Comrey & Soufi, 1960). What clearly emerges from these extensive investigations is that each of the MMPI scales is an amalgam of items psychometrically heterogeneous, not correlating well together, and not defining any kind of univocal trait or type. This great defect of the MMPI is due partly to the unreliability and heterogeneity of the criteria adopted (psychiatric diagnoses are notoriously unreliable, subjective, and idiosyncratic), and partly due to the lack of proper concern for psychometric procedures in the construction of the scales.

A number of factorial studies of the MMPI scales have been published and have usually yielded two major factors which have been interpreted in various ways. Welsh (1956) noted the marked consistency in several of these studies and interpreted the dimensions in terms of a general factor of maladjustment or anxiety and a second factor of repression or denial. He developed an anxiety (A) and a repression (R) scale from internal consistency criteria to measure these dimensions. Kassebaum, Couch, and Slater (1959) included many nonclinical scales in their factor analysis of the MMPI and obtained two factors, namely a general factor of neuroticism or ego-weakness and a second factor of introversion–extraversion. They noted that their results were consistent with Welsh's factor dimensions of A and R. Other research, for example, Hundleby and Connor (1968) and Goorney (1970), may also be mentioned in this connection, as may Blackburn (1968). For the most part these authors used both the MMPI and MPI and derived their conclusions from a comparison of the correlations between scales. In addition, Giedt and Downing (1961) and Drake (1946) developed social introversion scales on the basis of the MMPI factors.

A study by Corah (1964) is of particular interest because it was carried out to test several specific hypotheses. The first stated that the two recurring MMPI factors are indeed neuroticism and introversion–extraversion. A factor analysis of several MMPI scales clearly verified this hypothesis. Additional hypotheses related to external criteria to be used in identifying these factors. In order to do this, factor scores were obtained for seven groups of neurotics and sociopaths and two groups of normals. Most of the results are in line with prediction, the neurotic groups having significantly higher neuroticism scores than the two groups of normals, the hysterics, sociopaths, and antisocial reactions having higher extraversion scores than the groups showing dysthymic disorders (depressives, anxiety states, obsessives, phobics), although compared to the normals the hysterics are not distinguished by being more extraverted, a finding which is in line with previous research (H. J. Eysenck, 1959a). These results tend to confirm the identification of the factors.

H. J. Eysenck (1960a) had hypothesized, on the basis of some studies of symptom ratings, that extraverted neurotics would be characterized primarily by somatic manifestations of anxiety, whereas introverted neurotics would be characterized by psychic or cognitive manifestations of anxiety. This hypothesis has been verified several times, for instance, by Buss (1962), De Bonis (1968), Hamilton (1969), and Schalling, Cronholm, and Asberg (1973). Corah selected 25 extreme extraverts and 25 extreme introverts from a group of neurotic outpatients on the basis of their I-E factor scores, their N scores being not different statistically. Blind analysis by three raters of symptoms shown by these neurotics assigned the symptoms to either the somatic or the psychic anxiety pole. A "proportion of somatic anxiety" score was obtained for each patient by subtracting the number of cognitive symptoms from the number of somatic symptoms and dividing by the total number of symptoms. The means scores were .52 for the extraverts and .28 for the introverts, a difference significant beyond the .001 level.

Patients were divided into the respective categories of dysthymic neurotics and sociopathic neurotics, the prototypes of Eysenck's introverted and extraverted groups. Using a chi-square test for the fourfold table of MMPI score and diagnosis, a p value of .001 was obtained in the predicted direction. Thus, on all these tests using external criteria, the MMPI scales behaved as if they were indeed measuring neuroticism and extraversion–introversion. Rathus (1978) administered a shortened form of the MMPI to some 1,500 adolescent males and females and factor analyzed the intercorrelations between the K-corrected scales for the males. The first factor was labeled "psychoticism" and the other two factors "defensiveness" and "depression." The labels should not be taken too seriously, as they represent simply a verbal description. However, Rathus goes on to argue:

> The three factors emerging consistently in the present study may correspond to the three factors Eysenck ... states may largely account for the variance in human personality: psychoticism, extraversion, and neuroticism. The present pscyhoticism factor appears quite similar to Eysenck's psychoticism, which Feldman (1977) points out contains features that are as reminiscent of psychopathy as of psychotic distortion. What has here been labelled defensiveness bears resemblance to Eysenck's extraversion, and depression is similar to Eysenck's neuroticism. (p. 646)

Montag (1977), using a Hebrew version of the MMPI, which he applied simultaneously with the EPQ, found four major factors which he identified with P, E, N, and L. Varimax factor 1, he found,

> has high loadings on Eysenck's N and Pt, Sc, K, Hy (minus) and may be interpreted as the N factor.... Factor III is clearly the extraversion factor with a loading of .73 on Eysenck's E, $-.68$ on Si, .39 on Ma and $-.39$ on D. Factor IV is purely a lie factor since both lie scales have high loadings on it. Factor V is presumably Eysenck's P factor with relevant projection on the expected scales F .41, Ma .45, Sc .39, and Pd .27.... By oblique rotation the pattern is somewhat modified but essentially not different. (p. 48)

These findings agree well with those of Corah (1964), Platt, Pomeranz, and Eiseman (1971), and Wakefield et al. (1974).

The most interesting, as well as the most ingenious, study in the field of aligning the MMPI and the PEN system has been done by Wakefield and his colleagues (Wakefield et al., 1974, 1975). Wakefield started out by attempting to show in detail the relationship between the three dimensions of personality from the Eysenck system and the empirically developed MMPI. In order to demonstrate this relationship, he considered nine of the ten clinical scales as measures of points in Eysenck's three-dimensional framework. (The Mf scale was not considered in this connection as there was little to connect it with *P*, *E*, or *N*.) As Wakefield points out, MMPI scales that are commonly held to measure neuroticism are *Hs*, *D*, and *Hy* (Meehl, 1956; Gough, 1946). Indicators of psychoticism are *Pa*, *Pt*, and *Sc* (Ruesch & Bowman, 1945; Winter & Stortroen, 1963). The *Si* scale (Drake, 1946) originated as a measure of the nonclinical introversion–extraversion dimension of personality. The *Ma* scale has been considered an indicator of neurosis (e.g., Rousell & Edwards, 1971). Also, persons who score high on this scale tend to be outgoing and energetic (Carson, 1969). For these reasons Wakefield considered the *Ma* scale as a measure of both the neuroticism and extraversion dimensions. The *Pd* scale is associated with psychoticism; it supposedly measures psychopathic character disorders (Carson, 1969). Affleck and Garfield (1960) have noted the failure of the MMPI to discriminate between such character disorders and psychosis, and Carson (1969) has noted that high scores on this scale also indicate a tendency to "act out"; hence, Wakefield considered this scale to measure both psychoticism and extraversion.

Figure 18 shows the hypothetical placement of the nine scales in the space defined by *P*, *E*, and *N*. The three neurotic scales correspond to positive values on the *N* dimension. The three psychotic scales correspond to positive values on the *P* dimension. The *Si* scale corresponds to a negative value on the *E* dimension. *Ma* should have positive values on both the *E* and *N* dimensions. *Pd* should have positive values on both the *P* and *E* dimensions.

The correspondence between the empirical relationships among the MMPI scales and their theoretical placement in the *P*, *E*, and *N* system was tested by a procedure developed by Wakefield and Doughtie (1973). This procedure involves specifying the relative lengths of interpoint distances from the theory and comparing each theoretically ordered pair of distances with the empirically ordered magnitudes of the distances in three-dimensional factor space. This principle may be illustrated as follows: Among the three neurotic scales are three distances (*Hs-D*, *Hs-Hy*, and *D-Hy*.) Since the scales measure variations of the same construct, neuroticism, the distances among them should be relatively short. Thus three interpoint distances among the psychotic scales (*Pt*, *Pa*, and *Sc*) should also be short. These six distances are called *short* distances.

There are nine interpoint distances from the three neurotic scales to the three psychotic scales, three distances from the neurotic scales to *Si*, three from the psychotic scales to *Pd*, and one interpoint distance from *Ma* to *Pd*. These 22 distances should be longer than the short distances and are consequently called *middle* distances. Two remaining

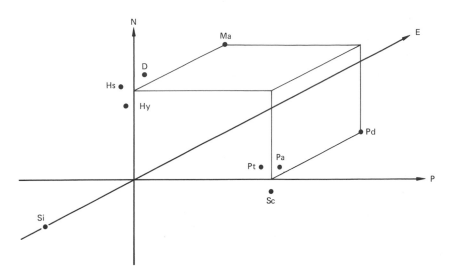

FIGURE 18. Conceptual placement of nine MMPI scales and Eysenck's three-dimensional personality theory. (From "Eysenck's Personality Dimensions: A Model for the MMPI" by J. A. Wakefield, B. H. L. Yom, P. E. Bradley, E. B. Doughtie, J. A. Cox, and I. A. Kraft, *British Journal of Social and Clinical Psychology*, 1974, *43*, 851–857. Copyright 1974 by the *British Journal of Social and Clinical Psychology*. Adapted by permission.)

interpoint distances are from *Si* to *Ma* and from *Si* to *Pd*. These should be the longest distances and are called *long* distances.

In Figure 18 there is a total of 176 ordered pairs of distances, where the theory predicts which should be longer than the other. A test was carried out to compare the prediction with reality by giving the MMPI to 205 married couples, intercorrelating the MMPI scales by principal component analysis, and locating each scale as a single point in the space created by the resulting components as coordinates on a set of orthogonal axes. Details of the statistics are given in the paper by Wakefield *et al.* (1974), but here it is only necessary to state that while 104 correct orders are required to support the theory at the .01 level, the actual number of correct orderings was 111 for the males and 134 for the females; this difference was in fact significant, but there is no explanation for the better agreement of theory with fact for the females. Thus the Wakefield study demonstrates fairly clearly that the hypothesis embodied in Figure 18 is in fact strongly borne out by the empirical data, and hence that there is considerable correspondence between the organization of the MMPI scales and the PEN system.

Wakefield *et al.* (1974) conclude their discussion by stating that an important aspect of the correspondence between the MMPI and the three personality dimensions is the support it gives to the dimensional system of personality description.

> The MMPI was not developed to correspond to Eysenck's personality theory. The actual geometric correspondence between the inventory and the theory suggests that the theory has a reality apart from the test construction skills of the theorist. (p. 419)

In their second paper, Wakefield *et al.* (1975) took up an interesting point stemming from the fact that many of the MMPI scales share identical items, a point which has given rise to much criticism and efforts on the part of several investigators (see Stein, 1968) to develop "pure" or nonoverlapping scales. Wakefield argues that if the overlapping items represent valid variances, they must represent variance that the scales containing them share. If the overlapping items of two scales represent a shared variance, the unique variance of each scale should be represented by the nonoverlapping items. By virtue of their appearing on only one scale, these items must not measure the shared aspects of a criterion employed to select items for the scales. For example, items that are scored for both the *Hy* and *D* scales measure some more general factor or trait that includes criteria for both scales, that is, neuroticism. Items that appear on only one of these scales do not measure neuroticism in general but rather a particular variation within the general trait.

Wakefield goes on to say that if the general traits neuroticism, psychoticism, and extraversion are validly measured by the *overlapping* items, and more specific traits measured by nonoverlapping items, then an analysis such as that of Wakefield *et al.* (1974) just described should yield quite different results for scales scored for the two types of items. That is to say, the scales scored for *overlapping* items only should be highly related to the theoretical model of the interrelations of scales, whereas the scales scored for the *nonoverlapping* items only should be unrelated to the theoretical model.

Using the ten clinical scales and the three validity scales of the MMPI on 100 adults, Wakefield scored the scales twice, producing a measure from the overlapping items and a measure from the nonoverlapping items of each scale. The nonoverlapping MMPI scales were obtained by eliminating all items scored in either direction for more than one scale. The findings of the distance analysis, carried out along the lines of the preceding study, were as follows: For the nonoverlapping scales, 93 of the 176 distance comparisons were in the correct theoretical order, a result not significantly different from chance. For the overlapping scales, however, 141 of the 176 distance comparisons were in the correct order, yielding a p value of $<.001$.

As Wakefield *et al.* (1975) point out:

> The results of the present study clearly show that the relationship between the MMPI and the three personality dimensions of Eysenck—neuroticism, psychoticism, and extraversion—is accounted for by the items that are scored on more than one scale. The items that were scored for only one scale showed almost no relationship to Eysenck's theoretical dimensions. (p. 855)

These highly original and ingenious comparisons made by Wakefield and his colleagues of the MMPI and the P, E, and N scales support the other work mentioned in this section, but take it a good deal further. It makes clear that indeed the structure of the MMPI can be represented in three-dimensional space, the dimension corresponding to the three personality factors of P, E, and N, with a certain amount of specific variance being left over for each of the scales, specific variance that is not related to P, E, or N. There seems to be no doubt that the major dimensions running through the MMPI are the same three dimensions we have found in our own and other people's questionnaires and that therefore research on the MMPI supports the universality of the P, E, and N dimensions.

Quite recently, Skinner and Jackson (1978) provided a model of psychopathology based on an integration of MMPI actuarial systems. They identified three superordinate types, which they called neurotic (Hs, D, Hy), psychotic (Sc, Pt), and sociopathic (Pd, Ma). This model clearly duplicates the P, E, and N model, particularly when it is recognized that sociopaths in Eysenck's original theory constitute the psychopathological form of extraversion.

This conclusion may seem to be contradicted by the fact that Cattell and Bolton (1969), Overall, Hunter, and Butcher (1973), and Hunter, Overall, and Butcher (1974) discovered primaries of varying kinds which do not bear much resemblance to P, E, or N, whether by themselves or in association; Koss (1979) has given a useful review of psychometric studies of the MMPI. What is noteworthy is that most of the factor analytic studies of the MMPI, when concentrating on primary factors, result in quite divergent trait conglomerates. For the most part this is due to the fact that the MMPI is not really suited to factorial analysis, for reasons stated by Kline and Barrett (1983) and by Jackson and Messick (1961). The approach pioneered by Wakefield and his colleagues seems to be much the soundest psychometrically, as far as the MMPI is concerned.

THE CALIFORNIA PSYCHOLOGICAL INVENTORY

The California Psychological Inventory, prepared by Gough (1957), is made up of 18 rationally (not factorially) constructed scales and has been widely used. Nichols and Schnell (1963) carried out a factor analysis of the intercorrelations between the scales and arrived at two major factors.

The first of these they call "value orientation," and the scales which load highly on this factor are those which according to Gough measure responsibility, self-control, and maturity, as well as a tendency to minimize complaints and a concern for the reactions of others. The items on this factor can be grouped into the following content categories: emotional stability and control, denial of impulsivity, even temperament and absence of anger and hostility, absence of anxiety and tension, physical health and well-being, dependability and sense of duty, happy home life, respect for others, and respect for rules and social customs. Correlations are presented with scales from the MMPI, the Guilford-Zimmerman Scale, and other scales; these make it clear that the scale is essentially a measure of *neuroticism* (or rather its obverse stability).

The second factor is labeled "person orientation," and the scales involved measure sociability and favorable interaction with others. The items can be grouped according to content into the following categories: comfortableness with others, joy in interpersonal interaction, dominance and leadership, absence of fear and embarrassment, and quickness of response. "This content and the scale correlates suggest that this scale is measuring the familiar extraversion-introversion dimension as reflected primarily in interpersonal interactions" (p. 232). Again, the interpretation is buttressed by correlation with scales from the MMPI, the Guilford-Zimmerman Scales, and other scales.

Goldberg and Hase (1967) included these two factor scales in a factor analysis of 68 scales developed from the CPI item pool. Only the first three rotated factors accounted for appreciable variance of the total variance; the value orientation or adjustment factor scale loaded .96 on the first factor, the person orientation or extraversion factor scale .96 on the second factor. The third factor appeared to be related mainly to the CPI flexibility scale.

Actually, it would be incorrect to assume that E and N are the only factors to emerge from an analysis of the CPI scales. Mitchell and Pierce-Jones (1960) extract four factors of which the first is called adjustment by social conformity; this appears to be the obverse of N. Factor 2 they call social poise or extraversion. Factor 3 they call superego strength, that is, the obverse of P, and the fourth factor appears to be an intellectual one relating to behaviors associated with intelligence. They label it "capacity for independent thought and action." A similar analysis by Crites, Bechtoldt, Goodstein, and Heilbrun (1961) gives similar results. Thus P and intelligence also emerge from the CPI scales, although not so clearly as do E and N.

Nichols and Schnell (1963) raised the question of the degree to which these two scales exhaust the meaningful and practically important variance of the CPI, but they did not answer it. In a later paper, Reynolds and Nichols (1977) formally asked, "How well [do] the two factor scales represent the information in the 18 scales CPI profile ... is the useful information in the scale contained mainly in this common factor portion of the variance, or is it contained mainly in the unique portion of the variance?" (p. 908). In order to answer this question, the writers regressed each of the CPI scales on the factor scales. The adjustment and extraversion factors were found to be quite independent in these data (r = .13), which were based on 763 subjects.

Having derived these equations, the authors predicted a score for each subject on each of the CPI scales from a knowledge of the person's scores on the two factors. These scores reflect the variance which the scales share with the common factors. The deviations of the actually obtained scores from the predicted scores were then calculated. These deviation scores reflected the scale variance which was unique, that is, not attributable to the two factor scales. Reynolds and Nichols were then able to compare the three scores (raw, predicted, and deviation) with respect to their correlation with selected criteria.

The data for the investigation were collected by questionnaire from a random sample of all the students who took the National Merit Scholarship qualifying test. The students supplied a great deal of information on personality inventories, vocational preference inventories, and objective behavior inventories, as well as other information about their personality, career plans, physical health and body statistics, religious commitments, life goals, time use, dating habits, perceived problems, relations with others, political and social opinions, social status, and school and nonschool achievements. The students' parents reported extensively on the child-rearing practices in the home and on the details of the child's development from infancy to adolescence. Two friends and two teachers of each student also provided estimates of the achievement potential of the student and rated him on a number of traits. Further information was gathered from the students in a follow-up study.

From this file of data, 178 variables were selected as being potentially related to personality. Included in this set of variables were measures of academic and extracurricular achievement, mental ability, mental and physical health, life goals, political opinions, religious beliefs, personal problems, interests, typical behaviors, dating behavior, child-rearing patterns, and parents' ages and education, as well as personality ratings obtained from the student himself, his parents, his friends, and his teachers.

As a next step, correlations between the 178 variables and the raw, predicted, and deviation CPI scales scores were calculated for all of the CPI scales. For each of the CPI scales, a number of criterion variables were selected from among the 178 variables in the file, these being chosen on the basis of the investigators' belief that they would be correlated with the CPI scales in question. From the very extensive statistical analysis of these data, the authors arrived at the following conclusion:

> For the most part, the factor scales do seem to capture the valid variance in the CPI scales. . . . In many instances a common factor portion of the scale's variance was actually more predictive of relevant criteria than was the total scale variance. These findings would suggest that users of the CPI might be better off to measure and interpret the two principal factors rather than attempt to derive meaning from a complex profile of scores. (p. 914)

This is an important result, which is certainly relevant to the controversy, to which we have alluded several times already, that has arisen between those who prefer reliance on primary factors and those who believe that most of the important variance is captured by superfactors. In a field wherein most opinions are based on preconceived notions of one kind or another, the Reynolds and Nichols study furnishes us with important factual information on this point.

THE EDWARDS PERSONAL PREFERENCE SCHEDULE AND THE JACKSON PERSONALITY RESEARCH FORM

Two personality inventories have been based on Murray's (Murray, 1938) Need Structure Theory of Personality, namely the Edwards (1957) Personal Preference Schedule and the Jackson (1967) Personality Research Form. Murray postulated a list of manifest needs, of which the EPPS attempted to measure 15 (achievement, deference, order, exhibition, autonomy, affiliation, intraception, succorance, dominance, abasement, nurturance, change, endurance, heterosexuality, and aggression), and the PRF 20 (abasement, achievement, affiliation, aggression, autonomy, change, cognitive structure, defendence, dominance, endurance, exhibition, harm-avoidance, impulsivity, nurturance, order, play, sentience, social recognition, succorance, and understanding). In addition, the PRF includes two stylistic scales, infrequency and desirability. The PRF appears in a true-false format; the EPPS uses the forced-choice item format, in which two statements with approximately the same social desirability scale values, but representing different needs, are paired and the subject's task is to select the alternative which best describes him (ipsative format).

Murray's original descriptions of these needs are far from clear, and having been translated into questionnaire form by Edwards and Jackson the relevance of the questions to the hypothetical needs becomes even less clear. Edwards, Abbott, and Klockars (1972) have correlated the scales in these two questionnaires with each other, and it is interesting to compare correlations of scales having similar names. Thus the two abasement scales only correlate .40; in other words, although based on the same trait description and sharing the same title, they only share 16% of the variance! Some scales in the one questionnaire have the highest correlation not with the scale in the other questionnaire bearing the same title, but with a different one. Thus the PRF *abasement* scale correlates .42 with the EPPS *endurance* scale. Some of the PRF trait scales which have no EPPS counterpart have higher correlations with differently named scales in the EPPS than many scales in the two questionnaires which have the same name. Some scales with different names but similar intention, such as the sentience and the intraception scales, have very low correlations (in this case, .25).

The picture is very confusing, and in an unpublished study we showed the items for the various scales to graduate psychology students with instructions to fit the correct title (from the Edwards and Jackson titles supplied) to the scales. They did better than chance, but there were so many errors that clearly the title does not always give a good indication to the content of the scale; this must be borne in mind in interpreting the factor analytic results to be described presently. It should also be borne in mind that most of the reports to be cited are so short and so lacking in psychometric detail that it is difficult or impossible to come to a reasonable decision about the adequacy of the work done.

In the factor analysis published by Edwards *et al.* (1972), three factors appear reasonably interpretable as P, E, and N. The authors extract 11 factors and do not give their intercorrelations; a clearer picture might have emerged had they carried out a second-order factor analysis. However, what we interpret as a P factor has a high loading on intraception $(-.86)$, sentience $(-.48)$, understanding $(-.69)$, and heterosexuality $(.31)$. The extraversion factor has loading on dominance $(.86$ and $.83$ for the two questionnaires), abasement ($-.64$), exhibition (.44), and around .4 for the social desirability scales used. Finally, the N factor has loadings on aggression and defence (all negative), affiliation (.38), and nurturance (.35 and .33 for the two questionnaires). There is also a high negative desirability loading for this factor.

Krug and Mayer (1961) used both the EPPS and the Guilford–Zimmerman Scales and arrived at three main factors which can plausibly be interpreted as P, E, and N. Thus their Factor 5 representing the opposite of P contains the Guilford–Zimmerman restraint and thoughtfulness scales and the intraception scale of the EPPS. Factor 3 resembles extraversion and is comprised of the ascendance scale of the Guilford–Zimmerman and the dominance, succourance, and abasement scales (negative) of the EPPS. Finally, the first factor appears the opposite of the neuroticism scale, being defined by the Guilford–Zimmerman variables of restraint, emotional stability, objectivity, friendliness, and personal relations and the EPPS variable of aggression, which has a negative loading.

Krug and Mayer (1961) also administered these two scales and obtained factors reasonably interpretable as P (their Factor 6), E (their Factor 2), and N (their Factor 3).

Other authors have used, in addition to the EPPS, the CPI and the Strong Vocational Interest Blank (Dunnette, Kirchner, & De Gidio, 1958), the PRF and various other variables (Stricker, 1974), and a combination of the EPPS, the Thurstone Temperament Schedule, the Guilford–Zimmerman Temperament Survey, and the Cattell 16PF (Borgatta, 1962). Borgatta, in his analysis, recognizes an extraversion cluster and an emotional stability cluster; in addition, there are a number of minor clusters, and the account is somewhat confusing. It is impossible to say that there is a clear-cut P factor; on the other hand, it would be equally impossible to deny that it may be hidden in the complexities of the analysis.

On the whole, results from the factor analytic study of the Edwards and Jackson Scales show clear-cut evidence for E and N, and putative evidence for P; given the incongruities appearing between the two scales ostensibly based on the same theoretical foundation, the results, although not so clear as one might have wished, are perhaps as closely in agreement with the theory of the three-factor solution as could be expected. What bearing the differential format (true–false versus ipsative) of the two questionnaires may have on the disconcerting lack of agreement between them is difficult to say; in our view the confusion is more in the interpretation of Murray's needs than in the format.

OTHER SYSTEMS

There is no space in this book to continue this detailed investigation and examination of the many personality inventories which have been used and which have been factor analyzed and/or correlated with the EPQ. Instead, we shall give in Table 14 a list of some of the more important personality scales used and the interpretation of the major factors emerging from factor analytic and correlation studies of these scales. In each case there is a reference to the author of the theory or the scales used, or both; in addition, there is a list of authors who have used the scales and on whose work our interpretation, given in the last column, is based.

It is obvious that the interpretation of all this work must, to some degree, be subjective, although we have checked our interpretation with independent experts in the field. Readers interested in specific personality inventories are urged to go back to the original studies to which we have referred and check the accuracy of our interpretation.

SUMMARY

There is such a wealth of information of a very detailed kind contained in this chapter that it would be very difficult indeed to formulate in summary form anything but the broadest conclusions that can be reached. We would suggest that on the whole the evidence supports the following conclusions:

1. Whenever a statistical analysis is carried out on intercorrelations between reasonably large and varied samples of items, factors similar to or identical with extraversion, neuroticism, and psychoticism emerge.
2. These factors take up the major part of the variance, and no other factor systematically produces a greater portion of the variance than P, E, and N.
3. These three factors also appear even in studies specifically devoted to alternative personality theories, such as the Freudian, the Pavlovian, and the Sheldonian.
4. In most analyses *primary* factors are discovered which tend to measure specific behaviors in addition to those contributing to P, E, and N.

TABLE 14. Eysenck Factors in Widely Used Polls

References	Eysenck factors identified
Comrey (1980) Personality Scales	
1. Vandenberg & Price (1978)	*E, N,*
2. Montag & Comrey (1982)	*E, N, L*
3. Comrey & Duffy (1968)	*E, N, P*
4. Barton & Cattell (1975)	Related to 16PF
5. Lorr, O'Connor, & Seifert (1977)	Related to EPPS and Jackson PRF
Howarth–Browne 20 Factor Scale (Browne & Howarth, 1977)	
1. H. J. Eysenck (1978b)	*P, E, N*
2. Barrett & Kline (1980a)	*P, E, N*
Freiburger Persönlichkeits Inventar (Fahrenberg & Selg, 1973)	
1. Hobi & Klar (1973)	Lack of reproducibility
2. Hobi (1973)	Fahrenberg & Selg 9-
3. Timm (1971)	Factor solution but presence of intercorrelations suggesting *E* and *N* superfactors
4. Hampel & Wittman (1973)	
5. Spiller & Guski (1975)	
6. Schenk (1974)	
7. Schenk, Rausche, & Steege (1977)	*P, E, N*
8. Amelang & Borkenau (1982)	*E, N*
Tellegen Differential Personality Questionnaire	
	P, E, N
Myers–Briggs Type Indicator (Myers, 1962)	
1. Stricker & Ross (1964a, b)	General criticism of rules, *E*
2. Steele & Kelly (1976)	*E*
3. Wakefield *et al.* (1976)	*E*
4. Carlyn (1977)	Review of scales
5. Carlson & Levy (1973)	Review of scales
Type A–Type B (Friedman & Rosenman, 1974)	
1. Steptoe (1981)	Review
2. Jenkins, Zyzanski, & Rosenman (1971)	Factor analysis
3. Lovalls & Pishkin (1980)	*N*
4. Irvine, Lyle, & Allen (1982)	*N*
5. Eysenck & Fulker (1983)	*N, E*
6. Furnham (1984)	*N, E*

TABLE 14. (*continued*)

References	Eysenck factors identified
Strelau Pavlovian Questionnaire (Strelau, 1970, 1972)	
1. Carlier (1982)	*E, N*
2. Strelau (1970)	*E, N*
Locus of Control (Rotter, 1960; Levenson, 1973)	
1. Kleiber, Veldman, & Menaker (1973)	*N*
2. Zuckerman & Gerbasi (1977)	Factor analysis
3. Reid & Ware (1973)	Factor analysis
4. Morelli, Krotinger, & Moore (1979)	*N*
5. Wambach & Panackal (1979)	*N*
6. Feather (1967)	*N*
7. Johnson, Ackerman, Frank, & Fionda (1968)	*N*
8. Lichtenstein & Kentzer (1967)	*N*
9. Platt *et al.* (1971)	*N*
Psychological Screening Inventory (Lanyan, 1970a)	
1. Lanyan (1970b)	*E*
2. Mehryar, Khayari, & Hebmat (1975)	*E*
3. McGurk & Bolton (1981)	*N, E, P*
Interpersonal Style Inventory (Lorr & Youniss, 1973; Nideffer, 1976)	
1. Lorr & Manning (1978)	*E, N, P*
2. Nideffer (1976)	*E, P*
Adjective Checklist (Gough & Heilbrun, 1965)	
1. Wakefield *et al.* (1976)	*E, N, P*
2. Brook & Johnson (1979)	*E*
3. Parker & Veldman, 1969	*P, E, N, L*
Sheldon's Typology (Sheldon, 1940, 1942; Cortes & Gatti, 1965)	
1. Metzner (1980)	*E, N*
Field Dependence–Independence (Witkin *et al.*, 1962)	
1. Bone & Eysenck (1972)	*E*
2. Evans (1967)	*E*
3. Loo (1976)	*E*

(Continued)

TABLE 14. (*continued*)

References	Eysenck factors identified
4. Loo (1978)	N
5. Loo & Townsend (1975)	E
6. Canter & Loo (1979)	E
7. Franks (1956)	E
8. Davidson & House (1978)	E
9. Goggin, Filemenbaum, & Anderson (1979)	E
10. Fine (1972)	E
11. Fine & Kobrick (1976, 1980)	E
Self-actualizations (Maslow, 1962; Shostrom, 1964)	
1. Shostrom & Knapp (1966)	E
2. Knapp (1965)	E, N
3. Doyle (1976)	E

Note. From *The Measurement of Personality* by H. J. Eysenck, Lancaster: Medical & Technical Publishers, 1976. Copyright 1976 by Medical & Technical Publishers. Adapted by permission.

5. The contribution made by these primary factors is usually relatively small, and they tend to differ from one investigation to another.

6. Where predictions are made as to external criteria, the major and possibly the total contribution to positive prediction is made by the three superfactors; there is little or no evidence that primary factors have much of a contribution to make.

7. It would appear to follow that there is much support for the suggestion that the three-dimensional hypothesis of personality structure in terms of P, E, and N has paradigmatic validity and covers a significant portion of the total variance attributed to temperament in human behavior.

CHAPTER FIVE

The Cognitive Dimension
Intelligence as a Component of Personality

GALTON VERSUS BINET: IQ AND REACTION TIME

Current terminology sometimes contrasts personality and intelligence and sometimes regards intelligence as part of personality. This is largely a semantic question; obviously we can define a term like *personality* so as either to include or exclude intelligence. H. J. Eysenck (1970c) has included it and would prefer to use the term *temperament* to denote those aspects of personality that are noncognitive. We would thus have a superordinate term, *personality*, subdivided into *temperament*, the noncognitive aspects of personality, and *intelligence*, the cognitive parts of personality. Not everyone would agree, and there may be more to this matter than a simple semantic issue. Binet (1911), for instance, treated emotion as integral to all thinking, considering that it reflected discharge of excitation; this, when powerful enough, was experienced as emotion, but some such discharge went on all the time. Hence, between an *intellectual* and an *emotional* attitude there was only a matter of degree. However, this is not an issue we can profitably discuss in this volume, and we will adopt the more common view that intellectual processes can be discriminated from emotional ones, a view which goes back to Plato at least.

We have already drawn attention to the fact that the term *intelligence* can be understood in three different ways: intelligence A, referring to the genetic basis of cognitive functioning and differences in ability; intelligence B, referring to intellectual and problem-solving behavior as observed in everyday life, determined by the interaction of genetic factors and educational, cultural, socioeconomic, and other environmental

159

determinants; and intelligence *C*, referring to IQ measurement of ability. These distinctions will be useful in looking at a fundamental dichotomy that has separated different approaches to the study of intelligence right from the very beginning of the experimental period around the turn of the century. The two major lines of thinking are identified with the names of Sir Francis Galton and Alfred Binet, one an English polymath, related to Charles Darwin, the other a French psychologist and educationalist. These two men differed, as far as conceptions of intelligence are concerned, in three major directions.

The first difference related to the very notion of intelligence. For Galton this was a general cognitive ability, which determined a person's success or failure at any kind of cognitive task, to varying degrees. Binet seemed to deny the existence of such a general quality; for him intelligence was simply the average of a number of separate abilities, such as verbal ability, numerical ability, suggestibility, and so on; in this conception, the very term *intelligence* is a misnomer, for nothing corresponding to it actually exists. It is an artifact, produced by averaging unlike abilities when what should really be done is to measure these abilities separately. Actually Binet (1903, 1907), whose views were very much influenced by Taine (1878), is not always consistent, but it is clear that on this point he differed very much from Galton.

The second point on which these two men differed relates to the importance of genetic factors. For Galton intelligence was a biological quality, and differences between people in intelligence were determined very largely by genetic factors. Binet, on the other hand, while not denying that heredity might play some part, continuously referred to social and other influences on a child's performance, and this difference in attitude toward genetic factors became very clear in the choice of tests for the measurement of intelligence adopted by Galton and Binet respectively.

Galton (1883, 1908), as one might have expected from his biological interpretation of intelligence, favored elementary and physiological measures, such as reaction times. He predicted that fast reaction times would be good measures of intelligence, a notion which seemed counterintuitive then and will still appear so to many people, although, as we shall see, it has now much empirical support.

Binet, on the other hand, looked for tests which would approximate intelligence *B*, that is, tests which in some way resemble ordinary life situations in which an individual could show intelligence, learning, problem solving, and other abilities. Thus he would incorporate tests which involved following complex instructions; he would use tests of practical ingenuity, such as how to set about finding a lost ball in a park; he would use simple scholastic problems or test achievements such as those involved in having a large vocabulary. It would probably be true to say that Binet's tests are geared to measure intelligence B, Galton's tests to measure intelligence A. Galton tried to develop tests that would come as close as possible to the genotype, whereas Binet tried to include as far as possible educational, cultural, and other environmental factors in his measurement.

The testing movement, in the United States and elsewhere, has completely followed the path outlined by Binet, and practically all existing tests nowadays are similar to those designed by him and Simon 80 years ago. It is interesting to speculate why Galton's suggestion was so coldly received and dismissed in such a wholesale fashion. The textbooks explain that this was due to empirical failure; Wissler (1901) reported a large-scale study in which he failed to find any correlation between intelligence and reaction time. This seems to have dissuaded psychologists from investigating the matter further, and had Wissler's study really been of high quality this might be understandable. However, this must be one of the worst experimental studies ever done, and the outcome is completely irrelevant to the question of reaction time as a measure of intelligence.

It is well known that in measuring reaction times we find that these are quite variable for any given individual, so that to achieve a meaningful average we need something like 100 or more measurements. Wissler used between 3 and 5! This was known already in his time to be quite inadequate and would rule out any meaningful interpretation of his data. A second criticism of his work is that he did not use an intelligence test, but correlated reaction times with grade point average; this is known to be a poor measure of intelligence, particularly in high-level university students, who furnish us with a very narrow range of ability. This indeed is the third criticism. Wissler did not use a normal sample including bright and dull individuals, but concentrated on highly intelligent university students at Columbia University, whose IQs would not show very much variation. For all these reasons Wissler's work is completely unacceptable; the reason why it was so influential can only be attributed to the Zeitgeist, which favored environmental and rejected genetic factors.

As regards this question of genetics, we have already pointed out that Galton was vindicated by ensuing research more than Binet, in spite of the fact that Binet-type tests were used by practically all investigators. On Binet-type tests we find that genetic factors contribute something like 80% to the variance, environmental factors something like 20% (Fulker & Eysenck, 1979). It seems likely that the heritability of Galton-type tests would be even higher, perhaps considerably so.

As regards the question of general intelligence, the crucial contribution here was made by Spearman (1927), who used factor analysis in order to settle this question once and for all. His argument in essence boils down to the rather simple statement that if tests of intelligence are made up in such a way that they do not resemble each other too closely and are then administered to large random samples of the population and intercorrelated, then the matrix of intercorrelations should show a rather interesting feature that mathematicians call forming a matrix of rank 1; this simply means that if the correlations are arranged from highest to lowest, they form perfect proportions. This is true and can be predicted if we can account for all the intercorrelations in terms of a general factor influencing each test to a different degree and specific factors which determine a person's ability to do a particular test but not any other. Spearman published evidence to show that most available matrices of intercorrelations formed a matrix of rank 1, at least roughly, and rested his case on these findings.

Thurstone (1938) took up the cudgels in support of Binet and published the intercorrelations between 56 different types of intelligence tests, claiming to find no evidence for a general factor of intelligence but rather to discover a number of separate abilities, such as verbal ability, numerical ability, visuospatial ability, memory and the like. H. J. Eysenck (1939) reanalyzed Thurstone's matrix and pointed out that alternative solutions were possible, suggesting a combination of a general factor with a number of group factors which resembled Thurstone's "primary abilities." He also pointed out that Thurstone's work had been done on university students, a sample wherein the range of intelligence was severely restricted, thus making it more difficult to discover a general factor.

Thurstone and Thurstone (1941) repeated the study on a fairly random sample of schoolchildren and found that in such a sample indeed there was no way of avoiding the postulation of a general factor of intelligence, as well as discovering the same group factors of primary abilities as before. Spearman also finally agreed to this compromise solution, and it is now fairly widely agreed that in the intellectual field we must postulate a strong general factor of intelligence as well as a number of independent special abilities (H. J. Eysenck, 1979).

The only well-known psychologist opposed to this solution and in favor of a Binet-type analysis is Guilford (Guilford & Hoepfner, 1971), whose wide-ranging structure-of-intellect model of intelligence postulates 120 independent abilities! In this system there are five different operations (evaluation, convergent production, divergent production, memory, and cognition) which can produce six different types of product (units, classes, relations, systems, transformations, or implications), differing in content (figural, symbolic, semantic, or behavioral). As H. J. Eysenck (1979) points out, this system is in fact psychometrically not viable because the intercorrelations between measures of these 120 abilities are not appropriately zero, but are in fact entirely positive, and often quite high. There are many other reasons why the Guilford system is unacceptable, but it would take us too far here to go into these (H. J. Eysenck, 1979).

We must now turn to the question of *measurement*. There is good evidence that IQ, as measured by Binet-type intelligence tests, correlates quite well with intelligence B, cognitive behavior in everyday life. There are the obvious and expected predictions and correlations between IQ and academic ability at school and at university; there are the expected differences between people in professions demanding high intelligence, like medicine, law, and engineering and those in jobs not requiring high intelligence, like dustbin collecting, laboring, and other sorts of manual occupation; there is the well-known success of IQ tests in officer selection; and there are many other positive associations which are predictable on the hypothesis that IQ tests measure intelligence as normally understood (H. J. Eysenck, 1979). There is no doubt that Binet-type tests have been extremely successful from the practical point of view; what is more doubtful is their contribution to a theoretical understanding of mental ability.

Most modern commentators on the nature of intelligence are concerned with the extreme complexity of the concept and the probably matching complexity of underlying processes. Such a conception, involving in particular, educational, cultural, socioeconomic, and other environmental factors as productive of differences in IQ, and also of different primary abilities or group factors of ability, would rule out any reasonably high correlations with such a very simple thing as reaction time. This makes the study of reaction times important, both from the theoretical and the practical points of view. What, then, is the present position?

Brand and Deary (1982) and Jensen (1982a,b) have given a thorough review of the literature, and in summary it is obvious that modern measures of reaction time correlate quite highly with IQ. Reported correlations differ, of course, with such factors as range of intelligence sampled, age, type of measurement of RT employed, and other factors, but given adequate measures of nearly random samples of the population, correlations up to .50 can be obtained; furthermore, when these are corrected for attenuation in both the IQ measure and the RT measure, the estimated true correlation may be well in excess of .70. Before turning to a discussion of the theoretical interpretation of these revolutionary find-ings, it may be worthwhile to look at some of the actual investigations carried out.

The first thing to note is that simple reaction times are not the only or even the most relevant types of measures to be taken. As H. J. Eysenck (1967b) has pointed out, basing his own work on some earlier work by Roth (1964), we must consider, as an additional alternative to simple reaction times, the slope of the regression line linking RT with the number of stimuli used in multiple choice RTs. This relationship is known as Hick's Law (Hick, 1952); he quantified the number of available alternative stimuli to which reaction time responses may be made in terms of information theory, that is, *bits* of information, these being defined as the $\log_2$ of the number of available choices. Thus for a single stimulus (simple reaction time) the number of bits of information is zero—because a single light goes on it does not tell us anything new; we already know that this is the only stimulus present. For two lights, one of which may constitute the stimulus when lit up, the number of bits of information is one. For four lights it is two, for eight lights it is three, and so forth. To a very close approximation there is no doubt that Hick's Law holds in that the increase in reaction time as more alternative stim-uli are added is indeed linear when we subdivide the baseline in terms of bits of information.

The hypothesis that it is the *slope* of this regression line which is correlated with intelligence, that in the dull person RTs increase more quickly with increase of bits of information than do RTs in bright sub-jects, was first verified by Roth (1964), and Jensen (1982a,b) cites much additional material from his own work and that of others (see Figure 19).

Another way of looking at the question of simple and choice reaction time as related to IQ is by correlating IQ with given paradigms involving different numbers of stimuli. Lally and Nettlebeck (1977) have published some data on a group of 48 subjects with Wechsler performance IQs ranging from 57–130. Simple RTs are correlated with IQ at below the $-.50$ level. With two stimuli the correlation rises to $-.55$, with four it rises to $-.64$, and with eight it rises to $-.74$. Thus, within limits, an increase in the number of bits of information, that is, the complexity of the problem, produces an increase in the correlation between RT and intelligence. Another variable in reaction time measurement which has been found to be closely related to intelligence is *variability*. To obtain a proper average of RT, it is necessary to carry out a large number of measures, preferably in excess of 100. These measures vary for each per-son around his mean, and the variability of these measures around the

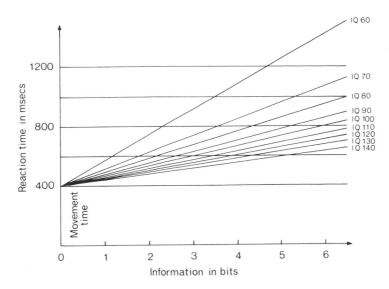

FIGURE 19. Relationship between information in bits, reaction time in msecs, and IQ. (From "Intelligenz, Informations Psychologische Grundlagen" by S. Lehrl, *Enzyklopädie Naturwissenschaft und Technik*, Jahresband 1983, Landsberg, West Germany: Moderne Industrie, 1983. Copyright 1983 by Moderne Industrie. Adapted by permission.)

mean is more highly correlated with IQ than is reaction time itself. Jensen (1982a) estimates a correlation of $-.70$ for random samples of the population when the results are corrected for attenuation; in other words, the higher the IQ, the lower the variability of RT. This is an important finding, as it is quite novel; previous work had not dealt with variability but only with simple and complex reaction times and the slope of the regression line.

In looking at the observed correlations, we should bear in mind that all the correlations with IQ as reported have been with measures of RT taken on a single day. It is well known that there are fluctuations in RT from day to day, lowering the reliability of the measure and suggesting that to get the best possible estimate of IQ we should repeat measures over several days, with the assurance that the averages so obtained would give significantly higher correlations with IQ. Our estimate is that if this were done, and if we based our estimate of IQ on a combination of simple and choice RTs, slope of the regression line, and variability, we would obtain correlations of around .80 between this compound of measures and IQ. The addition to this measurement of paradigms involving short-term memory (Sternberg, 1966) or long-term memory (Posner, 1969) does not appear to add significantly to the correlation between reaction time and IQ.

A rather different type of reaction time measurement has been developed recently by Lally and Nettlebeck (1977) under the name of "inspection time." In this type of work the subject is shown two lines on a tachistoscope, one much longer than the other. The longer line may appear on the right or on the left, and the subject is required to react as quickly as he can by pressing the appropriate button. Backward masking blots out the contours immediately after the two lines have disappeared from the screen, and the technique used involves finding the shortest inspection time which gives rise to 97.5% of accurate determinations. This very simple reaction time task has been found to give quite high correlations with IQ, particularly in subjects below average in intelligence (Brand & Deary, 1982). Presentation of stimuli does not have to be visual; it can also be auditory, having two separate sounds of differential length separated by white noise. Auditory inspection time, too, is quite highly correlated with IQ.

The possibility that reaction time and other very simple responses not involving learning or problem solving could be highly correlated with IQ was suggested to one of us (H. J. E.) originally in some collaborative work with W. D. Furneaux, dealing with attempts to break down the notion of the IQ into three major independent aspects, namely, mental speed, continuance or persistence, and error checking (Furneaux, 1961). (A history of the concepts and theories involved has been given by Berger, 1982, and the latest form of the paradigm developed by White, 1982.) Using mainly letter and number sequence tasks of the usual kind, Furneaux attempted to measure solution times for each problem for each person in a group-testing situation by mounting a cyclometer in front of the group. This cyclometer exhibited a three-digit number which would change once every second; subjects were instructed to write down the setting prior to attempting a problem and after solution. In this way it was thought it might be possible to time each person's solution.

Clearly the mechanics of noting the setting of the cyclometer and writing it down would figure as part of the solution time, when properly speaking it should be measured separately. In order to try and correct for this, Furneaux introduced a short session in which subjects did no problem solving but simply noted and wrote down settings of the cyclometer, so that individual differences in speed of noting and writing down these numbers could be subtracted from the total time taken for each problem. However, Furneaux also correlated the speed of carrying out this subsidiary task of noting and writing cyclometer settings with the IQ measures derived from the IQ task and found an astonishingly high correlation of .85! Given the reliabilities of the test in question, this suggests that there is an almost perfect true correlation between a purely mechanical task of noting and writing down numbers and the IQ test.

These data were never published, but quite recently Lehrl (1980; Lehrl, Straub & Straub, 1975; Lehrl & Erzigkeit, 1976; Lehrl, Gallwitz, & Blaha, 1980) has used a somewhat similar task (reading out letters aloud), finding very high correlations with orthodox IQ tests. Indeed, as long ago as 1909, Burt reported a similar measure of inspection time as having among the highest loadings on a factor of intelligence, otherwise defined by more orthodox types of IQ problem. Curiously enough, he never followed up this promising beginning. Spearman (1904) also used elementary sensory processes as indicators of intelligence with some success. However that may be, there does appear to be sufficient evidence to suggest that early negative results notwithstanding, reaction times, inspection times, and related measures of speed of elementary mental functioning might be a very important component of that mental ability which is traditionally measured by means of IQ tests.

THE PSYCHOPHYSIOLOGY OF INTELLIGENCE

Galton's hypothesis thus appears to have worked out extremely well, but of course reaction time measurement is not strictly speaking a *direct* physiological measure of the biological bases of intelligence. For this we may with advantage turn to the study of evoked potentials (H. J. Eysenck, 1982b). The nature of the evoked potential is illustrated in Figure 20. The baseline records time elapsed in milliseconds; the ordinate records EEG measures taken from the vortex of the skull. To begin with, we see the ordinary activity of the EEG, which has never been found to

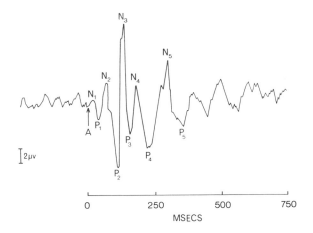

FIGURE 20. Diagrammatic representation of the average evoked potential.

give much in the way of correlations with IQ (until recently—see Gasser *et al.*, 1983a,b). At point *A* we introduce a stimulus, which could be either a flash of light, white noise, a click delivered over earphones, or indeed any other form of sensory stimulus. This produces a series of waves, going first negative (*N*), then positive (*P*), then again negative, and so on. This series of waves finally peters out after about 1,000 milliseconds. The idea of using the evoked potential for the measurement of intelligence apparently originated with Ertl (1971), who reported positive but not very high correlations of around .3 (Ertl & Schafer, 1969). Several attempts to replicate his work failed, but others succeeded; Shucard and Horn (1972) appear to have put the matter beyond doubt and to have demonstrated that correlations of between .2 and .3 could indeed be found between the evoked potential and several different varieties of IQ tests. The major research interest apparently was in the latency of the evoked potential waves, that is, the speed with which they succeed each other; the longer the latency, the lower the IQ. Amplitude also appears to be a promising candidate (Hendrickson, 1972), and by combining latency and amplitude and using auditory rather than visual stimuli (thus avoiding unnecessary artifacts) Hendrickson was able to achieve rather higher correlations than had previously been reported, amounting to something like .4 for latency and for amplitude. As these two measures are not themselves intercorrelated, the combined multiple *r* is something between .5 and .6 with intelligence (H. J. Eysenck, 1973b).

There are several reasons for being dissatisfied with these results. In the first place, there is no theoretical basis for the purely pragmatic correlations observed between latency and amplitude of evoked potential and IQ. In the second place, the correlations, although signficant and replicable, are too low to be of practical importance or to enable theoretical interpretations to be easily made. In the third place, the measurement of latency and amplitude is somewhat subjective, as the waves themselves are not always or even usually so clearly demarcated and set off from small squiggles and other effects to make clear just which is a particular wave and which is merely a small squiggle superimposed on that wave. A. Hendrickson (1972, 1982) and Hendrickson and Hendrickson (1980) proceeded to develop a general theory of neurological information processing in the brain and also to work out a new measurement paradigm which was derived directly from this theory.

The Hendricksons' theory deals in detail with synapse functioning at the biochemical level and also with neural transmission on the physiological level. It would be inappropriate here even to try to summarize the theory; instead, we will attempt simply to state those features that are relevant to the measurements subsequently made and the results achieved so far as a correlation with IQ is concerned.

The Hendricksons begin with a consideration of how information first enters the organism and how the specialized receptor cells encode the information as nerve impulses. In particular, the theory addresses the problem that a single nerve impulse is a binary event which either occurs or does not occur. Differences between nerve impulses in terms of their strength (voltage) are not meaningful; the only differentiation capable of transmitting information is variation in the frequency of firing. In this regard the neural pathways resemble the on/off nature of information carried in computers, and the theory developed by the Hendricksons in many ways resembles the ways in which computers work.

In their model they abandon the frequently used summation hypothesis of neural interconnection and instead postulate a "pulse-train hypothesis," which relies on the use of four discrete time intervals to encode information. These pulse-trains are transmitted from one neuron to another through the synapses, and *errors of transmission* inevitably occur. The Hendricksons posit the probability R to denote the probability of recognition of a pulse-train that has been transmitted without errors; the converse probability of failure is then $1 - R$. R is defined as the probability that any single recognition will succeed, that is, that any single pulse-train will occur without error in its propagation through the cortex. Given independent probabilities, and given the failure to recognize a pulse-train when it should be recognized as a random event, the probability of a longer chain of N event succeeding is simply R^N. The authors go into considerable detail regarding the statistical properties involved in these calculations and report computer simulations, but here we are concerned rather with the applications of this hypothesis of errors in transmission to the problem of the evoked potential and its relation to intelligence.

The Hendricksons postulate a reasonably close relationship between the averaged evoked potential and the pulse-trains that are occurring in the neighboring brain tissues. This immediately suggests that errors occurring in the pulse-train will be reflected in the evoked potential, and such a relationship should have important consequences from the point of view of measurement. It is well known that the evoked potential, superimposed as it is upon the ordinary EEG rhythm, has a rather poor signal-to-noise ratio, so that the recorded and published curves are in fact based on the average of a number of time-locked evocations of the response. (In our own work we have used 90 such replicated and averaged events and have found that a smaller number than this is not adequate for obtaining meaningful results.)

In other words, what we do is to superimpose 90 sets of waves on each other, in all their complexity, and record the average height of each wave at each of a large number of different data points (in our work we have used data points every 2 milliseconds for the total duration of the period we consider critical, the first 250 milliseconds after the presentation of the stimulus). Each individual wave is highly complex, but this complexity will be preserved in the average wave form only if all the waves are free of error. An error in a given pulse-train will be reflected as a change in the shape of the averaged evoked potential, and clearly if there are many errors occurring for a given individual the average wave form will lose its complexity and only record the broadest and most basic outline of the underlying averaged signal waves. Thus, we are led immediately to two measures that are directly derived from the theory and should correlate with intelligence, provided that intelligence is related to errorless transmission of information. The first measure would be the *complexity* of the wave form: the more intelligent the person, the more complex the wave form. The second measure would be the *variance* at each point across the 90 repetitions: the greater the variance, the less intelligent the person whose record is being analyzed. These two measures would be expected to be reasonably well correlated since they are both derived from the same fundamental property of the pulse-train and may thus be combined to form a single physiological measure of R. (The Hendricksons actually report quite high negative correlations between

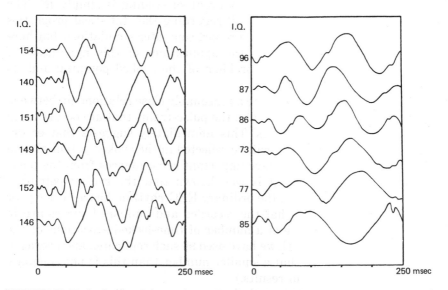

FIGURE 21. Evoked potential waveforms for six high and six low IQ subjects: auditory stimulation.

these two measures in their empirical work.) We would thus appear to have a rational measure that can be objectively quantified and correlated with IQ.

D. E. Hendrickson (1982) has reported a major study involving a reasonably random sample of some 219 male and female school children, aged 15–16, who were given the Wechsler scale and who also had their evoked potentials measured. Details of the procedure and results are given in the reference mentioned; here we will merely summarize the main results. Figures 21 and 22 show a comparison between six bright and six dull children (Wechsler IQs given in the figure) for both auditory and visual stimuli (both were actually used in the investigation, but we shall merely report on the auditory stimuli because these, as anticipated, gave better results). It will be clear from a simple inspection of these data, which are typical of the rest of the data, that the high-IQ children do in fact have evoked potential curves which are more complex in a way that is obvious even to the eye than do the low-IQ children. When a correlation was performed, it was found that the Wechsler IQ correlated .83 with our combined evoked potential measure (variance–complexity), a correlation very much higher than had been expected and one higher than that usually found between two different IQ tests of the Binet type.

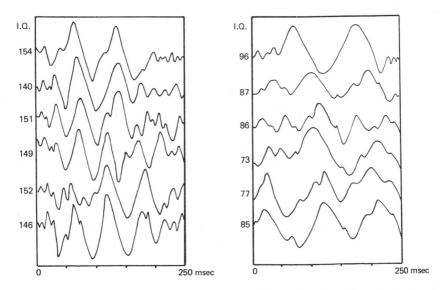

FIGURE 22. Evoked potential waveforms for six high and six low IQ subjects: visual stimulation.

It would thus appear that the evoked potential measure is a more relia-ble and valid index of intelligence than are current IQ tests. Figure 23 shows the actual scatter diagram which gave rise to this correlation.

This conclusion is strengthed by a factor analysis carried out on intercorrelations between the 11 subtests of the Wechsler plus the evoked potential measure. Only one general factor was extracted, as this would represent in a direct form the g factor common to all the tests.

If the general factor extracted from the intercorrelations between the Wechsler scales is the best available estimate of g, then the factor loadings of the different tests measure the degree to which they each reflect the general factor. If the evoked potential measure is now a true measure of intelligence, then the correlations of each of the 11 Wechsler scales with the evoked potential measure should be directly proportional to their factor loadings. This is indeed so: A correlation of .95 was in fact observed between the two sets of data. This again suggests that the two premises of the argument are probably along the right lines and that the evoked potential measure is a relatively pure and highly valid measure of intelligence (H. J. Eysenck & Barrett, 1984).

Before looking at the possible interpretations of these data, we may note a further study by Blinkhorn and Hendrickson (1982), which con-stitutes in a sense an independent replication of the earlier study in that the IQ measurement and the evoked potential measurement were carried out independently by the two authors on a different sample and using a different IQ test (Raven's Matrices) from that used previously. This study was carried out on university students, thus containing a severe

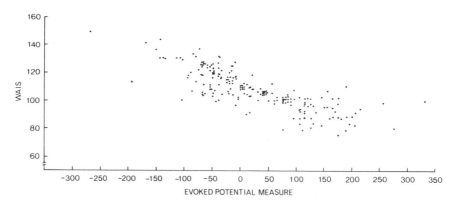

FIGURE 23. Scatter diagram showing relationship between WAIS IQ and evoked poten-tial measure.

restriction of range on the IQ measure adopted. The observed correlation between evoked potential and IQ was .45, which, when corrected for restriction of range, would give a corrected correlation of .84 assuming a standard deviation of 15 IQ points for this sample. As Blinkhorn and Hendrickson conclude:

> These results, and comparable results reported elsewhere using similar techniques on a large sample of school children, suggest the possibility of identifying sources of individual differences in measured intelligence distinct from the quality of cognitive strategies and processes involved in intelligent performance, which might ultimately resolve several of the outstanding issues in differential psychology. (p. 597)

There are two other rather interesting ways in which the evoked potential parallels findings from IQ tests. The first relates to sex differences. It is well known that on most IQ tests men and women have equal scores, but with men having a larger variance around the mean (H. J. Eysenck, 1979). It has been suggested that these data do not really support the view that men and women have equal intelligence on the average because psychologists constructing tests of intelligence take great care to eliminate items which discriminate between the sexes, or at least try to counterbalance them. The evoked potential does not suffer this drawback but does give equal scores to boys and girls, with the boys showing a significantly greater variance. Thus in this way the evoked potential closely parallels well known results from the IQ field.

Another comparison was made regarding differences in intelligence between children coming from high and low socioeconomic groups (H. J. Eysenck, 1982a). On the Wechsler test these two groups of children showed a difference of 23 points, or 1.64 standard deviations. It was predicted that on the evoked potential measure there would also be a significant difference between the two groups but that this would be smaller than on the Wechsler because 20% of the Wechsler score is contributed by environmental factors, which presumably would not be acting in the case of the evoked potential. This was actually found to be so; there is a significant difference between the two groups of children on the evoked potential, but it is just about 20% less than the difference on the Wechsler.

Very recently Haier, Robinson, Braden, and Williams (1984) have reported an investigation into the relationship between various AEP waveform measures, the Hendricksons' complexity measure, and Raven's Advanced Progressive Matrices, a nonverbal and culture-fair IQ test. Twenty-two nursing students took part in the study. Visual stimuli were used, and there were four levels of light intensity. The results indicated that intensity of stimulation is important in determining the size of the relationship between both the complexity and the amplitude measures and the IQ scores. Amplitude measures correlated highly with the complexity measure and both correlated quite highly with IQ, even though the restricted range of IQ measures suggest that higher correlations would have been obtained in samples resembling a more normal IQ distribution. Haier's observed correlation between AEP and IQ is .63, which, when suitably corrected, would again give us a correlation in the neighborhood of .80. For a critique of this study and for a review of various other less important studies, H. J. Eysenck and Barrett (1984) may be consulted.

The most recent study (I. Harris, personal communication) obtained a correlation of .85 between evoked potential and the AH5 IQ test. It is clear that the high relationship observed by Hendrickson is replicable and not a statistical artifact.

The Hendricksons' studies are not the only ones to show high correlations between intelligence and evoked potentials, and the paradigm which generated their particular types of measures is not the only one in the field. Schafer (1982) made use of the fact that unexpected stimuli produce AEPs of larger overall amplitude than those generated by stimuli the nature and timing of which are known to the individual. Schafer has extended the scope of this empirical phenomenon by hypothesizing that individual differences in the modulation of amplitude (cognitive neural adaptability) will relate to individual differences in intelligence. The physiological basis mediating this relationship is hypothesized to be neural energy as defined by the number of neurons firing in response to a stimulus. A functionally efficient brain will use fewer neurons to process a foreknown stimulus, whereas for a novel, unexpected stimulus the brain will commit large numbers of neurons.

Given the relationship between individual neuron firing patterns and observed cortical AEPs, the commitment of neural energy will be observed as amplitude differences between AEPs elicited from various stimulus presentation conditions. Schafer defines his operational measures as variations around the concept of an individual's "average amplitude." Thus individuals with high neural adaptability, characterized by AEPs with much smaller than average amplitude to expected stimuli and much larger than average amplitude to unexpected stimuli, should show high intelligence on test performance. Conversely, for individuals with low neural adaptability, the size of such AEP amplitude modulation should be diminished, with a corresponding low intelligence test score.

Schafer (1982) used two groups of subjects, a normal sample consisting of 63 female and 46 male adults and a retarded sample consisting of 32 male and 20 female subjects with a mean age of 30 years. The results indicated that the normal subjects gave a significantly smaller than average AEP to self-delivered stimuli (i.e., stimuli produced by the subject himself and hence completely expected), indicating the temporal expectancy effect. The retarded subjects showed no statistically significant difference. Correlating the neural adaptability score with the Wechsler on the normal group gave a correlation of .66, which, given the somewhat restricted range of the IQ scores, was corrected by Schafer to give a coefficient of .82. Thus the higher the IQ, the greater the amplitude differences between unexpected and expected AEP amplitude. This implies, as predicted by Schafer, greater flexibility in the attentional response to expected stimuli, that is, the higher-IQ subjects habituate to a greater extent, as indexed by amplitude, to regular, expected stimuli than do the lower-IQ subjects. The data are interesting and impressive, although, as H. J. Eysenck and Barrett (1984) have pointed out, there are criticisms to be made of methodology and analysis alike. These, however, relate more to the testing of the paradigm than to the size of the correlation found, which indicates again the close relationship between evoked potentials and measured intelligence.

THE THEORY OF INTELLIGENCE

We have now cited sufficient evidence to show that these data present a very great challenge to traditional theories of intelligence. Spearman (1927) tried to analyze thinking and problem-solving behavior as shown on IQ tests in terms of his "noegenetic laws." Noegenetic factors, that is, factors creating novel mental content, could, according to him, be accommodated by three major laws. The first of these was the law of *apprehension of experience;* before we can deal with a problem, we must apprehend the basic facts relating to it, that is, we must read the words describing the problem, or look at the figures constituting the problem, or listen to instructions, and understand them, or whatever. Next we have the *induction of relations;* we must see in what ways the different parts of the problems are related to each other. And finally comes the *eduction of correlates;* given the observed relations, we must extrapolate from these to the new items to be discovered, which stand to the items presented in some particular relationship. Thus if we take a very simple problem: A, C, F, J, O, ?, in which we have to find the missing letter, we have first to apprehend the actual content of the message, the letters presented. We then have to discover the relations in which these letters stand to each other, that for each step there is an increase of 1 in the number of letters omitted. We then have to educe the missing correlate by applying this relationship to the last gap and deduce that its missing letter is U. Sternberg (1977; Sternberg & Gardner, 1982) has taken this approach much further, but essentially it represents a typical cognitive approach to the problem of thinking, problem solving, and mental ability measurement in general.

Spearman and his successors devoted most of their attention to the eduction of relations and correlates, having very little to say about the apprehension of experience, although in his original theorizing Spearman saw the essence of intelligence in simple sensory processing. It was thought, quite rightly, that as everyone to whom an IQ test was presented could cope successfully with the intake of information, that is, the apprehension of experience, the true characteristic of differences in intelligence would not and could not be shown by this very simple mental process. Clearly this view is wrong. If reaction times, inspection times, evoked potentials, and other similar measures correlate as highly as they do with orthodox IQ tests, and if it is agreed that these physiological or semiphysiological measures do not involve an analysis of relations and correlates but are concerned rather with the simple apprehension of experience, then we would have to conclude that intelligence (at least as far as it is measured by orthodox IQ tests) is almost entirely concerned with the apprehension of experience. This would also appear to rule out at a stroke most of the environmentalist theories attributing great importance to education, cultural factors, socioeconomic status, and other such hypothetical determinants of thinking, reasoning, problem solving, and so forth. The same reasoning would apply to motivation; it is difficult to see how motivation would influence evoked potentials, when there is no task set to the subject who simply sits on his chair with the electrodes applied to his skull, listening to random tones without having to react to them in any way whatsoever. These results, therefore, present a very serious problem to those psychologists, constituting the great majority, who have followed the Binet line of measurement and have insisted on the complexity of the thought processes, motivational factors, and so on responsible for producing differences in IQ.

We may consider several theoretical possibilities that might account for the observed facts. One such, leaning very much on an interpretation of the data in terms of mental speed, has been given by Jensen (1982a). He posits that the conscious brain acts as a one-channel or *limited-capacity* information-processing system, which can deal simultaneously with only a very limited amount of information. The fact that capacity is limited also restricts the number of operations that can be performed simultaneously on the information that enters the system from external stimuli, or from retrieval of information stored in short-term or long-term memory (STM or LTM). It will be clear that rapidity of mental operations is advantageous in that more operations (e.g., eduction of relations and correlates) per unit of time can be executed without overloading the system.

Furthermore, because there is rapid decay of stimulus traces and information in the system, there is an advantage to speediness of any operation that must be performed on the information while it is still available (STM). Furthermore, to compensate for limited capacity and rapid decay of incoming information, the individual resorts to rehearsal and storage of the information into intermediate or long-term memory (LTM), which has relatively unlimited capacity. However, the process of storing information in LTM itself takes time and therefore uses up general capacity, so that we must postulate a trade-off between the storage and the processing of incoming information.

The important point is that the more complex the information and the operations required on it, the more time that is required, and consequently the greater the advantage of speediness in all the elemental processes involved. Loss of information due to overload interference and decay of traces that were inadequately encoded, or rehearsal for storage or retrieval from LTM, results in breakdown and failure to grasp all the essential relationships among the elements of a complex problem needed for its solution.

> Speediness of information processing, therefore, should be increasingly related to success in dealing with cognitive tasks to the extent that their information load strains the individual's limited channel capacity. The most discriminating test items thus would be those that "threaten" the information processing system at the threshold of "breakdown." In a series of items of graded complexity, this "breakdown" would occur at different points for various individuals. If individual differences in the speed of the elemental components of information processing could be measured in tasks that are so simple as to rule out "breakdown" failure, as in the several RT paradigms previously described, it should be possible to predict the individual differences in the point of "breakdown" for more complex tasks. I believe this is the basis for the observed correlations between RT variables and scores on complex g-loaded tests. (Jensen, 1982, p. 122)

These notions have to be taken together with Hick's (1952) model to account for the Hick paradigm; it essentially involves a type of search process which can be thought of as successive dichotomization of the total number (n) of stimulus elements to be searched, a type of central search process, which on average would take log nt amount of time where t is the time required for a single element. (This is equivalent to bits $\times$ t.) In the Hick paradigm the RT situation does not call for a search in the ordinary sense of the term; the search seems to consist in the resolution of uncertainty. The greater the uncertainty as to the particular stimulus, the greater the search (a central brain process) required for resolution, that is, reduction of the uncertainty to zero. Jensen uses this model in conjunction with the previously stated hypothesis about speed in order to give a realistic account of the relationship between Hick's Law and intelligence—the fact that the slope involved in Hick's law is related to IQ. It is interesting to note that Furneaux (1961) also involved a search process as an essential element in his own theory of intellectual functioning, and indeed it seems difficult to avoid postulation of such a search process, of a kind to be specified differentially in relation to different types of mental content, when talking about a scientific model of problem solving of any kind. It is here that we may have the basis for a relationship between simple RT measures, the other relatively straightforward "apprehension of experience" test, and more complex types of IQ measures.

Our own theory (H. J. Eysenck, 1982c) would differ from Jensen's only in substituting R for speed as being the underlying variable responsible for the observed correlations. Clearly individuals do not differ in the speed with which messages are relayed throughout the central nervous system; however, we can construct a model in which speed of reaction to given stimuli is monotonically related to the number of errors that occur in the processing of information. Such a scheme will here be outlined only very briefly, to give an indication of the kind of account we would give rather than to develop it in detail (H. J. Eysenck, 1982a, 1984).

From what is known about the functioning of the nervous system, it is clear that messages of a particular kind are not sent just once, but repeatedly. We may postulate a comparator which receives these messages and decides whether they are identical, or sufficiently similar, to accept them as accurate; only when this accuracy is accepted is a reaction performed. Let us assume, as a simple numerical model, that at least 10 messages are required, and that at least 9 out of these 10 have to be identical in order for the individual to act on the message. It will be clear that a person for whom no errors occur during transmission will only require 9 messages and will then be able to react. A person in whom several errors occur may require a larger number of messages in order to get 9 identical replicas and will then therefore react more slowly. A person in whom many errors occur will react even more slowly because it will take more redundant messages in order to obtain the required 9 correct ones. In other words, we make the number of errors occurring during transmission responsible for the need to send a larger number of messages, thus *delaying* the speed of reaction. Thus speed of reaction is not a primary cause of low IQ but is itself caused by R, the number of errors occurring during transmission. Given that important addition, our account would be similar to that given by Jensen and Furneaux.

It should of course be added that the messages involved in the solution of a problem do not come only from the outside, that is, do not concern only apprehension of experience gained through the eyes, ears, and so forth; what is also involved is accessing long-term memory and utilizing short-term memory in the process of problem solution. In accessing and using memory the same rules as stated before apply, that it is not speed of neural messages that is important, but error-free functioning. Errors in transmission may of course cause errors in apprehension and accessing when the comparator is set to a low figure of identical messages required; this would lead to errors in solution (White, 1982).

If some such theory as that suggested here is accepted, we still face a final problem, and that is the apparent absence in this scheme of the problem solver. Most cognitive theorists would postulate complex activity in the cortex to represent the actual *solution* of the problem involved in an IQ test. Quite frequently this "problem solver" appears to be some kind of homunculus, deposited in the center of the human brain, receiving messages and finally producing a solution which is then communicated to the motor cortex, the speech organs, and so on. This, of course, is a useless kind of postulation as it only refers the problem back to the homunculus itself, but nothing better seems to have been devised. In our analysis, the quantitative data leave very little if any space or time for the solution of the problem; it would appear that the correct transmission of impulses *by itself* is sufficient to generate the solution! This is a paradox for which at the moment we do not know the answer, but the quantitative data are quite insistent in making the simple information processing through limited capacity channels the major if not the only factor in the solution of IQ tests. Much future work will have to be done to solve this particular paradox.

Two further points should be made. In the first place, this scheme relates to the general factor of intelligence, as postulated by Galton and Spearman; it does not refer to any of the special *(primary)* abilities, such as verbal, numerical, and visuospatial, which are orthogonal to and independent of general intelligence. Thus we do not claim that this is a complete scheme for cognitive behavior; it is merely meant to refer to differences in general cognitive ability, *g*, or general intelligence. Obviously cognitive behavior goes well beyond this limited field, but it is important in science to solve specific problems rather than to attempt global solutions which in the nature of things cannot be tested.

Note further that the data reviewed in this chapter appear to indicate unambiguously that Galton was right and Binet wrong in relation to the postulate of a general factor of cognitive ability (g). The data are decisive, particularly those indicating that the 11 subtests of the Wechsler, although they touch on a great variety of different apparent abilities, are all highly and systematically correlated with the evoked potential measure, as required by Spearman's theorem. In other words, not only are there high and positive correlations, but these differ in size according to the actual factor loadings of these tests, showing very clear-cut proportionality. Such a finding is quite incompatible with the Binet-type hypothesis, and we may consequently claim with some justification that here as elsewhere Galton's theories have carried the day.

The objection has sometimes been raised to the argument that it assumes that IQ measures are a reasonable criterion of intelligence, as only then does the high correlation of reaction times or evoked potentials with IQ have any repercussions on the intelligence debate. If it is denied that IQ tests have any relation to intelligence, then the whole argument would seem to breakdown. There are two objections to this line of reasoning.

In the first place, the evidence is overwhelmingly strong that IQ and intelligence are indeed closely related, whatever definition of intelligence we may choose to give. This point has been fully discussed elsewhere (H. J. Eysenck, 1979), but an illustration may serve to show how close this relationship in fact is. Yule, Gold, and Busch (1982) tested all 5-year-old school children on the Isle of Wight on the Wechsler and then retested them 11 years later, again using the Wechsler and also noting their scholastic achievement. Very high correlations were observed between the two Wechsler tests, in spite of the long intervening period, and even the 5-year-olds' IQ test gave highly accurate predictions of scholastic achievement for these children. Unless we assume that intelligence is irrelevant to scholastic success (or to success at university, in business, in the army), we can hardly deny that IQ is quite intimately related with intelligence, however defined.

The other point to consider is the following: Typically, IQ tests contain problem to be solved, items to be learned and remembered, previous learning to be accessed (as in a vocabulary test), instructions to be remembered and carried out, and indeed a host of other types of activity which by general consent are cognitive and indicative of intelligence. What requires to be explained is why all these activities appear to be correlated so very closely with such very simple and obvious types of performance as reaction times and inspection times, or the Lehrl tasks, and with a simple neurological measure of information processing (if

that is what the evoked potential measures). Even if we never mentioned
the word *intelligence* in this connection, the problem would still remain,
consequently, and nothing would be gained by this purely semantic
subterfuge.

Psychologists are frequently suspicious of the notion (and the term)
intelligence because, as they rightly point out, there is no widely
accepted theory of intelligence, but rather many different theories, and
furthermore the way intelligence acts is completely unintelligible. However, the same might be said about gravitation (Roseveare, 1982). In a
review article Taylor (1876) described 21 theories purporting to provide
the inverse square law with a theoretical underpinning, and this list was
extended in 1881 by Stallo and again later by Zenneck (1903)! Newton
had emphatically dissociated himself from such efforts when he wrote in
the *Principia* the celebrated phrase, "Hypotheses non fingo" (1934, p.
547) about the cause of gravity. Clearly, advances in physics are not
dependent on such a deeper understanding of the observed relationships,
and it is not clear why a greater burden in this respect should be placed
on psychologists. As Mach said in 1872:

> The Newtonian theory of gravitation, on its appearance, disturbed almost
> all investigators of nature because it was founded on an uncommon unin-
> telligibility. People tried to reduce gravitation to pressure and impact. At
> the present day gravitation no longer disturbs anybody: it has become *com-*
> *mon* unintelligibility. (1911, p. 56)

Perhaps intelligence is destined to pass from a period of uncommon
unintelligibility to one of common unintelligibility too.

It may be useful to speculate a little about the practical and theoretical consequences of these findings. It is unlikely that reaction time
measurements or evoked potentials will displace the orthodox IQ tests in
education, industry, or the armed forces; the observed correlations are
so high and group IQ testing so cheap and convenient that it is doubtful
whether for most purposes much would be gained by using tests which,
although they measure the genotype more accurately, cannot in their
results be all that different from orthodox IQ tests. In particular, RT and
evoked potential measures require individual testing, laboratory conditions, skilled testers, apparatus, and other expensive paraphernalia; in
this they are clearly inferior from the practical point of view to orthodox
IQ tests.

On the practical side, we will have to look to certain unusual and
abnormal conditions to find the major usefulness for these new types of
intelligence tests. One obvious area is clinical and abnormal psychology,
where schizophrenics, depressives, very anxious patients, senile patients,

and others can be tested for intelligence only with great difficulty using the ordinary IQ tests, and where the results must always remain doubtful for reasons of insanity, a lack of motivation, and other factors. Here and in the case of brain damage much research will of course be necessary to see to what extent these new tests can be used in the clinical field, but they certainly appear to have a great deal of potential.

In other areas, such as educational, industrial, and military testing, ordinary IQ tests do not do justice to subjects of a different cultural background, subjects with language difficulties, or subjects growing up in very deprived conditions. Here the new tests should have a definite function to fulfill in giving an estimate of the subject's intelligence not much or not at all contaminated by cultural, educational, and socioeconomic factors. We have already noted the fact that the evoked potential gives more unequivocal support to the theory that men and women are equal in intelligence than do orthodox tests, and we have seen that socioeconomic differences are not so closely related to evoked potential as they are to Wechsler IQ measures. Again, a great deal of research remains to be done in this field, of course, but it looks a very promising area from the point of view of practical application.

On the theoretical side, the new measures should open up possibilities which have hitherto eluded the orthodox type of IQ test. One obvious possibility is the tracing of the development of intellectual powers from the neonate through childhood and adolescence to adulthood. Ordinary IQ tests are incapable of giving reasonable measures before the age of 5 or 6, baby tests showing very little and sometimes negative correlations with adult status (H. J. Eysenck, 1979). Neonates and very young children can, on the other hand, be tested for evoked potentials, and although there may be unforeseen difficulties in the way, the possibility does appear to exist of discovering a true zero point for the intelligence scale and of charting progress on a scale having equal intervals. This, of course, is a consequence of having a scale which is marked in physical units and has a genuine zero point, advantages not possessed by any type of IQ measurement, which must inevitably rely on comparisons between individuals and statistical manipulations and which in the nature of the case lacks a true zero point or equal intervals.

At the other end of the lifespan, there have been considerable difficulties in settling the question of whether and to what extent intelligence deteriorates with age; also, given that there is some deterioration, it is not clear to what extent the amount of deterioration is related to adult IQ (H. J. Eysenck, 1979). Different tests give different answers, with tests of fluid intelligence showing decline, tests of crystallized ability showing, if anything, an increase with advancing age. Evoked potential measurement should be able to settle the issue once and for all.

Many other questions arise which in the past have presented us with great difficulties but which now should be soluble. The reason for the difficulties, of course, has always been that the 20% of environmental determination of the IQ variance is sufficiently large to cover the total range of observed differences; this makes it very difficult, and subject to subjective argumentation, to decide between different causes and different effects. Possessing measures which apply at a more elementary, genotypic level, we should be able to dissect the genetic from the environmental and hence solve many of these hitherto insoluble problems. As always, statistical manipulation, through correlations and factor analysis, must be supplemented by direct experimental investigation of specific hypotheses in order to answer urgent theoretical and practical questions in the field of personality and individual differences. This clearly is true as much in the field of intelligence as in the field of temperament.

CHAPTER SIX

Summary and Conclusions

We are now in a position to summarize the major areas discussed so far and to draw certain conclusions that we believe to be justified by the data cited.

1. The most obvious conclusion, without which there would be no book at all, is that there is a remarkable degree of *consistency* in human conduct. It is not suggested that all types of behavior show such consistency, but the major and more important personality variables do. This consistency requires *description* and *explanation;* this is a task of personality theory.

2. The major areas in which the consistency of personality has been studied are (a) that of *temperament* and (b) that of *cognition.* There are other aspects of personality in its broadest sense, such as physique, but these areas are of interest mainly insofar as they correlate with temperamental or cognitive variables.

3. Descriptively, temperament can be analyzed in terms of *traits* and cognition in terms of *abilities.* It is in the main in terms of their traits and abilities that one person is distinguished from another, and the sum of these distinctions is conceived of as personality differences.

4. The concepts of traits and abilities are based ultimately on *correlated behaviors,* observed or self-rated.

5. The observed correlations are analyzed by means of techniques such as factor analysis; these enable us to impose order on large tables of intercorrelations and explain them in terms of a smaller number of factors or latent traits and abilities.

6. The resulting description, both on the temperament and the abilities side, is a hierarchical one, starting at the bottom, that is, with isolated observations of individual behaviors. These are grouped in habitual behavior patterns on the basis of correlations, thus giving rise to simple

traits. Traits themselves are found to be correlated and give rise to higher-order factors or superfactors, which are commonly called *types*.

7. Types are thus defined as *supraordinate concepts*, not in terms of discontinuous or bimodal distributions, as used to be the case in the centuries preceding empirical investigation.

8. Descriptively, traits can be measured by ratings or self-ratings (questionnaires). Such ratings and self-ratings must always be based on a large number of observations averaged over long periods of time.

9. The interactive influence of traits and situations produces transient internal conditions known as *states*.

10. Personality states, like traits, are equally measurable by means of ratings and self-ratings (questionnaires).

11. Traits and states are intervening variables or mediating variables that are useful in explaining individual differences in behavior to the extent that they are incorporated into an appropriate theoretical framework.

12. The relationship between traits or states and behavior is typically indirect, being affected or *moderated* by the interactions that exist among traits, states, and other salient factors.

13. Superfactors, identifying the major dimensions of variation in personality, are more easily replicable than are more elementary types of factors.

14. The three major superfactors identified in the description of human temperament are extraversion–introversion; emotional stability versus instability, or neuroticism; and psychoticism versus impulse control. Many other names have been used in describing these factors, and the naming itself, of course, is of no great importance. These three superfactors do not exhaust the field of personality description in terms of temperament; they simply account for a greater portion of the variance than other factors. The possibility of the existence of other superfactors cannot of course be ruled out, but hitherto none have emerged that cover such a wide area, are equally replicable, and have some kind of causal explanation in terms of the laws and concepts of academic psychology.

15. To fill in the descriptive picture of a person's temperament, a number of *primary traits* additional to the superfactors or major dimensions of personality will always be required.

16. Individual differences in personality are caused to a large extent by genetic factors, although of course environmental influences also play a prominent part.

17. Genetic influences are almost as strong in the field of temperament as in the field of ability, but there are also important differences. Nonadditive genetic factors, such as assortative mating and dominance,

play little part in the field of temperament but are very important in the ability field. On the environmental side, between-family influences are very important for intelligence but play little part in producing differences in temperament.

18. The fact that the major dimensions of personality have a genetic basis suggests that these dimensions must possess a certain amount of *universality*, such as historical universality. In other words, the same differences in behavior that produce our modern concepts should have been observable across historical times, and the evidence suggests that this is indeed so.

19. On a similar argument, the principle of universality suggests that the same major dimensions of personality should be observable among animals, and although research work here has not been very extensive, insofar as it exists it suggests that such factors as P, E, and N can be found in the animal field.

20. Another aspect of universality is that the major dimensions of personality should appear in cross-cultural studies, that is, among nations and cultures far removed from Western world, where most of the work on the description of temperament has been done. Again, the data appear to verify the hypothesis.

21. A fourth type of universality should emerge from longitudinal studies, in the sense that different age groups should show similar personality dimensions and a person should remain in a particular part of the three-dimensional temperament structure throughout life; both hypotheses appear to have found empirical support.

22. Finally, the principle of universality demands that the same dimensions of personality should emerge from all the different types of measures of personality that have been used. In a very extensive review of the evidence, we have shown that by and large this is true; the three major dimensions of personality are truly universal in the sense that they emerge from all types of different personality inventories and ratings.

23. Validity of scores of P, E, or N can be established in several different ways, the most convincing of which is probably consensual validation using the multitrait–multimethod matrix. Experiments show that these three dimensions of personality and the methods of measuring used possess sufficient validity to pass this test.

24. The descriptive analysis of personality must be supplemented by a causal analysis, taking seriously the findings about the genetic basis of differences in temperament and looking for psychophysiological, hormonal, and other biological mechanisms which could be responsible for these differences. Such theories would then be testable in terms of psy-

chophysiological, laboratory experimental, and social experiments; no theory would be considered valid that did not make testable and verified predictions in these three fields.

25. Cognitive abilities, having a strong genetic basis, must also be presumed to have a physiological mechanism underlying their functioning and responsible for differences in individual performance. Such a mechanism appears to be related to the averaged evoked potentials that follow auditory or visual stimulation.

26. Individuals with high IQs tend to show *complex* evoked potential patterns, whereas those with low IQs tend to show very *simple* evoked potential patterns.

27. The observed facts were predicated on the hypothesis that information transmission through the cortex is subject to *errors*, probably occurring at the synapse, and that propensity to commit such errors is likely to lead to low IQ and to simple evoked potential patterns. Low probability of commission of errors would lead to complex evoked potential patterns and high intelligence.

28. The theory also predicts that high-IQ subjects would have short reaction times, that their reaction times would increase less than those of dull subjects when choice among stimuli is involved, and that they would have less *variability* in reaction times.

29. Equally, the theory predicts that high-IQ subjects would have shorter inspection times. All these predictions have been found to be empirically verified.

30. It is not suggested that general intelligence, as so measured and so grounded in theory, is the only factor in cognitive achievement; however, it is the most fundamental, the most general, and the most important.

These are the major conclusions derived from the work so far reviewed. The second part of this book will deal with causal factors and experimental analyses of predictions made from these theories.

PART TWO

CAUSAL

CHAPTER SEVEN

Theories of Personality and Performance

Our emphasis so far in this book has been on the attempt to discover the major dimensions of personality. Once that task of *describing* the basic structure of personality had been accomplished, the next obvious step was to propose *explanatory* theories that provide a systematic account of the personality dimensions that had been discovered. Since the most progress has been made with the extraversion dimension, much of our discussion will focus on that dimension.

The distinction between description and explanation is usually regarded as being one of degree rather than absolute, but it is important in the present instance. At a descriptive level, it is known that extraverts tend to be both more impulsive and more sociable than introverts. That is fine as far as it goes but obviously leaves many major questions unanswered. In particular, we want to know *why* extraverted individuals tend to be impulsive and sociable whereas introverted individuals tend not to be. It would also be of great value to have available an explanatory theory of extraversion that provided a firm basis for predicting behavioral differences between introverts and extraverts across a wide range of situations.

What might such an explanatory theory look like? One probable ingredient is an account of the underlying physiological differences between introverts and extraverts. It is worth distinguishing between two different kinds of physiological theories of extraversion. One kind invokes quasi-physiological constructs and processes but fails to relate them to what is known about human physiology. The other kind also specifies various underlying physiological processes but is more precise because the putative physiological mechanisms accounting for behavioral differences between introverts and extraverts are specified.

In this chapter, we consider four explanatory theories concerned with extraversion. Two of these theories (those of H. J. Eysenck, 1957, and Brebner) refer to physiological processes without implicating any specific parts of the physiological system, whereas the other two (those of H. J. Eysenck, 1967a, and Gray, 1973) go further and specify actual physiological structures and processes. Other things being equal, theories that relate extraversion to definite components of the physiological system have certain clear advantages over those that do not. For example, such theories are likely to generate more specific predictions because knowledge about the functioning of the specified physiological structures is available. As a consequence, such theories are likely to be more testable.

Other theories attempting to describe the physiological bases of personality have been put forward. For example, Claridge (1967) identified two components of psychophysiological activity, one of which was an arousal factor and the other of which was concerned with the modulation of sensory input into the nervous system and also with attentional selectivity. The dynamic interrelationships between these two mechanisms were assumed to underlie neurosis and psychosis. Zuckerman (e.g., 1979b) has investigated the personality dimension of sensation seeking and made significant progress in uncovering the physiological basis of individual differences in sensation seeking. However, these two theories will not be discussed in detail here because Claridge's theory is primarily designed to account for abnormal behavior and has not generated a substantial amount of experimental research, and much of Zuckerman's contribution is compatible with H. J. Eysenck's (1967a) theory.

Before proceeding to a more detailed analysis of the four main theories, it is perhaps worth providing the reader with some orientation concerning the interrelationships of these theories. The original theory was proposed by H. J. Eysenck (1957) and then substantially modified (H. J. Eysenck, 1967a). Gray's (1973) theory is, in essence, a modification of H. J. Eysenck's (1967a) theory and shares many features with that theory. Finally, the theory developed by Brebner and his associates owes much to the theory of H. J. Eysenck (1957) as well as to that of H. J. Eysenck (1967a). Thus it would be quite inappropriate to regard the four theories discussed in this chapter as entirely separate and distinct from each other. The true state of affairs is that the various theories represent the gradual evolution of theoretical ideas stemming from H. J. Eysenck (1957).

H. J. EYSENCK (1957)

H. J. Eysenck has proposed two related but conceptually distinct explanatory theories. The first of these was put forward in 1957 and will henceforth be referred to as the inhibition theory. As a result of increasing evidence that this theory was inadequate in some respects, Eysenck (1967a) suggested a modified theoretical conceptualization; for convenience, this will be referred to as the arousal theory.

Although the arousal theory largely superseded the earlier inhibition theory, the ways in which thinking and research have developed owe much to the impetus provided by the inhibition theory. Accordingly, a relatively brief account of that original theory is in order at this point. One of the major aims of the inhibition theory was to provide some kind of theoretical understanding of the substantial differences between introverts and extraverts. More specifically, Eysenck asked, in effect, whether it was possible to account for the myriad behavioral differences between introverts and extraverts in terms of an extremely small number of fundamental physiological or quasi-physiological differences. To cut a long story short, he answered this question in the affirmative, and his major theoretical statement relating to the personality dimension of extraversion was contained in the following typological postulate:

> Individuals in whom excitatory potential is generated slowly and in whom excitatory potentials so generated are relatively weak, are thereby predisposed to develop extraverted patterns of behaviour and to develop hysterical-psychopathic disorders in cases of neurotic breakdown; individuals in whom excitatory potential is generated quickly and in whom excitatory potentials so generated are strong, are thereby predisposed to develop introverted patterns of behaviour and to develop dysthymic disorders in case of neurotic breakdown. Similarly, individuals in whom reactive inhibition is developed quickly, in whom strong reactive inhibitions are generated, and in whom reactive inhibition is dissipated slowly, are thereby predisposed to develop extraverted patterns of behaviour and to develop hysterical-psychopathic disorders in case of neurotic breakdown; conversely, individuals in whom reactive inhibition is developed slowly, in whom weak reactive inhibitions are generated, and in whom reactive inhibition is dissipated quickly, are thereby predisposed to develop introverted patterns of behaviour and to develop dysthymic disorders in case of neurotic breakdown. (p. 114)

This typological postulate refers to two different explanatory constructs (i.e., excitation and inhibition). However, for most purposes the excitation–inhibition balance was treated as a unidimensional construct. The problem of disentangling the separate influences of excitation and inhibition can be seen if we note that, according to the typological postulate, those who generate excitatory potentials with ease also generate inhibitory potentials with difficulty, whereas those who generate excitatory potentials with difficulty also generate inhibitory potentials with ease. Accordingly, it is often possible to interpret performance differences between introverts and extraverts either in terms of greater excitation in introverts than in extraverts or in terms of greater inhibition in extraverts than in introverts. It is worth pointing out that the concept of *inhibition* as used in the typological postulate refers to some central process of unknown physiological origin and should not be confused with inhibition of behavior. Very often, the greater central inhibition of the extravert results in his behaving in an uninhibited way.

How can the typological postulate be used to account for the effects of extraversion on behavior such as the performance displayed on laboratory tasks? In practice, the strategy used by Eysenck and his associates was to attempt to locate tasks for which there was good reason to suppose that inhibition (or excitation) made a substantial contribution to performance. If some kind of performance effect on a laboratory task was produced by inhibition, then extraverts (being more prone than introverts to inhibition) should produce a greater effect. Perhaps the clearest example of this strategy in operation concerns the reminiscence effect on the pursuit-rotor task. This task usually involves a metal disc embedded in a rotating turntable and a metal stylus. The subject holds the stylus and attempts to keep it in contact with the metal disc. If someone performs this task continuously for several minutes and then has a rest for a few minutes, the level of performance upon resumption of the task is usually considerably higher than at any time before the rest. This surprising spontaneous improvement is what is known as the reminiscence effect.

From a theoretical perspective, it is tempting to assume that this reminiscence effect occurs because inhibitory processes occurring prior to the rest interval depress performance. This inhibition is dissipated during the rest interval, and so performance improves. If one combines this theoretical analysis of the reminiscence effect with the typological postulate, then it follows that extraverts should show more reminiscence than introverts. This prediction has been confirmed repeatedly, but unfortunately the original interpretation of the data can no longer be sustained (see Chapter 9 for details). For present purposes, however, the important point is that research on the reminiscence effect exemplifies the approach taken to investigate the typological postulate.

We have discussed in general terms the notion that inhibition can impair performance. More specifically, the idea was that sustained activity of any kind leads to the development of inhibition. When a high level of inhibition has been created, there is a block or involuntary rest pause in performance that allows some of the inhibition to dissipate. It is the existence of these involuntary rest pauses that is directly responsible for the disruptive effects of inhibition on performance. Once again, the empirical evidence does not provide strong support for this theoretical contention (see Chapter 9).

One of the problematical aspects of the research strategy outlined so far is that it does not permit experimental manipulation of what is allegedly the crucial causal factor (i.e., the excitation–inhibition balance). As a consequence, it is possible only to obtain rather indirect evidence concerning the behavioral impact of variations in that balance. Eysenck argued that it was, in fact, possible to manipulate the excitation–inhibition balance; his theoretical position on this point was encapsulated in the drug postulate:

> Depressant drugs increase cortical inhibition, decrease cortical excitation and thereby produce extraverted behaviour patterns. Stimulant drugs decrease cortical inhibition, increase cortical excitation and thereby produce introverted behaviour patterns. (p. 229)

Informal evidence consistent with the drug postulate can be obtained any evening at your local pub or hostelry. Alcohol is basically a depressant, and it certainly makes many introverts much livelier, more garrulous, and generally more extraverted than normal. (A review of the more formal evidence appears in H. J. Eysenck, 1983c).

The use of depressant and stimulant drugs in research not only provides a powerful way of altering the excitation–inhibition balance; it also broadens considerably the range of laboratory tasks that can be used to test Eysenck's typological postulate. In a nutshell, the expectation is that the effects of stimulant drugs and of introversion on performance should be broadly comparable, and the same should be true of depressant drugs and extraversion. If these expectations were to be confirmed, this would provide rather strong support for the notion that differences between introverts and extraverts are determined in large measure by their different locations along the excitation–inhibition balance.

What we have here is a kind of wedding of insufficiencies. Any behavioral differences between introverts and extraverts may be due to the excitation–inhibition balance, but they may also be due to a variety of other factors over which we have no control. In similar vein, effects of depressant and stimulant drugs on performance may reflect their inhibitory and excitatory effects respectively, but they may also be due to the side-effects produced by the drugs. However, if we conduct two parallel studies, one investigating the effects of extraversion on a task and the other the effects of stimulant and depressant drugs on the same task, then any behavioral equivalence of the effects of personality and drugs is most likely to be due to the influence of the excitation–inhibition balance.

In view of the fact that Eysenck had already identified the three orthogonal personality dimensions of extraversion, neuroticism, and psychoticism, it may seem strange to the reader that our discussion of his theory as it relates to performance has had little to say about any personality dimension other than extraversion. His research strategy has, for the most part, involved an attempt to elucidate the fundamental characteristics of one personality dimension before attention is turned to a different dimension. The historical sequence has been as follows: Extraversion was the first personality dimension to be explored systematically, followed by neuroticism (H. J. Eysenck, 1967a) and then by psychoticism (H. J. Eysenck & S. B. G. Eysenck, 1976).

Why has the inhibition theory put forward by Eysenck become considerably less influential than it once was? Far and away the most important reason is the success of the subsequent arousal theory (H. J. Eysenck, 1967a). With relatively few exceptions, the arousal theory is able to handle the findings the inhibition theory accounted for and also many of the findings that appeared anomalous from the perspective of the inhibition theory. In addition, the arousal theory has the advantage of identifying the physiological systems underlying individual differences in extraversion and neuroticism, whereas the earlier inhibition theory was silent on these matters.

A further complication with the inhibition theory concerns the concept of *inhibition* itself. In the first place, the Eysenckian notion of inhibition represents a complex amalgam of the Hullian concepts of reactive and conditioned inhibition, Koehler's stimulus satiation, and Pavlov's internal inhibition. Second, it seems clear that the inhibitory state produced by, say, massed practice on the pursuit rotor is by no means synonymous with the inhibitory state produced by a depressant drug. Inhibition of the former type is thought to be specific to a particular task, whereas inhibition of the latter type is much more general and would presumably affect the performance of most tasks.

H. J. EYSENCK (1967a)

Increasing evidence that his inhibition theory was inadequate led H. J. Eysenck (1967a) to replace it with the arousal theory. This theory was considerably more specific than the earlier one in its description of the physiological mechanisms underlying individual differences in personality. The basic physiological assumptions made in this theory can be seen with reference to Figure 24. The ascending reticular activating system (ARAS) was initially investigated by Moruzzi and Magoun (1949). They discovered that electrical stimulation of parts of the ARAS elicited a general activation pattern in the cortical EEG. What happens is that collaterals from the ascending sensory pathways produce activity in the ARAS, which subsequently relays the excitation to numerous sites in the cerebral cortex. It was this excitation which produced the EEG desynchronization observed by Moruzzi and Magoun.

Since the pioneering efforts of Moruzzi and Magoun, our knowledge of the workings of the ARAS has increased considerably. In essence, it appears that the ARAS is involved in a wide range of psychological processes, as was made clear by Stelmack (1981) in a review chapter:

> The reticular formation is implicated in the initiation and maintenance of motivation, emotion and conditioning by way of excitatory and inhibitory control of autonomic and postural adjustments and by way of cortical coordination of activity serving attention, arousal, and orienting behaviour. (p. 40)

According to H. J. Eysenck (1967a), the extraversion dimension is identified largely with differences in levels of activity in the corticoreticular loop. Introverts are characterized by higher levels of activity than extraverts and so are chronically more cortically aroused than extraverts. It should be noted that this theoretical view of the underlying differences between introverts and extraverts represents an extension and modification of the previous inhibition approach rather than a totally

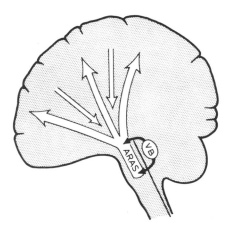

FIGURE 24. Interrelationships between the visceral brain (VB) and the ascending reticular activating system (ARAS) and the cortex. Also shown are the ascending afferent pathways (AAP). (From *The Biological Basis of Personality* by H. J. Eysenck, Springfield, Ill.: Charles C Thomas, 1967. Copyright 1967 by Charles C Thomas, Publishers. Adapted by permission.)

new theory. Thus, introverts in the earlier theory were those with high excitation and low inhibition and now are those with high arousal level, and extraverts have become those individuals having low arousal level rather than those with low excitation and high inhibition. It would be tempting (but simplistic and erroneous) to argue that the two theories are essentially the same, except that the emphasis has shifted from the inhibition component of the excitation–inhibition balance to the excitation or arousal component. We will return to the similarities and differences of the two theoretical accounts of extraversion a little later.

Reference back to Figure 24 will show the location of the physiological system known as the visceral brain, which consists of the hippocampus, amygdala, cingulum, septum, and hypothalamus. The visceral brain appears to be largely concerned with emotion. As the arrows in the figure make clear, the visceral brain and the ARAS are only partially independent of each other. Indeed, one of the ways in which cortical arousal can be produced is through activity in the visceral brain which reaches the reticular formation through collaterals. Activity in the visceral brain produces autonomic arousal, and Eysenck used the term *activation* to distinguish this form of arousal from that produced by reticular activity. The relevance of the visceral brain to personality theory, according to Eysenck, is that individual differences in neuroticism depend upon its functioning. More specifically, people who are high in neuroticism produce activity in the visceral brain (i.e., activation) more readily than those low in neuroticism.

Now that we have outlined the physiological mechanisms underlying the personality dimensions of extraversion and neuroticism, it is appropriate to return to the issue of the exact relationship between the inhibition and arousal interpretations of the crucial differences between introverts and extraverts. As Gray (1981) pointed out in his perceptive discussion of this issue, the proof of the pudding is in the eating; if the two theories generate different predictions, then they cannot be the same.

A clear example of a difference between the two theories concerns the critical flicker fusion threshold, which is obtained by increasing the rate at which a light flashes until it appears to stop flickering and becomes continuous. According to the inhibition theory (1957), extraverts should have higher critical flicker fusion thresholds than introverts. The argument was that two separate flashes can most readily be distinguished if the perceptual effects of the first flash are curtailed or inhibited, and extraverts are more susceptible to inhibition. In contrast, the arousal theory (1967a) makes the opposite prediction. Introverts are more cortically aroused than extraverts and so augment incoming stimulation. This makes the light flashes subjectively more intense, and it is known that critical flicker fusion thresholds increase in line with stimulus intensity (Gray, 1964). As a result, introverts should have higher thresholds than extraverts.

The actual effects of extraversion on critical flicker fusion thresholds are discussed in some detail in Chapter 9. Suffice it to say at this point that most of the data are more in line with the prediction from arousal theory than with that of inhibition theory. Thus, the case of critical flicker fusion illustrates both the differences between the inhibition and arousal theories and the superiority of the arousal theory.

Another difference between the two theoretical approaches is that the predictions of the inhibition theory are often rather more detailed than those of the arousal theory. Consider the likely effects of extraversion on vigilance tasks in which subjects spend a long time attempting to detect infrequent signal stimuli. According to the inhibition theory, perception is a response, and the repeated production of responses generates inhibition. When sufficient inhibition has been generated, involuntary rest pauses occur in order to permit some of the inhibition to dissipate, and performance is impaired. This leads to the prediction that extraverts, who generate more inhibition than introverts, should show a greater decrement in performance over time. In contrast, arousal theory merely predicts that some intermediate level of arousal is optimal for performance (Yerkes & Dodson, 1908). Depending on the exact locations of introverts and extraverts with respect to the optimal level of arousal at the start of the vigilance task and the precise changes in arousal level during the experimental session, it is possible for the arousal theory to accommodate the finding that the performance of extraverts declines more rapidly over time than that of introverts, and also the opposite finding. Both findings have, in fact, been reported (Bakan, 1959; Bakan, Belton, & Toth, 1963; Keister & McLaughlin, 1972).

A final but important difference between the inhibition and arousal theories concerns the range of applicability of each theory. Since inhibition is extremely difficult to measure, the major way in which the inhibition theory was tested was by using tasks on which massed practice was assumed to produce a progressive increase in the amount of inhibition produced. In contrast, since the level of arousal can be indexed at least approximately by various physiological measures, predictions can be generated for a much greater variety of situations, factors, and tasks. For example, consider the discovery by Colquhoun (1960) that the effects of extraversion on vigilance performance were systematically affected by the time of day at which testing occurred. More specifically, he looked at the results from 17 vigilance studies and found that the morning experiments gave a negative correlation between extraversion and efficiency, whereas the afternoon experiments produced a positive correlation. Such a pattern of findings is rather mysterious from the perspective of inhibition theory, but can be handled by arousal theory. The key fact is that physiological arousal appears to increase progressively over much of the waking day (M. W. Eysenck, 1982). It can thus be surmised that extraverts are suboptimally aroused in the morning, and introverts are supraoptimally aroused in the afternoon.

We have so far glossed over some of the complexities associated with using the arousal theory of H. J. Eysenck (1967a) to predict behavioral differences between introverts and extraverts. While some theoretical progress was made by arguing that introverts are more cortically aroused than extraverts, it is obviously essential to couple this hypothesis with theoretical statements about the behavioral consequences of high and low arousal on various tasks. The earliest of such theoretical statements (and still influential today) was put forward by Yerkes and Dodson (1908). Their theoretical position was outlined in what is generally called the Yerkes–Dodson Law; it comprises two major assumptions. The first assumption is that there is an inverted-U relationship between the level of arousal, tension, or motivation and performance quality. In other words, performance is at its most effective when an individual is at an intermediate level of arousal. The second assumption is that the optimal level of arousal is inversely related to task difficulty, that is, easy tasks are associated with an optimal level of arousal that is higher than that for difficult tasks. These various predicted relationships are shown in Fig. 25.

It is usually assumed that Yerkes and Dodson (1908) successfully identified two of the major generalizations relating arousal to performance. There is strong evidence that the optimal level of arousal is inversely related to task difficulty (see M. W. Eysenck, 1982, for details), but there is little consensus about the best way of assessing task difficulty. It is less clear that there is necessarily an inverted-U relationship between arousal and performance. In many studies, only three levels of arousal were compared, so that there was a total of six possible orderings of these three levels with respect to the quality of performance. Since only the two orderings in which the intermediate level of arousal produces the worst performance are inconsistent with the Yerkes–Dodson Law, it follows that two-thirds of studies with only three levels of arousal would provide apparent support for the Yerkes–Dodson law by chance alone!

The most serious inadequacy of the Yerkes–Dodson Law is the fact that it is descriptive rather than explanatory. What the law signally fails

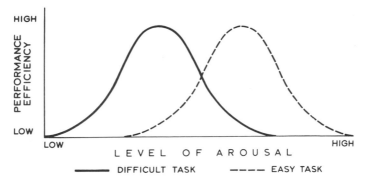

FIGURE 25. Performance efficiency as a function of arousal level and task difficulty according to the Yerkes–Dodson law. (From "The Relation of Strength of Stimulus to Rapidity of Habit-formation" by R. M. Yerkes and J. D. Dodson, *Journal of Comparative and Neurological Psychology*, 1908, *18*, 459–482.)

to do is to tell us why there should be a curvilinear relationship between arousal and performance and why the nature of this relationship is affected by task difficulty. The best-known and most influential attempt to provide this missing theoretical underpinning for the Yerkes–Dodson Law was that of Easterbrook (1959). He assumed that increasing arousal, anxiety, or emotionality all produce a progressive reduction in the range of environmental cues used, which

> will reduce the proportion of irrelevant cues employed, and so improve performance. When all irrelevant cues have been excluded, however,. . . further reduction in the number of cues employed can only affect relevant cues, and proficiency will fall. (p. 193)

Thus, Easterbrook's notion that arousal causes attentional narrowing provides a direct explanation of the curvilinear relationship between arousal and performance.

Easterbrook's (1959) hypothesis also provides a potential explanation of the fact that the optimal level of arousal is lower for difficult than for easy tasks. If we assume that there are more cues associated with difficult than with easy tasks, then task-relevant cues should begin to be ignored at lower arousal levels on the more difficult tasks. As a consequence, increasing arousal should produce a deterioration on difficult tasks earlier than easy ones.

How adequate is Easterbrook's hypothesis? It has generally been tested by using a paradigm in which subjects are given a primary task and a subsidiary task at the same time. The expectation is that high arousal will be more likely to disrupt performance on the subsidiary than on the primary task. The reason for this is that the reduced range of cue utilization under high arousal excludes cues from the subsidiary task before those from the primary task. Most of the data are consistent with this expectation (M. W. Eysenck, 1982). However, Easterbrook's hypothesis is clearly inadequate in some ways. It assumes that high arousal impairs performance because it leads to intense concentration on some, rather than all, of the task stimuli. This implies that high arousal should reduce distractibility and therefore introverts should be less distractible than extraverts. In fact, most of the available evidence indicates that introverts are significantly *more* distractible than extraverts (Howarth, 1969b; Morgenstern, Hodgson, & Law, 1974; Shanmugan & Santhanam, 1964).

Although one can test the arousal theory of extraversion by investigating the extent to which the performance of introverts and extraverts conforms to various theoretical predictions, it is also possible (and perhaps preferable) to adopt a somewhat more empirical approach (Broadbent, 1971; Duffy, 1962). In this approach, it is assumed that an individual's level of arousal can be altered experimentally by manipulations such as intense noise, failure feedback, incentive, sleep deprivations, time of day, electric shocks, and drugs. The central contention of the arousal-theory approach is that the behavioral effects of most, or all, of the above factors are mediated by a common arousal mechanism. To the extent that introverts are actually more aroused than extraverts, then the performance differences between introverts and extraverts should resemble those between people exposed versus not exposed to arousal manipulations (e.g., intense noise).

Within this general approach, it is worth distinguishing between a *strong* and a *weak* version of arousal theory (M. W. Eysenck, 1982). The strong version assumes that virtually all of the effects of different arousers are mediated by a single arousal system, whereas the weak version of arousal theory claims only that there is a partial similarity in the performance patterns produced by the various arousal manipulations. The similarities among arousers occur because they all affect a single arousal system, and the dissimilarities occur because each arouser has its own specific or idiosyncratic effects.

The strong arousal theory is easier to test than the weak theory. The reason for this is that findings that are awkward for the weak theory can always be explained away by postulating additional idiosyncratic effects. So far as the strong theory is concerned, the crucial prediction is that of behavioral equivalence, that is, the various arousers should all produce comparable performance patterns.

We will provide a more detailed analysis of the evidence relating to behavioral equivalence a little later in the chapter. However, it is worth noting here that there is more empirical support for the notion of behavioral equivalence at a relatively molar level of analysis than at a more molecular one. For example, it is well established that different arousers usually enhance long-term memory, but fine-grain analysis indicates that everything in the garden is not lovely. As M. W. Eysenck (1982) pointed out:

> The available evidence suggests that long-term retention is better in the afternoon than in the morning because high arousal leads to deeper or more semantic processing. On the other hand, intense white noise enhances long-term retention in spite of the fact that the arousal produced by noise leads to shallower or less semantic processing. Incentive seems to improve long-term retention by producing increased elaboration of processing without any discernible effect on the depth of processing. (p. 185)

A second important prediction relates to what will happen when two arousers are used either separately or together during the performance of a task. If the two arousers actually affect different internal mechanisms, then their effects on performance when used together should simply correspond to the sum of their effects when used separately. In other words, they should have *independent* effects on performance. On the other hand, if the two arousers both affect the same internal arousal mechanisms in the same way (as the strong version of arousal theory proposes), then their combined effects may differ substantially from those predicted by simply adding together their effects when used singly. Thus, two arousers should frequently have *interactive* effects on performance. It is possible to go further and to predict the form of the interaction, provided that we are willing to make one or two additional assumptions. The most popular of such notions is that an intermediate level of arousal is optimal for performance (Yerkes & Dodson, 1908).

We can flesh out the discussion at this point by considering some concrete examples. The effects of two arousers (incentive in the form of knowledge of results and white noise, a meaningless blend of sound frequencies) and one de-arouser (sleep deprivation) have been assessed on the continuous serial reaction task. The various resultant interaction effects can be seen in Figure 26. As is shown in Panel A, white noise improved performance for sleep-deprived subjects, perhaps because it prevented their arousal level from being suboptimal. However, it impaired performance for those subjects who had slept normally, presumably because it produced a supraoptimal level of arousal. A similar explanation is applicable to the data shown in Panel B, where arousal in the form of knowledge of results or incentive enhanced the performance

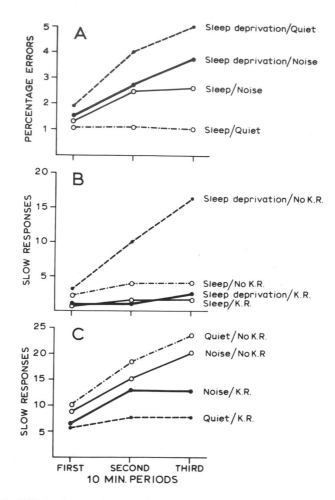

FIGURE 26. Effects of arousal on continuous serial reaction. A. White noise and sleep deprivation. B. Incentive (knowledge of results) and sleep deprivation. C. White noise and incentive (knowledge of results). (Panels A and C from "Interaction of Noise with Knowledge of Results and Sleep Deprivation" by R. T. Wilkinson, *Journal of Experimental Psychology*, 1963, *66*, 332–337. Copyright 1963 by the American Psychological Association. Panel B from "Interaction of Lack of Sleep with Knowledge of Results, Repeated Testing and Individual Differences" by R. T. Wilkinson, *Journal of Experimental Psychology*, 1961, *62*, 263–271. Copyright 1961 by the American Psychological Association. Figures reprinted by permission.)

of the underaroused subjects deprived of sleep more than that of subjects who had not suffered sleep loss. Finally, in Panel C, an arousing stimulus (noise) produced a decrement in performance among subjects who would have been aroused in any case (i.e., those receiving incentive) because the overall level of arousal became too high, but it had the opposite effect on underaroused subjects who were not given feedback.

The detailed interpretation of the data shown in Figure 26 may be somewhat speculative. However, it seems probable that there is some important similarity among incentive, noise, and sleep deprivation. Whatever the deficiencies of the arousal-theory approach, it is worth noting that the fact that such apparently disparate factors have something in common is of considerable interest and has only been accounted for in arousal-theory terms.

The strategy of comparing the effects of two different arousers when used singly and together has been used to investigate the arousal theory of extraversion. It has been assumed that individual differences in extraversion reflect naturally occurring differences in arousal level, and so this constitutes one of the arousal factors. In other words, the conventional approach of manipulating arousal in two different ways has been modified slightly so that a dimension of personality (i.e., extraversion) replaces one of the arousal manipulations. The results from studies using this approach have been fairly encouraging and are discussed in Chapter 9.

A third prediction of the strong arousal theory is relevant if arousal is construed in physiological terms as H. J. Eysenck (1967a) has done. The prediction is simply that different arousers should produce similar patterns of physiological responses indicative of an aroused system. Although it may appear that physiological measures provide a straightforward way of assessing many of the contentions of arousal theory, matters are, in fact, quite complex. First, the various physiological measures of arousal (such as heart rate, EEG, and skin conductance) typically produce only modest intercorrelations of approximately $+.2$ to $+.3$. Second, there are some situations in which one measure indicates increased arousal whereas a second measure suggests the opposite. Lacey (1967) used the term *directional fractionation* to refer to this state of affairs. An example of directional fractionation occurs in reaction-time studies. When the subject is waiting for the imminent arrival of the stimulus, some of the components of the EEG indicate high arousal, but at the same time heart rate decreases. Third, account has to be taken of autonomic response stereotypy (Lacey, 1950, 1967). Stress typically produces a pattern of great activation on some autonomic measures and a much smaller activation on other measures. This pattern remains fairly constant for any given individual from one stressor to another, but there are pronounced individual differences in the precise pattern of autonomic activation.

There are additional complications with physiological data when one is attempting to use them to discriminate between cortical arousal and autonomic activation (H. J. Eysenck, 1967a). It will be remembered that theoretically the personality dimension of extraversion is linked to cortical arousal and neuroticism is linked to autonomic activation. However, autonomic activation can lead in an indirect fashion to cortical arousal. Indeed, as H. J. Eysenck (1967a) pointed out, "when highly emotional people are involved, i.e., people for whom even quite mild stimuli are emotionally activating, then the distinction tends to break down, and activation = arousal" (p. 233).

In an attempt to clarify matters, H. J. Eysenck (1967a) proposed that "skin conductance and alpha activity are measures of extraversion" (p. 170). The relevant evidence is discussed in detail in Chapter 8. Suffice it at this point to say that the precise findings obtained depend critically on the nature of the experimental situation, the instructions given to the subject, and several other factors.

With respect to the personality dimension of neuroticism, H. J. Eysenck (1967a) argued that differences between people high and low in neuroticism may be interpreted "in terms of differential thresholds for hypothalamic activity" (p. 237), and to differences in responsivity of the sympathetic nervous system, "with high neuroticism scores associated with greater responsivity." Accordingly, a wide range of measures indicative of autonomic activation should discriminate between those high and low on neuroticism. Appropriate measures include skin conductance, muscular tension, heart rate, blood pressure, EEG, and breathing rate.

It does not follow from this hypothesis that high and low neuroticism scorers will always differ with respect to autonomic activation. In particular, it is important to note that autonomic activation typically reflects an emotional reaction to a given situation. As a consequence, there will probably be no effect of neuroticism on measures of autonomic activation obtained in relaxed conditions. Thus, any adequate test of the hypothesis would seem to require the use of relatively stressful conditions in which differences in responsivity of the sympathetic nervous system as a function of neuroticism would have a reasonable chance of manifesting themselves.

We have now discussed various ways in which arousal theory can be tested. How adequate does that theory appear to be in light of the available evidence? In a nutshell, there is much stronger support for the weak version of arousal theory than for the strong version. In other words, different arousal manipulations do manifest certain similarities in terms of their effects on physiological functioning and behavior, but various dissimilarities are also evident. This evaluation of the evidence can be given more concrete form if we turn to the research summarized by M. W. Eysenck (1982). He considered in some detail the effects of time of day, white noise, incentive, anxiety, extraversion, and sleep deprivation on performance. The aspects of performance that have received most attention in the literature include attentional selectivity (Easterbrook, 1959), the efficiency and/or capacity of the short-term storage system, long-term memory, speed and accuracy of performance, susceptibility to distraction, and efficiency of retrieval.

The relevant data are summarized in Table 15. In all cases, the effects of high arousal on performance are shown. Thus, for example, the data for noise indicate the effects of intense noise compared with either weak noise or no noise and those for time of day represent performance later in the day (when arousal is assumed to be highest) compared with performance earlier in the day. In the case of sleep deprivation, which is assumed to reduce the level of arousal, the actual effects have been reversed so as to produce a better basis for comparison with the effects of manipulations that enhance arousal level.

It is clear by inspection of Table 15 that there is some evidence for a common behavioral pattern associated with high arousal. It consists of increased attentional selectivity, increased distractibility, greater speed, increased proneness to errors, reduced short-term storage capacity, and improved long-term memory. However, although a common behavioral pattern can be identified, there are various discrepant findings. In the first place, there is not a single arouser that produces a behavioral pat-

TABLE 15. Effects of High Arousal in Various Forms on Aspects of Performance.[a]

Performance	Time of day[b]	White noise	Incentive	Anxiety	Introversion	Sleep deprivation[c]
Attentional selectivity	?	+	+	+	+	+
Short-term storage	−	−	+	−	0	+
Long-term memory	+	+	0	−	+	+
Speed	+	+	+	0	−	+
Accuracy	−	−	−	−	+	0
Distractibility	?	?	+	+	+	?
Retrieval efficiency	+	+	0	−	−	?

[a]+ = improved performance; − = impaired performance; 0 = no effect; ? = effect unknown.
[b]Later in the day is assumed to represent high arousal.
[c]The findings are reversed for sleep deprivation (which is a de-arouser) to facilitate comparison.

tern that coincides in all details with the one claimed to be characteristic of high arousal. Second, there is no aspect of performance that is affected by each and every arouser in a uniform way. In view of these complexities, it seems likely that we may ultimately need a number of interrelated arousal theories rather than just one.

As we have seen, arousal theory provides a rather imprecise and oversimplified perspective. This has led some researchers to argue that each arouser should be considered separately at a theoretical level. This seems to us very much like throwing out the baby with the bathwater, because it ignores the crucial insight incorporated into arousal theory that there are important similarities among a gallimaufry of experimental variables that appear on the surface to have nothing in common.

An alternative and more reasonable approach is to consider replacing the unitary concept of arousal with a more complex conceptualization incorporating two or more relatively specific arousal concepts. This line of argument has been followed by Broadbent (1971), M. W. Eysenck (1982), and Kahneman (1973). For example, Broadbent postulated two interrelated arousal mechanisms. The Lower mechanism is involved mainly in carrying out well-established decision processes, whereas the Upper mechanism monitors and changes the parameters of the Lower mechanism in order to maintain the level of performance. According to Broadbent, the factors affecting the Lower mechanism directly include sleeplessness, amphetamine, noise, and chlorpromazine, whereas the Upper mechanism is affected by extraversion, time of day, alcohol, and task duration.

A related distinction between two arousal systems was proposed by M. W. Eysenck (1982). The first arousal system resembles Broadbent's lower mechanism and consists of a passive arousal state which is rather general and undifferentiated physiologically. The effects of activity in this arousal system depend upon the extent to which the induced arousal state is appropriate for the task in hand. In contrast, the second arousal system consists of an active and effortful reaction to the effects of activity in the first arousal system on performance. There will usually be greater compensatory activity in the second arousal system when there are clearly adverse effects on task performance stemming from the first arousal system.

Do these kinds of modifications to arousal theory have any important implications? At the most general level, the crucial difference between the early versions of arousal theory and the theoretical views of Broadbent (1971), M. W. Eysenck (1982), and others is simply that it was originally assumed that an individual's level of arousal determined his behavior in a direct fashion, whereas it is believed now that the behav-

ioral effects of arousal are usually indirect. The reason for this indirect-
ness is the intervention of cognitive control systems, which are probably
responsive to feedback about performance under different levels of
arousal. In other words, we are not so completely at the mercy of our
internal physiological states as was suggested by the original proponents
of arousal theory.

If we relate some of these notions to predictions about the effects of
extraversion on performance, then we have an interesting contrast
between H. J. Eysenck's (1967a) theory and those of Broadbent (1971)
and M. W. Eysenck (1982). H. J. Eysenck argued that introverts are more
affected physiologically than extraverts by arousing stimuli, and this
implies that arousing stimuli will usually alter the task performance of
introverts to a greater extent than that of extraverts. In contrast, Broadbent and M. W. Eysenck argued that the cognitive control exerted by the
Upper mechanism or second arousal system is much greater in introverts than in extraverts. As a consequence, we are left with the prediction opposite to that of H. J. Eysenck (1967a). In the words of M. W.
Eysenck (1982):

> The strongly functioning upper mechanism in introverts tends to prevent
> changing levels of arousal from manifesting themselves in performance,
> whereas the weakly functioning upper mechanism in extraverts means that
> their performance is fairly directly determined by the prevailing level of
> arousal in the lower mechanism. (p. 138)

It is not clear as yet which theory provides the better interpretation
of the data. However, it is worth emphasizing the similarities between
the two theoretical positions. They both start from the initial assumption that introverts are more aroused than extraverts and differ merely
in the details of how this difference is reflected in performance.

GRAY'S THEORY

Gray (1970, 1972, 1973, 1981) has proposed a theory of personality
that bears a close resemblance to that of H. J. Eysenck (1967a). Gray
agreed with Eysenck that the two-dimensional space formed by the
orthogonal personality dimensions of extraversion and neuroticism was
of major significance. However, he disputed the notion that the primary
lines of causal influence within this space were identifiable with the
Eysenckian dimensions of personality. Instead, as is shown in Figure 27,
Gray argued that two different personality dimensions (anxiety and
impulsivity) were of primary importance. Gray located his anxiety and

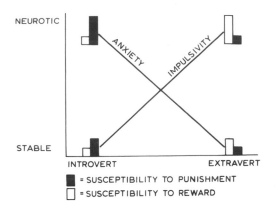

FIGURE 27. The interrelationships between the personality dimensions of anxiety and impulsivity and those of extraversion and neuroticism according to Gray. (From "Causal Theories of Personality and How to Test Them" by J. A. Gray. In J. R. Royce (Ed.), *Multivariate Analysis of Psychological Theory*, New York: Academic Press, 1973. Copyright 1973 by Pergamon Press. Adapted by permission.)

impulsivity dimensions 45° away from the Eysenck dimensions of neuroticism and extraversion, but it would seem preferable to locate the anxiety dimension closer to the neuroticism dimension than to that of extraversion. The reason is that questionnaire measures of anxiety (e.g., the Manifest Anxiety Scale) correlate more highly with neuroticism than with extraversion (approximately +.7 and −.3, respectively). If the location of the anxiety dimension is altered in this way, then impulsivity would have to be moved closer to the dimension of extraversion so that it would remain orthogonal to the dimension of anxiety. S. B. G. Eysenck and H. J. Eysenck (1977) reported questionnaire data that are consistent with this expectation.

What are the potential advantages of Gray's reformulation of the Eysenckian position? Gray (1973, 1981) argued that his theory was in better agreement with research on animal learning and physiology, which has led to the discovery of two major systems. One system is responsive to signals of punishment or frustrative nonreward, whereas the other system responds to signals of reward or nondelivery of anticipated punishment. According to Gray, individual differences in susceptibility to punishment lie along the anxiety dimension, and at a physiological level the behavioral inhibition system is involved. It comprises the septo-hippocampal system, its monoaminergic afferents from the brain stem, and its neocortical projection in the frontal lobe. In contrast, individual differences in susceptibility to reward are determined by the impulsivity dimension and relate physiologically to the approach system, which involves the medial forebrain bundle and the lateral hypothalamus.

At the behavioral level, one of the major predictions from Gray's theoretical approach is that individual differences in anxiety and impulsivity should by and large prove to be more powerfully related to performance than individual differences in extraversion and neuroticism. Unfortunately, the available evidence is rather sketchy. Revelle, Humphreys, Simon, and Gilliland (1980) investigated the effects of extraversion, the stimulant drug caffeine, and time of day on performance of an academic test resembling the American Graduate Record Examination. They found in each of several experiments that there were consistent interaction effects involving all three independent variables. Of greatest interest, however, was their discovery that the crucial aspect of personality in producing these interactions was impulsivity rather than extraversion.

A similar result in the field of conditioning was obtained by H. J. Eysenck and Levey (1972). The basic assumption of H. J. Eysenck (1967a) was that introverts are more conditionable than extraverts (at least within certain limits), but H. J. Eysenck and Levey discovered that the effects of extraversion on conditioning performance were mediated entirely by impulsivity.

It has not invariably been found that performance is better predicted by impulsivity than by extraversion, however. H. J. Eysenck and S. B. G. Eysenck (1969) considered the effects of personality on salivation in response to lemon juice. The amount of salivation correlated more highly with extraversion than with impulsivity.

It is entirely possible, of course, that Nature has not arranged matters so tidily that impulsivity (or extraversion) consistently predicts all kinds of performance data better than its related personality dimension. It would obviously be valuable to know which aspects of performance are better predicted by which personality dimension, but nothing like enough data have been collected to establish this information.

Let us now turn to the issue of whether anxiety should be regarded as a single entity (Gray) or as an amalgam of introversion and neuroticism (H. J. Eysenck). Gray (1970) attached considerable importance to the fact that various physiological treatments successfully reduced dysthymic symptoms (phobias, obsessional-compulsive rituals, anxiety states, and neurotic depression). Since H. J. Eysenck regards dysthymics as individuals who are high in neuroticism and low in extraversion, it follows that successful treatments must have two effects, reducing neuroticism and increasing extraversion. Although this may be a reasonable description of what happens, it is more parsimonious to argue with Gray that dysthymics are high in anxiety and that effective forms of treatment simply reduce anxiety.

In sum, it is perhaps worth emphasizing that Gray's theory is very similar to that of H. J. Eysenck in several important ways. In the first place, their theories both focus on the same two-dimensional personality space. However, Gray has clarified the reasons *why* this particular two-dimensional space is so significant; in essence, his argument is that individual differences in susceptibility to reward and punishment are centered within this space (see Figure 27). Second, many experimental findings can readily be interpreted within both theoretical frameworks. If, for example, it is found that high-anxiety subjects perform some task less efficiently than low-anxiety subjects, H. J. Eysenck can simply redescribe the findings as revealing the inferiority of neurotic introverts to stable extraverts. In similar fashion, Gray's theory is able to account for performance differences between introverts and extraverts:

> We may regard the dimension of introversion-extraversion as a dimension
> of susceptibility to punishment and non-reward: the greater the degree of
> introversion, the greater is this susceptibility. (Gray, 1972, p. 194)

Third, introverts exposed to punishing stimuli are predicted to be more aroused than extraverts, either because they are chronically more aroused and more affected by stimulation (H. J. Eysenck, 1967a) or because they are more susceptible to punishment (Gray, 1970).

In spite of these similarities between the two theories, they do make some different predictions. Consider the case of conditioning, where it has been assumed (H. J. Eysenck, 1967a) that high arousal usually enhances performance. While both theorists therefore predict that introverts should typically condition better than extraverts when aversive unconditioned stimuli are used, they disagree about the effects of introversion-extraversion on conditioning when appetitive or rewarding unconditioned stimuli are used. H. J. Eysenck continues to predict superior conditioning in introverts, whereas Gray predicts the opposite, because of the extravert's greater susceptibility to reward. As we will see later, the empirical evidence is rather equivocal.

BREBNER'S THEORY

Brebner and his colleagues (Brebner & Cooper, 1974, 1978; Brebner & Flavel, 1978; Katsikitis & Brebner, 1981) have put forward a theoretical model that represents in essence an amalgam of H. J. Eysenck's inhibition and arousal theories. They argue that it is important to determine whether the effects of extraversion on performance of any task are stimulus effects or response effects. Thus, for example, the poor performance of extraverts on vigilance tasks may reflect the need to engage in continuous stimulus analysis, or alternatively it may be due to the relative lack of opportunity to respond.

According to Brebner, extraversion affects both stimulus analysis and response organization. In a nutshell, he proposed that the introvert is "geared to inspect," whereas the extravert is "geared to respond." These hypotheses were derived from the following line of reasoning. Brebner and Cooper (1974) suggested that there were central mechanisms that could be in either of two states (excitation or inhibition). Either of these two states can be produced as a consequence of the demands for stimulus analysis (stimulus excitation or stimulus inhibition) or for response organization (response excitation or response inhibition). The meanings of the terms *excitation* and *inhibition* were clarified by Katsikitis and Brebner (1981). Excitation refers to the tendency to continue the same behavior when the opportunities to vary that behavior are constrained, whereas inhibition refers to the tendency to discontinue the same behavior in the same circumstances.

Personality enters the picture as follows: Introverts tend to derive excitation from stimulus analysis and inhibition from response organization, whereas extraverts are more likely to derive inhibition from stimulus analysis and excitation from response organization. This provides the theoretical underpinning for the notion that introverts are "geared to inspect" and extraverts "geared to respond." However, there is a slight complication when attempting to use this theoretical model to predict performance data. According to Brebner and Cooper (1974), the perception of a stimulus is itself a response, and the act of responding generates stimuli which have an excitatory effect on introverts but an inhibitory effect on extraverts.

Let us now turn to the experimental evidence relating to this theoretical position. Since the model espoused by Brebner and his colleagues incorporates theoretical ideas taken from H. J. Eysenck's 1957 inhibition theory and his 1967 arousal theory, it can safely be assumed that most of the research supportive of Eysenck's theorizing is also supportive of Brebner's model. This research is discussed in detail in Chapter 9. In addition, Brebner has carried out a number of studies specifically designed to test his theory, and it is appropriate to consider them at this point.

The theoretical prediction that introverts derive excitation from stimulus analysis whereas extraverts derive inhibition has been tested by Brebner and Cooper (1974) and by Katsikitis and Brebner (1981). Brebner and Cooper used a reaction-time task in which visual signals were presented at a slow, regular rate of one every 18 seconds. In their second experiment, extraversion correlated +.69 with the number of missed signals. An analysis of the reaction-time data revealed that only the extraverts slowed down during the course of the experiment, with the result that they had much slower reaction times than introverts during the second half of the trials.

Brebner and Cooper (1974) chose to interpret their data in terms of stimulus inhibition, arguing that extraverts are more likely than introverts to respond to stimulus analysis with inhibition. However, it is not clear that much stimulus analysis was actually required in view of the regular intervals between signals. They claimed that response inhibition was not a relevant factor because little response inhibition would be generated with the low response rate required; by the same logic, little stimulus inhibition would be generated with the low regular signal rate that was used.

Katsikitis and Brebner (1981) tried to manipulate the amount of stimulus analysis that was required in two different ways. They used a letter-cancellation task and varied the number of letters to be crossed out (one or four). Some of the subjects performed the task in crowded conditions. The key finding was that performance of extraverts relative to introverts was especially poor when the demands for stimulus analysis were allegedly maximal (i.e., the more difficult version of the cancellation task performed under crowded conditions). It was argued that these conditions produced the greatest amount of stimulus inhibition in extraverts. It seems strange that Katsikitis and Brebner predicted that extraverts would be at a disadvantage to introverts only when considerable stimulus analysis was required, whereas Brebner and Cooper (1974) predicted (and found) that extraverts were considerably inferior to introverts on a task making relatively modest demands on stimulus analysis.

We have concentrated so far on the effects of extraversion on stimulus excitation and inhibition. Brebner and Flavel (1978) considered response excitation and hypothesized that extraverts should be more likely than introverts to respond on a reaction-time task in the absence of the appropriate stimulus. The argument was that extraverts are more reliant on response excitation to maintain an adequate level of arousal. The number of erroneous responses produced was greater than in most reaction-time studies because of the introduction of numerous catch trials on which the warning signal was not followed by the stimulus requiring response. As expected, extraverts made considerably more commission errors than introverts on catch trials. Although the findings are consistent with the hypothesis, the problem is that the greater cautiousness of responding found in introverts may be due to various factors that have nothing to do with response excitation.

Tiggemann, Winefield, and Brebner (1982) considered some of the possible effects of extraversion on response inhibition by investigating the learned helplessness effect. This effect involves a lack of responsiveness and a performance decrement following exposure to events which the subject cannot control. In line with the notion that introverts tend to develop response inhibition, whereas extraverts show response excitation, they predicted that introverts should be more susceptible than extraverts to the helplessness effect.

According to Brebner's theory, this prediction is most likely to be supported when the task used imposes heavy demands on response organization. As a result, Tiggemann *et al.* (1982) followed a task in which button pressing failed to control the offset of a buzzer with a second task on which a buzzer could either be avoided or turned off by a relatively complex series of button presses (one press on the left button and two on the right button). As expected, there was a considerably stronger helplessness effect in introverts than in extraverts. What is perhaps surprising, however, is that there was practically no effect of extraversion when it was preceded by a button-pressing task on which the subject could control the offset of a buzzer. Why did the introverts' tendency toward response inhibition not impair their performance under these conditions as well?

Perhaps the most direct investigation of the Brebner formulation was undertaken by Brebner and Cooper (1978). They used a task in which subjects had the choice of inspecting a visual stimulus or of responding in an attempt to change it. More specifically, key-press responding produced reinforcement in the form of stimulus change on a variable ratio schedule. In line with the notion that introverts are "geared to inspect," introverts inspected the stimuli much longer than extraverts. On the other hand, extraverts, who are putatively "geared to respond," responded more than twice as often as introverts and almost three times as fast.

How adequate is the theory put forward by Brebner and his colleagues? It contains a grain of truth but appears to suffer from a lack of testability. In order to make accurate predictions of the relative performance levels of introverts and extraverts, we must have information about four separate factors: (1) the stimulus excitation or inhibition produced by stimulus analysis, (2) the response excitation or inhibition produced by response organization, (3) the response excitation or inhibition produced by perceptual responses, and (4) the stimulus excitation or inhibition produced by the stimuli generated by responding. In practice, we usually have extremely sketchy information about most of these factors. As a consequence, it would typically be possible to account for any set of data in a *post hoc* fashion.

A further complication was added by Brebner (1983). Previous statements of the theory had clearly proposed that the more excitation that was generated in any particular situation, the greater would be the tendency to continue with the stimulus analysis and/or response organization that was producing the excitation. As Paisey and Mangan (1982) pointed out, it surely makes some difference whether these increases in excitation are accompanied by positive or negative hedonic tone. In particular, extremely high levels of excitation would probably be aversive and thus avoided. It could therefore be the case that introverts, who derive excitation from stimulus analysis and are "geared to inspect," might actually be less likely than extraverts to inspect relatively intense stimuli. Thus, Weisen (1965) found that introverts worked harder than extraverts to produce the temporary removal of loud music and colored lights. In view of these considerations, Brebner (1983) agreed that the notion that excitation is invariably rewarding is not generally true:

> When the organism's top capacity for excitation is reached, demands for stimulus-analysis or response-organization which add further loading will produce an inhibitory effect. This is how the model tries to incorporate the concepts of overarousal or the induction of transmarginal inhibition. (p. 229)

In other words, successful prediction of performance differences between introverts and extraverts requires some assessment of the point at which increasing excitation fails to facilitate performance. Thus, we require *in toto* accurate information about five separate factors in order to predict task performance. Worse still, the only evidence that is typically available comes from performance on the task itself, so that there is a very real danger of circular reasoning. Until such time as measures of the various factors can be obtained that are independent of the performance data that are to be predicted, this theory will prove extremely difficult to test in an adequate fashion.

CHAPTER EIGHT

The Psychophysiology of Personality

According to H. J. Eysenck (1967a), individual differences in personality depend ultimately on underlying physiological processes. As a consequence, although the behavioral data discussed in Chapters 9 and 10 provide indirect support for Eysenckian theory, it is apparently at the physiological level that the most direct evidence can be obtained. However, as we will see shortly, it would be overly optimistic to assume that psychophysiological measures provide irrefutable evidence about individual differences in physiological functioning. Thus, for example, tempting as it is to believe that arousal level can be measured more precisely physiologically than behaviorally, the temptation should probably be resisted. The plain fact is that both physiological and psychological data require interpretation in order to make theoretical sense.

What, then, is the value of the psychophysiological approach to personality? In essence, it is of great importance to broaden the data base when attempting to unravel the complexities of individual differences in personality. It is perhaps possible to dismiss the behavioral evidence relating to the arousal theory of extraversion when considered on its own, and the same is true of the psychophysiological evidence taken in isolation. However, if both the psychophysiological and psychological evidence point to the same conclusion, then that conclusion must be taken seriously.

As with the behavioral data, the physiological data collected so far relate mainly to the extraversion dimension. However, a number of studies have dealt with the physiological correlates of neuroticism or anxiety, leaving the psychoticism dimension largely unresearched. Our coverage of the psychophysiology of personality will reflect these research preferences.

THEORETICAL BACKGROUND

The theory of personality proposed by H. J. Eysenck (1967a) spelled out in some detail the biological basis of the extraversion and neuroticism dimensions. Extraversion was related to the ascending reticular activating system (ARAS), which is located in the brain-stem reticular formation. According to H. J. Eysenck (1967a), collaterals from the ascending sensory pathways excite cells within the ARAS, which then sends the excitation to various sites in the cerebral cortex. The ARAS was first directly associated with arousal by Moruzzi and Magoun (1949), who discovered that electrical stimulation of the ARAS elicited an activation pattern in the cortical EEG. The general significance of the ARAS was expressed in the following way by Stelmack (1981):

> The reticular formation is implicated in the initiation and maintenance of motivation, emotion and conditioning by way of excitatory and inhibitory control of autonomic and postural adjustments and by way of cortical coordination of activity serving attention, arousal and orienting behavior. (p. 40)

The physiological structure alleged to underlie the neuroticism dimension is the visceral brain, which comprises the amygdala, hippocampus, septum, cingulum, and hypothalamus. There is a loop consisting of the visceral brain and the reticular formation. Messages from the visceral brain reach the reticular formation and then proceed to the cortex, where they have an arousing effect. The physiological structures relating to neuroticism are mainly concerned with emotion.

It should be clear that the structures underlying the extraversion and neuroticism dimensions are only partially independent in their functioning. For example, cortical arousal is associated with the corticoreticular loop underlying extraversion and with the loop underlying neuroticism. H. J. Eysenck (1967a) attempted to clarify matters by proposing a conceptual distinction between arousal and activation, with arousal referring to reticular activity and activation referring to autonomic activity. However, we are still left with various uncertainties, since psychophysiological measures do not reflect directly the activity of either the reticular formation or the visceral brain. For instance, EEG desynchronization occurs as a consequence of either arousal or activation. This means that it is difficult to provide a satisfactory empirical test of the notion that arousal and activation are related but separate.

EXTRAVERSION

THE ORIENTING REACTION

The general notion that introverts are more cortically aroused than extraverts can be subdivided into two more specific predictions: (1) introverts will produce a greater increment in arousal than extraverts in response to stimulation; and (2) introverts will characteristically be more aroused than extraverts. The first prediction has most frequently been tested by investigating the effects of extraversion on the orienting reaction. The orienting reaction or "what is it?" response occurs in response to novel stimuli and indicates that the individual concerned is responsive to some environmental change. According to Sokolov (1963), information about any given stimulus is stored in a neuronal model. An orienting reaction is elicited when there is a mismatch between the neuronal models of current and previous stimulation. Habituation occurs when the current stimulus matches a stored neuronal model and as a result there is no orienting reaction. There are several physiological components of the orienting reaction, including increased skin conductance, digital vasoconstriction, cephalic vasodilation, heart rate deceleration, and EEG desynchronization.

The orienting reaction seems especially relevant to the personality dimension of extraversion because of the central role apparently played by cortical excitatory and inhibitory activity in its elicitation and subsequent habituation. It appears that nonspecific activating effects of the stimulus act through the reticular formation to activate the hypothalamic sites that produce the autonomic components of the orienting reaction. If a stimulus is repeated several times, then cortical inhibitory impulses affect the collaterals transmitting impulses from the ascending sensory tracts to the reticular formation. This in turn produces a reduction in autonomic activity and habituation occurs.

It has usually been predicted (e.g., H. J. Eysenck, 1967a) that introverts should have larger orienting reactions than extraverts and should take longer to habituate. These predictions can be based on the putative high level of arousal in introverts or on the notion that extraverts generate more cortical inhibition than introverts. The theoretical problem is that the orienting reaction and its habituation depend on a complex mixture of arousal, inhibition, and neuronal model formation, and it is not easy to decide which of these three processes is most important in producing any observed effects of extraversion.

The usual experimental procedure in studies of habituation is to present the same stimulus (visual or auditory) several times to the subject and measure his reactions (usually in the form of electrodermal responses). What are the findings with respect to habituation of the orienting reaction in introverts and extraverts? A thorough review of the literature was provided by O'Gorman (1977). Of the 20 studies he considered, 8 provided significant support for the hypothesis that habituation occurs more rapidly in extraverts than in introverts, and the remaining 12 reported nonsignificant effects of extraversion. Since that review was published, H. J. Eysenck's hypothesis has been supported at a significant level in at least five other studies (Gange, Geen, & Harkins, 1979; Smith, Rypma, & Wilson, 1981; Smith & Wigglesworth, 1978; Stelmack, Bourgeois, Chian, & Pickard, 1979; Venturini, De Pascalis, Imperiali, & San Martini, 1981).

Although such a head count serves to indicate that introverts do often take longer than extraverts to habituate to repeated stimulation, it ignores the quality of the individual studies and provides relatively little insight into the processes involved. However, it is noteworthy that most of the studies reporting significant effects of extraversion are reasonably sound methodologically. One such study reporting unusually consistent findings was that of Stelmack et al. (1979). In one of their experiments, they found that introverts showed less habituation than extraverts to chromatic stimuli as revealed by cardiac, skin resistance response, and vasomotor indices of the orienting reaction. In a further experiment, introverts produced more orienting reactions than extraverts to visually presented affective and neutral words as evidenced by electrodermal and vasomotor measures. In a third experiment, introverts produced more orienting reactions than extraverts to chromatic stimuli, and introverts also showed greater initial amplitude of the skin conductance response.

One way of clarifying matters is to consider some of the studies in which interactions were obtained between extraversion and some aspect of the experimental situation. If it proves possible to delineate the conditions under which extraverts habituate more quickly than introverts, then this would represent significant progress. One of the factors influencing differences between introverts and extraverts is the intensity of the stimulus that is presented repeatedly. Wigglesworth and Smith (1976) found in their first experiment that there were no differences in habituation rate to a tone between introverts and extraverts. However, introverts initially produced greater skin conductance responses than extraverts to 80 dB intensity tones, whereas exactly the opposite happened with 100 dB tones. The greater responsiveness of extraverts than introverts to the more intense tone was accounted for by reference to the

Pavlovian law of transmarginal inhibition. According to this law, the usual tendency for increased stimulation to produce increased responding breaks down when stimulation is very intense, and instead transmarginal inhibition or response decrement occurs. The level of stimulus intensity at which this is found is thought to be lower for introverts than for extraverts. Unfortunately, although their second experiment was very similar to the first, Wigglesworth and Smith failed to discover any effects of extraversion on initial skin conductance responses, and extraverts habituated more rapidly than introverts only with the most intense tone.

The law of transmarginal inhibition also appears to be applicable to a series of experiments reported by Fowles, Roberts, and Nagel (1977). They used a fairly typical orienting reaction paradigm but reported only skin conductance level rather than the skin conductance responses to each stimulus. A series of tones was presented following a stressful (i.e., difficult task) or a nonstressful (i.e., easy task) experience. Extraverts had higher skin conductance levels than introverts under the most arousing conditions (i.e., most intense tones following stress). In contrast, when no task was given prior to the tone presentations, introverts had higher skin conductance levels than extraverts with tones of moderate intensity. In other words, as Fowles et al. remarked, introverts showed "greater responsiveness at low stimulus intensities and the decline in responsiveness at high stimulus intensities as a result of transmarginal inhibition" (p. 142).

The most interesting study in which transmarginal inhibition occurred was that of Smith et al. (1981). The subjects were given either the stimulant caffeine or a placebo at the start of the experiment, and habituation, dishabituation, and spontaneous recovery of the electrodermal orienting reaction were measured. In general, introverts had higher tonic levels and larger phasic responses than extraverts under placebo conditions, whereas extraverts showed greater electrodermal responsiveness than introverts after caffeine administration. These interactions suggest strongly that differences between introverts and extraverts in the orienting reaction are attributable to the higher level of arousal in introverts. A final point of interest is that virtually all of the effects of extraversion on electrodermal responses were due to the impulsivity component of extraversion rather than the sociability component.

Rather similar findings were obtained by Smith, Wilson, and Jones (1983). They also found that skin conductance responses were greater in introverts than in extraverts under placebo conditions, but the opposite was the case under caffeine. If we consider all of the relevant evidence, it seems reasonably well established that decrements in arousal as measured by electrodermal activity occur at lower levels of stimulation for introverts than for extraverts (Smith, 1983).

Stimulus intensity is one of the factors determining whether or not introverts and extraverts differ in the rate of habituation, but it is highly improbable that the use of an inappropriate stimulus explains all, or even most, of the nonsignificant effects in the literature. Part of the reason is the existence of autonomic response specificity (Lacey & Lacey, 1958), which is the tendency for individuals to differ with respect to that autonomic response most affected by stimulation. The value of paying heed to autonomic response specificity was shown by Stelmack *et al.* (1979) in an experiment on habituation to neutral and affective words in which the cardiac, vasomotor, and electrodermal components of the orienting reaction were all recorded. No single component accounted for more than 24% of the variation in extraversion, but the conjoint influence of all three components accounted for 54%.

In addition to the complexities associated with autonomic response specificity, there is the related problem that different physiological measures of the orienting reaction usually show disappointingly low intercorrelations. This is true even within a single response system. For example, Bull and Gale (1973) obtained a correlation of only +.27 between mean response levels of skin conductance responses and skin resistance responses. Apart from the noisy data that result from such specificity, there is the further problem that it is dangerous to extrapolate from the experimental findings. Even if extraversion has a clear-cut effect on the electrodermal component of the orienting reaction, it cannot safely be assumed that the same effect would be found with other components.

In sum, there are two major reasons why it is of importance to investigate the effects of extraversion on the orienting reaction and on habituation. The first reason is that the theoretical constructs (e.g., arousal, inhibition) that have been used to explain the occurrence and disappearance of the orienting reaction seem remarkably similar to those used to describe the basic physiological differences between introverts and extraverts. The second reason is that the orienting reaction is often regarded as an attention-like process of general significance. As such, it is thought to facilitate sensory intake, to lower sensory thresholds, and to enhance sensory discrimination and conditioning. For example, some evidence that the introvert superiority in vigilance performance may be due, at least in part, to the fact that introverts produce more orienting reactions than extraverts was obtained by Gange *et al.*

(1979). They discovered that introverts displayed more electrodermal responses than extraverts to signal stimuli during a visual vigilance task.

It is reasonably well established that introverts show more persistent electrodermal responses to repetitive stimulation than extraverts, except with intense stimuli, and this suggests that extraversion is related to electrodermal habituation. However, there is less evidence available with respect to other physiological response systems. The proper interpretation of these findings is still unclear. Since introverts often have higher skin conductance levels and produce more nonspecific responses than extraverts, it is likely that part of the slowness to habituate shown by introverts reflects their high level of arousal rather than simply their reactions to stimuli. Whether the effects of extraversion on habituation are due primarily to individual differences in excitatory activity or in the buildup of inhibition remains unclear. In other words, we know roughly what is happening, but not why it is happening.

The EEG

The notion (H. J. Eysenck, 1967a) that introverts are characterized by higher levels of cortical arousal than extraverts has been investigated several times by means of EEG recordings. It has typically been assumed that high levels of arousal are associated with low amplitude, high frequency activity in the alpha frequency range of 8–13 Hz. However, serious doubts have been expressed about the value of such measures. Although the EEG is regarded as an indicator of brain activity, it is recorded from the outside of the skull. A more consequential objection is that the EEG represents a kind of composite or amalgam of electrical energy generated from different parts of the cortex and may thus provide a misleading impression of the actual activity in any specific area of the brain.

An inchoate picture confronts anyone who considers the published studies. In his review, Stelmack (1981) discusses 11 studies. In four studies there were nonsignificant effects of extraversion on EEG activity; in five studies introverts had higher levels of cortical arousal than extraverts; and in the remaining two studies extraverts appeared to be more aroused cortically than introverts.

The most thorough review of the literature relating extraversion and the EEG was provided by Gale (1983). He considered 33 studies containing a total of 38 experimental comparisons. Extraverts were less aroused than introverts on the EEG in 22 comparisons; introverts were less aroused than extraverts in 5 comparisons; and nonsignificant effects of extraversion were reported in the remaining comparisons.

There are a number of possible reasons for this apparent inconsistency. First, there are considerable variations from one study to another with respect to technical details such as electrode placements and the ways in which alpha activity was defined. Second, hand-scoring techniques, which may be unreliable and prone to systematic error, were used in some of the studies. Third, and perhaps most obviously, the task performed by the subject while the EEG recordings were being made varied considerably. In some studies the subject reclined in a semisomnolent state with his eyes closed, whereas in others the subject sat upright and attempted to solve complex arithmetic problems.

Is there any way of reconciling the various findings? A valiant attempt to do so was made by Gale (1973), who argued that the effects of extraversion on the EEG were influenced by the level of arousal induced by the experimental conditions. More specifically, he suggested that introverts are most likely to be more aroused than extraverts in moderately arousing conditions, with the differences between introverts and extraverts either disappearing or being reversed with conditions producing either low or high levels of arousal.

Additional support for this conclusion was reported by Gale (1983). He classified all of the relevant EEG studies according to whether the test conditions were minimally, moderately, or highly arousing. Introverts appeared to be more aroused than extraverts in all 8 of the studies using moderately arousing conditions that reported significant effects of extraversion, but the expected result was found in only 9 out of 12 significant studies using low-arousal conditions and 5 out of 7 using high-arousal conditions.

Why does the extent to which the experimental conditions induce arousal affect the nature of the effect of extraversion on the EEG? In low-arousal conditions, extraverts may be more likely than introverts to disregard instructions to sit still and think about nothing (stimulus hunger), and in high-arousal conditions introverts may attempt to relax under sensory bombardment (stimulus avoidance).

The fairest conclusion is probably that introverts have higher levels of cortical activity than extraverts as measured by the EEG only under certain conditions. As yet, the nature of those limiting conditions has not been clearly established. However, the extent to which the experimental conditions induce arousal is probably one of the crucial factors.

A rather different method of investigating the relationship between personality and EEG activity has become increasingly popular in recent years. A stimulus is presented several times, and the cortical potentials evoked by that stimulus are averaged so that stimulus-related cortical activity can be distinguished from other EEG activity. It has usually been found that there is a positive relationship between the intensity of the repeated stimulus and the amplitude of the resultant evoked potential, and this has led to suggestions that cortical evoked potentials provide a useful measure of sensory sensitivity and attention.

If introverts are characteristically more cortically aroused than extraverts, and if they also tend to augment incoming stimulation (H. J. Eysenck, 1967a), then the natural prediction is that the amplitude of cortical evoked potentials should be greater in introverts than in extraverts. The findings are rather inconsistent. Shagass and Schwartz (1965) recorded somatosensory evoked responses to stimulation of the right median nerve at the wrist. Extraverts had greater amplitude of evoked responses than introverts among subjects under the age of 20, whereas the opposite was the case among those over 40 years of age. There was no evidence of greater amplitude of evoked potentials in introverts in the studies by Burgess (1973) and Häseth, Shagass, and Straumanis (1969). However, in both studies different levels of stimulus intensity were given to different individuals, with the actual level of stimulation being determined by each subject's absolute threshold. This procedure probably prevented the discovery of any effect of extraversion on evoked potential amplitude. Further negative findings were reported by Rust (1975) in two studies investigating the auditory evoked response to tones.

More promising results with the auditory evoked response were reported by Hendrickson (1972) and by Stelmack, Achorn, and Michaud (1977). Hendrickson obtained a significant negative correlation between extraversion and the amplitude of the auditory evoked response to low frequency tones at 60 dB. Stelmack *et al.* also discovered that introverts produced evoked responses of greater amplitude than extraverts to low frequency tones (55 db and 80dB), but there was no effect of extraversion with high frequency tones. While the reason for this discrepancy between the effects of low and high frequency tones is not known, it may be relevant that introverts demonstrate greater auditory sensitivity than extraverts on a signal-detection task only when low frequency tones are presented (Stelmack & Campbell, 1974).

A rather more complex analysis of evoked potential data was employed by Haier, Robinson, Braden, and Williams (1984). Subjects viewed light flashes of four intensities, and the difference between the first positive wave and the first negative wave was calculated. Augmenters were defined as those in whom increased stimulation produced increased amplitude of this component of the evoked potential, and reducers as those showing the opposite effect. Although it might have been expected that augmenters would tend to be extraverted and reducers introverted, the actual finding was that the augmenters were significantly more extraverted than the reducers. Since relatively low stimulus intensities were used, it seems improbable that the results can be explained in terms of the law of transmarginal inhibition.

A complex theoretical interpretation of the above findings was offered by Robinson, Haier, Braden, and Krengel (in press). In essence, they argued that the diffuse thalamocortical system is more strongly activated by stimulus input in introverts than in extraverts and that this produces inhibiton of the brain-stem reticular formation. As a consequence, certain components of the evoked potential in introverts indicate a reducing effect.

If we may tentatively conclude that there are at least some circumstances in which introverts produce cortical evoked potentials of greater amplitude than extraverts, then the theoretically interesting question is the level within the processing system at which the effects of extraversion are occurring. One possibility is that extraversion affects the initial brain-stem response to stimulation; an alternative view is that extraversion affects later aspects of processing relating to attention. The most plausible interpretation is that extraversion affects attentional processes. This could help to account for some of the differences between studies. For example, extraversion affected the auditory evoked potential in the study by Stelmack et al. (1977) but did not in that of Rust (1975), and the level of attention required by the subject was greater in the former than in the latter study.

An appropriate way of resolving some of the interpretative ambiguities is to consider the effects of extraversion on the brain-stem evoked potentials occurring very shortly after stimulation and before any effects of attention can be identified. Campbell, Baribeau-Bräun, and Bräun (1981) measured the auditory brain-stem evoked potential in two separate experiments and failed to discover any effects of extraversion. These findings suggested that extraversion does not affect neuronal transmission at the receptor and brain-stem levels and that attentional factors may be responsible for the observed cortical evoked response differences. However, a recent study by Stelmack and Wilson (1982) produced very different findings. They also investigated the auditory brain-stem evoked response and found that extraverts had longer latencies than introverts for some of its waves. This result implies that there are effects of extraversion on the initial peripheral action of the auditory nerve. If this is confirmed, then physiological differences between introverts and extraverts may extend beyond the corticoreticular loop and encompass axonal or synaptic transmission.

An exciting recent development in research relating personality and the EEG has been reported by Robinson (1982). Visual stimuli were sinusoidally modulated, and the relative amplitudes of the sinusoidal EEG responses were measured for each stimulus frequency. The EEG data were analyzed in a complex manner that produced measures of inductance (L) and capacitance (C). The C measure was regarded as an index of the inhibitory process. Individuals in whom the excitatory and inhibitory processes were of comparable strength were regarded as balanced. By combining Eysenckian and Pavlovian theoretical ideas, Robinson argued that stable extraverts are strong balanced individuals whereas neurotic introverts are weakly balanced individuals. When only those individuals whose stability and extraversion scores were comparable and whose C and L scores were also similar were considered, a remarkably high correlation of $+.95$ between personality and an EEG measure was obtained. Even when the total sample was included in the correlation, a correlation coefficient of $+.63$ was still obtained.

These are the strongest relationships between personality and the EEG ever obtained, perhaps because EEG data have rarely been analyzed in such a sophisticated manner and in accordance with a theoretical model of physiological functioning. If these findings can be replicated, then we are well on the way to understanding the underlying physiological basis of individual differences in extraversion.

PUPILLOMETRY

There has been increasing interest in recent years in the use of the pupillary response as a psychophysiological measure relating to personality. The iris muscle around the pupillary aperture is innervated reciprocally by the autonomic nervous system. Pupillary dilation is due primarily to sympathetic activity, whereas constriction reflects parasympathetic activity. Pupillometry can be used to measure individual differences in responsiveness to stimulation, and tonic pupil size in the absence of specific stimulation can provide an index of the general level of autonomic arousal. In spite of the potential usefulness of pupillometry, it is often difficult to interpret pupillary response data in an unequivocal fashion. Part of the reason is that the pupillary response is affected by several different factors, some of which are of little interest to the personality theorist.

In the first study to investigate the effects of extraversion on pupillary responses, Holmes (1967) measured speed of pupillary dilation to the offset of a light and pupillary constriction to the onset of a light. The fast dilators tended to be extraverted, whereas the fast constrictors were introverted. The proper interpretation of these findings is unclear, but Holmes argued that the rapid pupillary constriction of introverts indicated that they had greater amounts of acetylcholine at cholinergic synapses than extraverts.

Frith (1977) confirmed some of Holmes's findings. He reported that high scorers on the impulsivity component of extraversion showed less pupillary constriction than low impulsives in response to a light flash, perhaps because they were less reactive to stimulation. Of greater interest, Frith also discovered that impulsivity was negatively correlated with pupil size during an initial interval of no stimulation. This suggests that the more impulsive subjects were less aroused than the less impulsive subjects.

Stelmack and Mandelzys (1975) also found that introverts had larger pupils than extraverts in the absence of specific stimulation, suggesting that the introverted subjects were more aroused throughout the experiment. They also investigated phasic pupillary responses to auditorily presented neutral, affective, and taboo words. Ther results are shown in Figure 28. Introverts showed significantly more pupillary dilation to these stimuli than extraverts, especially in response to the taboo words. In other words, introverts responded more strongly than extraverts to the auditory stimuli.

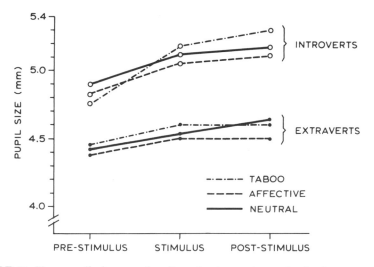

FIGURE 28. Mean pupil size as a function of extraversion and stimulus type. (From "Extraversion and Pupillary Response to Affective and Taboo Words" by R. M. Stelmack and N. Mandelzys, *Psychophysiology*, 1975, *12*, 536–540. Copyright 1975 by the Society for Psychophysiological Research. Adapted by permission of author.)

The limited pupillometric data are generally consistent with the notion that introverts have a higher general or tonic level of arousal than extraverts. There is also some support for the hypothesis that introverts are more responsive to stimulation.

SEDATION THRESHOLD

We have concentrated so far on the ways in which relatively peripheral psychophysiological measures can be used to provide indirect evidence concerning the underlying biological basis of personality. An alternative approach, and one that seems more direct in some ways, is to examine individual differences in the reactions to centrally acting drugs. If the drugs affect the brain mechanisms responsible for producing individual differences along a given dimension of personality, then there should be predictable differences in drug sensitivity between high and low scorers on that dimension.

The most popular embodiment of this research strategy has involved assessment of the sedation threshold. A sedative drug (usually one of the barbiturates) is administered by intravenous injection until the sedation threshold is reached. This has been determined in several different ways, including physiological changes and unresponsiveness to verbal stimuli.

An attractively simple prediction concerning individual differences in the sedation threshold is that introverts, starting from a high level of cortical arousal, should require more sedative drug than extraverts to reach the sedation threshold. The predicted result has been obtained a number of times. Laverty (1958) found that the amount of sodium amytal (a depressant) needed to produce slurred speech was greater for introverts than for extraverts. In a series of studies, Shagass and his colleagues (Krishnamoorti & Shagass, 1964; Shagass & Jones, 1958; Shagass & Kerenyi, 1958) discovered that the sedation threshold tended to be higher in introverts than in extraverts. However, they also found that manifest anxiety was associated with increased tolerance of barbiturates.

There appear to be fairly complex interactions between different personality dimensions in the determination of the sedation threshold. Rodnight and Gooch (1963) found that tolerance of nitrous oxide gas was not related to either introversion or neuroticism considered separately. However, reanalysis of their data revealed that extraversion was negatively correlated with tolerance among subjects high in neuroticism, whereas there was a positive correlation between extraversion and tolerance among those low in neuroticism. Very similar results were

reported by Claridge and Ross (1973) in a study of tolerance of amylo-
barbitone sodium and by Claridge, Donald, and Birchall (1981) using
thiopentone.

Futher complexities were revealed by Claridge *et al.* (1981) when
they combined their thiopentone data with earlier data collected by Claridge (1967) and by Claridge and Ross (1973). This produced a total of 126
subjects who were assigned to nine groups on the basis of low, medium,
or high extraversion and low, medium, or high neuroticism. The results
are shown in Figure 29. There was a highly significant interaction
between extraversion and neuroticism; in this interaction, the highest
drug tolerance was shown by introverts with moderate neuroticism, and
the lowest tolerance occurred among neurotic extraverts. Thus, the
hypothesis that introverts should have higher sedation thresholds than
extraverts was supported among those of medium neuroticism, whereas
the opposite tendency was present among those of low neuroticism.

What are we to make of these findings? The most troublesome
aspect of the research in this area is the fact that administering large
amounts of a sedative drug is analogous to firing a blunderbuss: You may
hit the desired target (e.g., the corticoreticular loop) but you probably
also hit other unwanted targets. In other words, the process of achieving
sedation through drug administration involves myriad effects on all of
the various excitatory and inhibitory activities that maintain conscious-
ness. It has been clearly established that the sedation threshold is

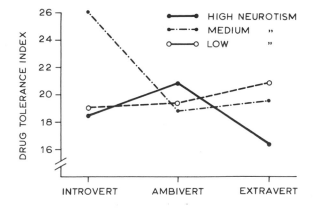

FIGURE 29. Mean drug tolerance index scored as a function of extraversion and neuroticism. (From "Drug Tolerance and Personality: Some Implications for Eysenck's Theory"
by G. S. Claridge, J. R. Donald, and P. M. Birchall, *Personality and Individual Differences,* 1981, *2,* 153–166. Copyright 1981 by Pergamon Press, Inc. Adapted by permission.)

affected by extraversion and neuroticism in interaction, but the theoretical significance of that interaction will remain obscure until more is known of the ways in which sedative drugs affect physiological functioning.

CONCLUSIONS

Attempts have been made to discover physiological differences between introverts and extraverts in a variety of experimental paradigms and using a number of psychophysiological measures (e.g., the EEG, electrodermal responses, pupillary responses). It is perhaps reasonable to conclude that introverts usually show greater physiological responsiveness than extraverts to stimulation, with the most consistent findings being obtained with electrodermal measures. This conclusion is very much in line with H. J. Eysenck's (1967a) contention that introverts tend to augment incoming stimulation, and it also accords well with the psychological data discussed in Chapter 9. However, there is as yet inadequate understanding of the precise conditions in which introverts will be more responsive than extraverts, as is indicated by the several studies reporting nonsignificant effects of extraversion. One relevant factor appears to be the intensity of stimulation, since at high levels of stimulus intensity the normal effect of extraversion is sometimes reversed.

When we turn to the issue of whether introverts have a higher characteristic or tonic level of arousal than extraverts, the evidence is less satisfactory. There are supporting results based on the EEG and pupil size, but there are several nonsignificant findings and even some in which extraverts were more aroused than introverts. Some, but not all, of these apparent inconsistencies can be explained if we assume that the predicted effect of extraversion is most likely to occur under moderately arousing conditions. It is rather surprising that so little attention has been paid to the possibility that differences in tonic arousal between introverts and extraverts may be affected by time of day, although there are one or two notable exceptions (e.g., Gale, Harpham, & Lucas, 1972). The extremely limited evidence that is available suggests that introverts may be more aroused than extraverts in the morning but that the opposite is the case in the evening. The body temperature of introverts is higher than that of extraverts early in the day but lower in the evening (Blake, 1967; M. W. Eysenck & Folkard, 1980), and the same is true of ratings of subjective alertness (Folkard, 1975). It is entirely possible that more systematic effects of extraversion on levels of tonic arousal will be found if individual differences in the diurnal rhythm of arousal are taken into account.

NEUROTICISM

H. J. Eysenck (1967a) argued that the personality dimension of neuroticism relates to individual differences in excitability and emotional responsiveness, which are reflected in autonomic activation. In view of the nature of the neuroticism dimension, it is likely that large differences in autonomic activation between high and low neuroticism scorers will only occur under relatively stressful conditions. It is unfortunate that nearly all of the research on psychophysiological differences as a function of neuroticism has used conditions that cannot realistically be described as stressful. It could thus be claimed that H. J. Eysenck's (1967a) prediction that those high in neuroticism should have greater autonomic activation than those low in neuroticism has not been put to an adequate test.

There has been a certain amount of interest in the effects of neuroticism on the rate of habituation of the orienting reaction to repeated stimuli. Of 11 studies reviewed by O'Gorman (1977), 5 reported a significant relationship between neuroticism and speed of habituation and 6 did not. However, there was little consistency among the significant results. High neuroticism was associated with fast habituation in two studies but with slow habituation in a third. Both directions of relationship were reported by Mangan and O'Gorman (1969) and by Marchman and Fowles (1973), with neuroticism interacting with extraversion in the former study and with stimulus intensity in the latter. If we replace mere head counting by considering the technical merit of the various studies, then two of the best studies (Coles, Gale, & Kline, 1971; Nielsen & Petersen, 1976) reported that high neuroticism slowed down the rate of habituation.

There are rather more interpretable effects of anxiety (which correlates highly with neuroticism) on habituation rate. Anxiety reduced the speed of habituation in three studies and had no effect in nine others (O'Gorman, 1977). The psychophysiological measure of habituation appears to have been of critical importance, because significant findings were obtained in all of the studies in which finger-volume responses were considered, whereas nonsignificant findings were reported in all of the studies examining habituation of electrodermal responses.

In some cases the amplitude of the initial orienting reaction has been assessed, usually by means of electrodermal measures. High neuroticism has been found both to increase the amplitude of the orienting reaction (Siddle, 1971) and to decrease its amplitude (Mangan, 1974; Mangan & O'Gorman, 1969). It is likely that neuroticism would be found to be positively related to the amplitude of the orienting reaction if stress-evoking stimuli were presented.

A similarly confusing state of affairs is present in studies of EEG measures of arousal. High neuroticism produced higher levels of arousal in one study (Winter, Broadhurst, & Glass, 1972) but was associated with lower levels of arousal in another (Coles *et al.*, 1971). There may have been important differences between the two studies in the extent to which the experimental conditions were stressful.

One of the few studies in which the degree of environmental stress was manipulated was reported by Plouffe and Stelmack (1979). They recorded the pupillary light reflex under conditons of stress produced by the cold pressor test and under control conditions. Stress led to a decrease in the extent of pupillary constriction to light stimulation and to a greater amount of redilation following the end of stimulation. These findings may have occurred because stress produced sympathetic excitation of the reticular formation. There were negligible effects of neuroticism on the pupillary response, and the theoretically predicted enhanced level of autonomic activation among stressed subjects high in neuroticism was certainly not manifest in enhanced pupil diameter.

There is an alternative way of investigating the effects of neuroticism or anxiety on autonomic activation, and that is based on comparisons of physiological activity between normal and neurotic patient groups. According to H. J. Eysenck (1967a), individuals who have experienced powerful emotions over long periods of time may show a breakdown of the usual distinction between activation and arousal, since for them "quite mild stimuli are emotionally activating" (p. 233).

Kelly and Martin (1969) compared patient and control groups differing in neuroticism and discovered that the patients had higher tonic levels of heart rate, blood flow, and blood pressure than the controls under nonstressful conditions. However, the two groups did not differ with respect to the same physiological measures during a stressful mental arithmetic task, possibly due to a "ceiling" effect. Lader and Wing (1966) found that neurotic patients showed slower habituation of the orienting reaction than did controls and also showed generally heightened activity on various psychophysiological indices of arousal.

In general, anxiety patient groups tend to be chronically highly activated at the physiological level, but they are often less autonomically reactive to stimulation than controls, perhaps because their prestimulus level of activation is so high. These findings further illustrate the value of distinguishing between long-lasting or tonic physiological effects and the phasic reactions to specific stimulation.

The data are undoubtedly disappointing, and it is hard to disagree with Stelmack's (1981) pessimistic conclusion that "correlations between neuroticism and psychophysiological responsiveness have not been reported with sufficient consistency to permit inferences of the physiological determinants" (p. 61). The problem may lie in the persistent use of insufficiently stressful conditions. This problem is obviated to some extent by the use of anxiety neurotics who are very susceptible to environmental stress, and it may be for this reason that more clear-cut findings have emerged from comparisons between patients and normals than between normals high and low in neuroticism.

THEORETICAL IMPLICATIONS

The theoretical implications of the psychophysiological research can be seen most clearly in the context of the formulations of H. J. Eysenck (1967a) and Gray (1973, 1981). There are three differences between the two theories that are of crucial importance in this connection:

1. Eysenck argues that orthogonal dimensions of extraversion and neuroticism should be postulated, whereas Gray prefers to rotate these dimensions in order to produce orthogonal dimensions of impulsivity and anxiety.

2. In line with the disagreement over the precise identification of the major personality factors, there are also differences between Eysenck and Gray with respect to the underlying physiological structures. Gray (1981) argued that the behavioral inhibition system determines trait anxiety; it consists of the septo-hippocampal system, its neocortical projection in the frontal lobe, and its monoaminergic afferents from the brain stem. Gray admitted with respect to the physiological system underlying the impulsivity dimension that "little progress has been made in describing the structures that go to make up this system" (p. 261). The contrasting views of Eysenck were discussed both in Chapter 7 and earlier in this chapter.

3. In terms of predicting psychophysiological responses, Eysenck argued that introverts have a chronically higher level of cortical activity than extraverts and that introverts were more responsive to stimulation. In contrast, Gray (1981) claimed that impulsive individuals are especially susceptible to reward, whereas anxious individuals are particularly susceptible to punishment.

Since Eysenck and Gray both assume that the same two-dimensional personality space is important, it follows that any set of data can be described by both of them. Thus, for example, results that Gray would ascribe to the effects of anxiety would be attributed by Eysenck to the conjoint influence of introversion and neuroticism. However, to the extent that the personality dimensions favored by each theorist are assumed to reflect causal lines of influence, Gray must predict that psychophysiological data should usually be determined interactively by extraversion and neuroticism. In fact, main effects of either extraversion or neuroticism are far more prevalent than interactive effects in the psychophysiological findings. An exception to the general trend is to be found in research on the sedation threshold. However, when the exact form of the interactions between extraversion and neuroticism is examined, it turns out that they do not provide strong support for Gray's theory. High sedation thresholds for neurotic introverts and low sedation thresholds for stable extraverts would indicate the involvement of the anxiety dimension, but the actual data are usually more complex (e.g., Claridge et al., 1981).

At the level of the physiological systems involved, Eysenck emphasized the role of the hypothalamus and of autonomic nervous system activity in accounting for individual differences in neuroticism. On the other hand, Gray argued that rather different physiological structures were involved. His theoretical position reduces the relevance of those physiological systems controlling autonomic response measures to the basis of neuroticism. As a consequence, many of the failures of autonomic responses to relate to neuroticism can be simply explained by arguing that the physiological responses being measured are inappropriate. However, the general conclusion must be that the psychophysiological data collected so far are not sufficiently sensitive or detailed to permit precise identification of the underlying physiological structures.

Two of the implications of Gray's theory are that introverts are more physiologically responsive than extraverts to aversive stimuli but that the opposite is the case with appetitive stimuli. These hypotheses have rarely been tested but are supported by some evidence. Stelmack et al. (1979) discovered that introverts habituated less than extraverts to taboo words as measured by electrodermal and vasomotor responses. Stelmack and Mandelzys (1975) investigated pupillary dilation to auditorily presented taboo words and found that it was greater for introverts than for extraverts. When interesting appetitive stimuli in the form of erotic nude pictures were presented, the amplitude of the electrodermal response was positively (but not significantly) correlated with extraversion.

It is also possible to make use of Gray's theory to account for the numerous failures of extraversion and neuroticism to relate to physiological measures. According to Gray, individual differences in physiological responsiveness should be greatest when definitely aversive or appetitive stimuli are presented, but the usual practice has been to use rather neutral stimuli (e.g., tones, light flashes). However, this line of reasoning encounters great difficulties when it comes to explaining the frequent discovery of significant individual differences in responsiveness to neutral stimuli in habituation. Gray's theory also fails to account satisfactorily for some of the EEG studies in which introverts appeared to be more aroused than extraverts in the absence of any specific stimulation (e.g., Gale *et al.*, 1969).

It thus appears that physiological differences between introverts and extraverts occur in various situations in which neither aversive nor appetitive stimuli are presented. This suggests that Gray's theory is too narrow at this point. In contrast, Eysenck claimed that introverts are chronically more physiologically responsive to stimulation, and these theoretical assumptions enable one to interpret most of the significant findings.

Eysenck's theory also makes sense of the otherwise puzzling tendency for physiological measures to correlate more consistently with extraversion than with neuroticism. High and low scorers on neuroticism differ physiologically only when emotionally threatening stimuli are presented, whereas the differences between introverts and extraverts with respect to chronic arousal level and responsiveness to stimulation ensure that introverts and extraverts differ physiologically in a wide range of situations.

CHAPTER NINE

Extraversion, Arousal, and Performance

Other things being equal, it is fairly obvious that a personality theory's range of applicability is relevant to any assessment of its importance. If a theory of personality includes within its compass an account of the impact of personality on basic processes such as perception, learning, thinking, and memory, then it deserves to be taken more seriously than one that is concerned only with a description of personality *per se*.

It is one of the most noteworthy features of the theory of personality proposed by H. J. Eysenck (1957, 1967a) that it has an unusually broad range of applicability. As was pointed out in Chapter 7, this is especially true with respect to the personality dimension of extraversion, where a huge volume of research has successfully identified many significant behavioral differences between introverts and extraverts. As Eysenck demonstrated, extraversion has been found to be related to the performance of a very large number of tasks, only some of which are included in the following list: classical conditioning, operant conditioning, sensory thresholds, pain tolerance, vigilance, sensory deprivation, time estimation, perceptual defense, critical flicker fusion, visual constancies, sleep-wakefulness patterns, verbal learning, figural aftereffects, visual masking, rest pauses in tapping, speech patterns, expressive behavior, and reminiscence.

There has been less experimental interest in the effects of neuroticism on performance. However, there is a reasonable body of research on anxiety and performance, and this is of relevance because personality measures of anxiety typically correlate approximately $+.6-+.7$ with measures of neuroticism. Theoretical and experimental contributions concerned with the ways in which performance is affected by individual differences in anxiety are discussed in Chapter 10. There has been relatively little research addressed to the issue of the behavioral correlates of psychoticism, and so it would be premature to attempt to draw any general conclusions.

To return to work on the effects of extraversion on performance, there is no way in which a complete account of the relevant research could possibly be given within the confines of a single chapter. Accordingly, the approach adopted is a highly selective one. Upon what criteria has the selection been based? First, some kinds of performance are of much more central importance to the theory than others. Since the theory is primarily a biological one, it tends to be those performance aspects that seem to have fairly direct biological significance (e.g., basic reactions to stimuli of varying intensity) that are emphasized. Second, there has been some preference for tasks on which the effects of arousal manipulations such as intense noise or stimulant drugs have already been established. The reason is that this additional information makes it easier to decide whether performance differences between introverts and extraverts can plausibly be ascribed to the higher level of cortical arousal of introverts. Third, the process of selection was inevitably influenced by the amount of research that has been done on various performance tasks. In order to avoid excessive speculation, it was desirable to concentrate on tasks where the effects of extraversion are reasonably clear.

Before we proceed to the research itself, it is worth discussing some of the complexities associated with the task of discovering meaningful behavioral differences between introverts and extraverts. The laboratory situation differs in many ways from everyday situations. Of particular significance, in everyday life an individual's personality characteristics play a part in determining both the situations in which he finds himself and the ways in which he reacts to those situations. Evidence for the assertion that personality determines the situations selected by individuals was obtained by Furnham (1981). He asked people to indicate the amount of time they had spent in various leisure situations over the previous week and discovered that introverts and extraverts had significantly different patterns of activity preferences. In contrast, experimental studies do not usually permit subjects much control over the situation, since that is controlled by the experimenter. As a result, the impact of personality on behavior is likely to be greater in everyday life than in the laboratory because a major way in which personality affects behavior is through its influence on situation selection.

A further difficulty is that successful prediction of the effects of extraversion on task performance can be achieved consistently only when there is a good understanding of the nature of extraversion and of the task in question. It has turned out repeatedly that reality is much more complex than had been envisaged. Although it is perhaps natural to predict simply that extraverts will outperform introverts on a task (or vice versa), it is usually the case that performance is determined interactively by extraversion and the precise parameters of the task. For example, it has been found with eye-blink conditioning that introverts condition better than extraverts under some conditions but that the opposite occurs under other conditions (H. J. Eysenck & Levey, 1972). Unless careful attention is paid to the details of an experimental task, there is the ever-present danger that the effects of extraversion on performance will appear to be inconsistent and unpredictable.

It follows from what has been said that it is often extremely difficult to assign an unequivocal interpretation to a failure to obtain a predicted effect of extraversion on performance. Obviously something is at fault, but there are many possible sources of error, including the theoretical analysis of extraversion, the measure of extraversion used, the analysis of the task, or the task parameters that were selected. Only a long program of research can enable the researcher to adjudicate among these possibilities. In contrast, life is much simpler if the predicted effect is obtained. When this desired state of affairs is achieved, it provides some support for the entire set of theoretical assumptions involved in generating the original prediction. However, there is the complicating factor that although the results obtained may be consistent with one theoretical position, they may also be consistent with an alternative formulation. In such cases, it is necessary to compare the predictions of the two theories in some situation for which they make different predictions.

CONDITIONING

There has been a considerable amount of research into the ways in which individual differences in personality influence the process of conditioning. It is customary to distinguish between classical and operant conditioning, and this distinction makes sense at a heuristic level. Some of the earliest work on classical conditioning was undertaken by Pavlov (1927). In a typical experiment, a bell was sounded immediately prior to the presentation of food. The subject of the experiment (usually a hungry dog) would eventually learn to associate bell and food, so that the sound of the bell would produce salivation even in the absence of food. This kind of experiment is illustrative of work on classical conditioning in that the subject is relatively passive, and a simple autonomic response (e.g., salivation) is conditioned.

In contrast, operant conditioning usually involves a more active subject, and a much greater variety of responses can be conditioned. The best-known exponent of work on operant conditioning is Skinner (1938). In many of his studies, rats learned to press a lever in a so-called Skinner box in order to obtain food. In this situation, the conditioned response is the lever press, and the reward or reinforcement consists of the food that follows lever pressing.

While we retain the conventional distinction between these two kinds of conditioning, it is perhaps worth mentioning that there has been much theoretical controversy about the similarities and differences between classical and operant conditioning (Walker, 1975). A complicating factor is the suspicion that it is extremely difficult to carry out an experiment that involves only one form of conditioning. If the norm is for conditioning situations to comprise an amalgam of classical and operant conditioning, then this poses obvious interpretative problems.

CLASSICAL CONDITIONING

Interest in the possibility that classical conditioning might be affected by individual differences in temperament was initiated by Pavlov. According to legend, Pavlov became convinced that the dogs in his laboratory differed importantly in temperament during the stressful period of the Leningrad flood. Some of the dogs became excitable, whereas others reacted in a rather inert way. Pavlov ascribed these behavioral variations to individual differences in the excitability of nervous tissue in the cerebral cortex and argued for an underlying dimension of *strength* of the nervous system.

Initially, Pavlov claimed that *weak* dogs were timid and difficult to condition whereas *strong* dogs were bold and conditioned readily. Subsequently, as Levey and Martin (1981) pointed out, he identified three major properties of the nervous system based very largely on conditioning data. In this conceptualization, strength was reflected in the speed of conditioning, equilibrium was based on the relative ease of formation of excitatory rather than inhibitory conditioned responses, and mobility referred to the ease with which an individual could switch from positive to negative conditioned response formation.

The theoretical approach adopted by H. J. Eysenck (1957) was in part an extension of some of these Pavlovian notions to the relationship between human personality and conditioning. In particular, the concept of the balance of inhibition–excitation was retained, and it was assumed that introverts were high in excitation and low in inhibition, whereas extraverts were the opposite. The general expectation was that the greater susceptibility of extraverts to inhibition would mean that they would condition less well than introverts.

In the revised version of the theory (H. J. Eysenck, 1967a), the same prediction was obtained from somewhat different theoretical assumptions. It was argued that introverts are more cortically aroused than extraverts; when this assumption is linked with the notion that conditioning is usually facilitated by high arousal, it is plain that introverts are still expected to show better conditioning performance than extraverts in most circumstances.

The assumption that high arousal enhances conditioning and low arousal reduces it has been tested a number of times by means of drug manipulations. Franks and Laverty (1955) found that intravenous sodium amytal (which is a depressant drug that reduces arousal) reduced the number of conditioned responses produced during conditioning, and similar results with two further depressant drugs (Doriden and meprobamate) were obtained by Willett (1960). Franks and Trouton (1958) found that a stimulant drug (dexamphetamine sulfate) enhanced conditioning, whereas a depressant drug (amobarbital sodium) reduced conditioning. All in all, the drug studies provide quite strong support for the predicted effects of arousal on conditioning performance.

A final point that must be made about the theoretical approach of H. J. Eysenck (1957, 1967a) is that differences in conditionability between introverts and extraverts were regarded as of fundamental importance. In the first place, the theory is primarily a biological one, and so conditioning performance provides a more direct test of its adequacy than could be obtained from most laboratory tasks. Second, Eysenck assumed that the differences between extraverts and introverts in degree of socialization and in susceptibility to various psychiatric disorders were attributable in large measure to the greater conditionability of introverts.

The most popular way of investigating the effects of extraversion on classical conditioning has been to use the eye-blink conditioning paradigm, and this was also the paradigm used in the drug studies discussed above. In this paradigm, a tone or other neutral stimulus is used as the conditioned stimulus, and a puff of air to the eye is the unconditioned stimulus. Conditioning occurs when the conditioned stimulus is sufficient to produce the eye-blink response. The only other paradigm that has been examined extensively by personality researchers involves conditioning of the skin resistance response.

Much of the work carried out in the 1950s and 1960s revolved around the controversy of the relative importance of extraversion and anxiety as determinants of conditioning performance. The relevant evidence was reviewed by Spence (1964), who concluded that 64% of the studies indicated that anxious subjects conditioned more rapidly than nonanxious subjects. In contrast, H. J. Eysenck (1965) also reviewed the evidence and came to the conclusion that 55% of the studies pointed to superior conditioning in introverts than in extraverts. There have even been studies (e.g., Piers & Kirchner, 1969) in which conditioning was found to be related to both anxiety and introversion.

Not surprisingly, it turns out that the effects of personality on conditioning are determined by various aspects of the experimental situation and of the task. Anxiety is most likely to influence the rate of conditioning when the situation is somewhat stressful (e.g., there is an impressive array of equipment or the subject is isolated in an experimental cubicle). In contrast, the effects of extraversion on conditioning depend in large measure on the precise parameters of the conditioning task. A major attempt to identify some of these parameters was made by H. J. Eysenck and Levey (1967). They argued that partial reinforcement (i.e., presentation of the UCS only on some of the trials) should have a greater detrimental effect on the conditioning of extraverts than on that of introverts. The argument was that unreinforced trials produce inhibition, and extraverts are more susceptible than introverts to inhibition. They also suggested that a weak unconditioned stimulus (i.e., puff of air) would be perceived as subjectively stronger by introverts because they have lower sensory thresholds. If a strong unconditioned stimulus were used, however, introverts would be more likely than extraverts to produce "protective inhibition" which would disrupt conditioning. Finally, H. J. Eysenck and Levey claimed that introverts have faster reaction times than extraverts and so should show better conditioning than extraverts with short time intervals between the conditioned stimulus and the unconditioned stimulus.

The results supported these predictions, except the one relating to partial reinforcement. The impact of the precise nature of task conditions in modifying the effects of extraversion on conditioning can be seen most clearly if we consider only those combinations of conditions that are theoretically most and least likely to reveal an introvert superiority. The optimal combination should involve a weak unconditioned stimulus, a short interval between the conditioned stimulus and the unconditioned stimulus, and partial reinforcement, whereas the worst combination should consist of a strong unconditioned stimulus, a long interval between the conditioned stimulus and the unconditioned stimulus, and continuous reinforcement. The findings from these two task conditions are shown in Figure 30. The effects of extraversion are opposite in the two cases; in the optimal condition, there was a correlation of +.40

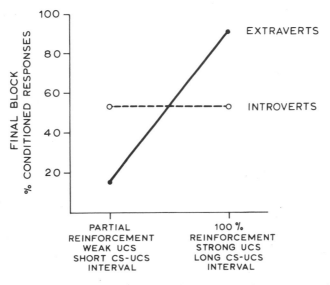

FIGURE 30. Eyeblink conditioned responses of introverts and extraverts under extreme conditions. (From "Konditionierung, Introversion-Extraversion und die Starke des Nervensystems" by H. J. Eysenck and A. B. Levey, *Zeitschrift für Psychologie*, 1967, *174*, 96–106. Copyright 1967 by the *Zeitschrift für Psychologie*. Adapted by permission.)

between introversion and conditioning, whereas the correlation was −.31 for the worst condition.

It is of some theoretical interest to note that H. J. Eysenck and Levey (1972) reanalyzed the data of their earlier study in terms of the two major components of extraversion (i.e., sociability and impulsiveness). The results were striking. Sociability had a negligible effect on performance, and all of the effects of extraversion on conditioning were attributable entirely to impulsiveness. Further evidence of the importance of impulsiveness was obtained by Barratt (1971). He divided his subjects into four groups on the basis of their scores on his own impulsiveness questionnaire and the Manifest Anxiety Scale. The high anxious, low impulsive subjects gave the most conditioned responses overall. Furthermore, it was one of the extraverted groups (the low anxious, high impulsive subjects) that was least aroused at the time of presentation of the conditioned stimulus, according to a measure of alpha abundance.

These findings should be borne in mind when interpreting the results of other conditioning studies. In particular, it is usually assumed that the crucial personality variable responsible for many of the observed individual differences in conditioning performance is extraversion. It is at least possible that impulsiveness is actually the key factor (cf. Revelle et al., 1980).

Jones, Eysenck, Martin, and Levey (1981) argued that most of the research on personality and conditioning was limited because the only performance measure taken was the number of conditioned responses produced. In their study of eye-blink conditioning, they measured several different aspects of conditioning performance. Their findings were somewhat complex, but they discovered that introverts were superior to extraverts on nearly all of the measures of conditioning they considered. However, although extraverts were found to condition much more poorly than introverts with the less intense of two unconditioned stimuli, they did show considerably better placement of the conditioned response in relation to the unconditioned stimulus than did introverts under the same conditions. The proper interpretation of this finding is not clear, but it may be that the extraverted subjects made optimal use of their inferior conditioned responses by placing them in such a way as to produce the greatest possible avoidance of the air-puff. Whatever the merits of that speculation, the use of such fine-grained analyses of conditioning performance appears to offer considerable potential for clarifying the effects of personality on conditioning.

One of the major characteristics of the conditioning studies discussed so far is that they all involved the use of unpleasant or aversive unconditioned stimuli such as a puff of air to the eye. What would we expect if appetitive unconditioned stimuli were used in a classical conditioning paradigm? According to H. J. Eysenck (1967a), introverts are generally more conditionable than extraverts, and this should be the case whether or not the unconditioned stimulus is aversive. In contrast, Gray (1970, 1973) argued that introverts condition better than extraverts with aversive unconditioned stimuli because they are more susceptible to punishment but that extraverts condition better than introverts with appetitive unconditioned stimuli because of their greater sensitivity to reward.

Unfortunately, relatively few studies have dealt with the effects of personality on appetitive conditioning. One of the few was carried out by Kantorowitz (1978). He discovered that extraverts showed greater sexual appetitive conditioning than introverts in terms of tumescence to slides of attractive female nudes associated with orgasm. Indeed, extraversion correlated +.88 with appetitive conditioning. In contrast, introverts showed superior conditioning of detumescence to nude slides associated with the period immediately after orgasm. This situation may be somewhat aversive, and here extraversion correlated −.76 with conditioning.

It might be supposed that these findings provide decisive support for Gray's theoretical conceptualization, but this is not in fact the case. It seems probable that the subjects in the appetitive conditioning situation were more aroused than those in Kantorowitz's aversive conditioning task. In other words, there was a confounding between the type of conditioning (appetitive or aversive) and the level of arousal. This may well be of importance, since introverts appear to condition better than extraverts only under relatively nonarousing conditions (H. J. Eysenck & Levey, 1972).

Attempts to demonstrate a general factor of conditionability, the existence of which was postulated by H. J. Eysenck (1967a), have by and large proved unsuccessful. An exception is the work of Barr and McConaghy (1972). They conditioned penile volume and galvanic skin responses (GSRs) to both aversive (shock) and appetitive (slides of nude females) unconditioned stimuli. When individual differences in reflex sensitivity to the aversive and appetitive unconditioned stimuli were controlled, most of the correlations between response systems (i.e., GSR and penile volume) and across type of unconditioned stimulus (i.e., aversive and appetitive) were statistically significant.

Somewhat similar results were reported by Paisey and Mangan (in press). They looked at both appetitive and aversive conditioning in classical and instrumental paradigms. Differential sensitivity to the unconditioned stimuli that were used in this study was partialled out statistically, and there then appeared to be a general factor of conditionability running across both aversive and appetitive classical conditioning.

The issue of whether or not there is a general factor of conditionability is of considerable importance. Most of the researchers who have discovered that, for example, introverts show better conditioning of the eye-blink response to air puffs than extraverts have extrapolated from this finding to the conclusion that introverts are more conditionable than extraverts. A *sine qua non* for this conclusion to be correct is that there is a general factor of conditionability. However, there are formidable problems in arriving at an empirical resolution of this issue, as Levey and Martin (1981) made clear:

> The assessment of conditionability is no easy matter and the obstacles cannot be underestimated [*sic*]. Conditioning for any particular sensory modality and for any somatic or autonomic response system is likely to have its own temporal parameters in terms of optimum interstimulus and intertrial intervals....Levels of stimulation which are arousing in one CS-UCS modality may not be the same in another, and the difficulty of equating this factor has never been satisfactorily resolved. (p. 159)

In sum, introverts often show better conditioning performance than extraverts in studies of aversive classical conditioning, but this main effect of extraversion must be interpreted in light of the interactions between extraversion and task conditions. As a first approximation, introverts condition better than extraverts under relatively unarousing conditions, whereas the opposite happens under stimulating or arousing conditions. The position with respect to appetitive classical conditioning is less clear. Extraverts tend, if anything, to condition better, but possible interactions with task conditions have not been explored systematically. It remains a matter of controversy whether there is a general factor of conditionability.

In general terms, most of the findings are consistent with H. J. Eysenck's (1967a) arousal theory. It is particularly noteworthy that the effects of stimulant drugs resemble those of introversion, whereas the effects of depressant drugs resemble those of extraversion. It is also relevant that the extent to which the conditioning situation is arousing interacts with extraversion to determine performance. What remains to be done is to determine whether extraversion *per se* or one of its components (e.g., some aspect of impulsiveness) is more responsible for the observed effects, and also to explore in more detail the conditions under which introvert and extravert superiority in conditioning are likely to be found.

OPERANT CONDITIONING

The theoretical positions of H. J. Eysenck (1967a) and Gray (1970, 1973) are intended to be relevant to operant or instrumental conditioning as well as to classical conditioning. In other words, Eysenck predicts that introverts will manifest better operant conditioning than extraverts unless the conditions are so arousing that transmarginal or protective inhibition is produced in the introverted subjects, whereas Gray assumes that introverts will condition better than extraverts when negative reinforcement is used but that the opposite will happen with positive reinforcement.

Far and away the most popular experimental approach has been to use a verbal operant paradigm in which sentences are constructed by the subjects. Reinforcement (e.g., "good," "not so good") is provided if the subjects' utterances conform, or fail to conform, to certain experimenter-determined specifications (e.g., use of a plural noun). The intention is that the subjects should not be consciously aware of the factors determining whether their responses are reinforced or not. If they are aware, then we are dealing with problem solving rather than operant conditioning *per se.*

The importance of taking account of awareness of reinforcement contingencies was demonstrated clearly by Gidwani (1971). He used the standard verbal operant paradigm with sweets as positive reinforcers. There was no overall effect of extraversion on conditioning performance. However, there was a highly significant interaction between personality and awareness. Among those subjects who were unaware of the reinforcement contingencies, introverts conditioned better than extraverts, whereas extraverts produced more operant conditioning than introverts among the aware subjects.

The studies concerned with the effects of extraversion on verbal operant conditioning under reward or positive reinforcement conditions were reviewed by Mangan (1982). The modal finding (e.g., Mohan & Dharmani, 1976) is that introverts condition more readily than extraverts, a result that accords better with the theory of H. J. Eysenck (1967a) than with that of Gray (1970, 1973). Furthermore, comparable findings have been obtained when arousal level has been manipulated. Gupta (1973) examined the effects of two depressant drugs (chlorpromazine and phenobarbitone) and two stimulant drugs (Dexedrine and ephedrine) on verbal operant conditioning with positive reinforcement. He discovered that the depressant drugs inhibited conditioning, whereas the stimulant drugs enhanced conditioning. Since he also found that introverts conditioned better than extraverts, the results were in line with the notion that high arousal improves operant conditioning.

The rather consistent pattern of findings described above has been disturbed by more recent work of Gupta and his associates. Gupta (1976) investigated verbal operant conditioning under various conditions of positive and negative reinforcement. With negative reinforcement, there was a clear conditioning superiority for the introverted subjects. In contrast, there was some evidence that extraverts conditioned better than introverts when positive reinforcement was provided. These findings were replicated by Gupta and Nagpal (1978), who also tested Gray's (1973) theory more directly by carrying out additional analyses on the impulsiveness and sociability components of extraversion. According to Gray's theory, the effects of positive reinforcement should be greater for impulsiveness than for sociability, whereas the opposite should be the case with negative reinforcement. In fact, impulsiveness and sociability were comparably affected by the different reinforcement conditions.

Rather stronger support for Gray's (1973) theory was obtained by Nagpal and Gupta (1979). As predicted by that theory, it was the neurotic introverts (high-anxiety subjects) who showed the greatest amount of conditioning with negative reinforcement, whereas it was the neurotic extraverts (high impulsives) who conditioned best with positive reinforcement.

In sum, while introverts usually show more operant conditioning than extraverts when negative reinforcement is used, the effects of extraversion on operant conditioning are much less consistent when positive reinforcement is used. The fact that introverts sometimes condition significantly better than extraverts with positive reinforcement, whereas precisely the opposite has been found in other studies, suggests that we have an incomplete understanding of the factors involved in operant conditioning. The degree of involvement of conscious awareness of reinforcement contingencies is certainly one factor that is commonly overlooked, but there may well be others.

SENSITIVITY TO STIMULATION

The theoretical position espoused by H. J. Eysenck (1967a) assumes that there are a number of important differences between introverts and extraverts. An especially crucial difference is thought to be that introverts are more sensitive to stimulation than are extraverts. Why should this be so? According to H. J. Eysenck (1976b), it is possible to account for this differential sensitivity on the basis of the fact that introverts have a chronically higher level of cortical arousal than extraverts:

> If sensory stimulation is registered in the cortex to a degree which is a joint function of the objective level of intensity of the stimulation and of the arousal existing in the cortex at the time of arrival of the neural message, then identical intensities of input will be experienced as stronger by introverts than by extraverts, and as weaker by extraverts than by ambiverts. (p. 113)

Some of the potential consequences of this assumption that the high cortical arousal of introverts acts as "an amplifying valve for incoming sensory stimulation" (H. J. Eysenck, 1971c) are shown in Figure 31. The notion that any given level of stimulation is experienced as higher by introverts than by extraverts is incorporated into the figure, as are two further reasonable assumptions: (1) very low and very high levels of stimulation produce negative hedonic tone, and (2) positive hedonic tone is found only at intermediate levels of sensory stimulation. It follows from this theoretical formulation that the preferred level of stimulation should be lower for introverts than for extraverts. It also follows that

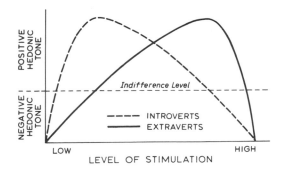

FIGURE 31. The relationship between level of stimulation and hedonic tone as a function of extraversion. (From *Readings in Extraversion-Introversion: 3. Bearings on Basic Psychological Processes*, London: Staples, 1971. Copyright 1971 by Pergamon Press, Inc. Adapted by permission.)

introverts should be better able than extraverts to cope with extremely low levels of stimulation, whereas at high levels of sensory stimulation it should be extraverts who are less adversely affected.

These predictions are intuitively reasonable. Extraverts are people who enjoy a varied and stimulating life-style which includes such activities as going to parties, extensive socializing, and playing at sports. It could thus be said that extraverts are characterized by "stimulus hunger" (H. J. Eysenck, 1967a). In contrast, introverts typically favor activities of a relatively unstimulating nature (e.g., reading), and thus their behavior is suggestive of "stimulus aversion" (H. J. Eysenck, 1967a).

Let us begin our discussion of the empirical evidence with the effects of very low levels of sensory stimulation on the behavior of introverts and extraverts. The usual paradigm is that of sensory deprivation, in which the subjects rest quietly in a room with minimal visual and auditory stimulation or are immersed in a tank of water. Sensory deprivation studies typically last for several days, and tolerance of the unstimulating conditions is assessed by observing how long each individual is able to endure the situation.

The prediction that introverts should tolerate sensory deprivation better than extraverts has sometimes been confirmed. Francis (1969) discovered that those subjects with the longest toleration times for immersion in water were significantly more introverted than those who showed the poorest toleration. He also reported that people who claimed to be well able to withstand isolation in everyday situations (e.g., being confined to bed, sitting in waiting rooms) were mostly very introverted. In another study (Hull & Zubek, 1962), subjects were asked to remain in an isolation chamber in constant darkness and silence for seven days. Those who managed to complete the seven days were less impulsive than the unsuccessful subjects on the Thurstone Temperament Schedule.

Unfortunately, the opposite findings have been obtained in other studies. Arnhoff and Leon (1963) compared those who lasted to the end of a sensory deprivation experiment with those who were unsuccessful on Cattell's 16PF test. Far more of the unsuccessful subjects were found to be desurgent (i.e., introverted). Tranel (1962) also discovered that extraverts showed more toleration of sensory deprivation than introverts, at least when toleration was measured in terms of the amount of time spent in the experimental room. However, a more detailed analysis of the subjects' behavior under sensory deprivation conditions pointed to a rather different conclusion. Tranel instructed his subjects at the beginning of the experiment to lie quietly on their couches and not to fall asleep, but the extraverts tended to disregard these instructions. The extraverted subjects experienced more difficulty than the introverts, and they also made many more movements per minute on average (38 versus 23, respectively). It could quite plausibly be argued that the extraverts actually tolerated sensory deprivation less well than the introverts, and this led them to adopt the coping strategy of increasing their level of arousal by means of self-stimulation.

In sum, the sensory deprivation paradigm does not provide a straightforward way of testing the notion that introverts are better able to tolerate very low levels of sensory stimulation. It is not clear that the time for which sensory deprivation can be endured is an adequate measure of deprivation tolerance in view of the tendency of some subjects to resort to daydreaming and irrelevant motor behavior in order to increase arousal. As a consequence, it is by no means always the case that the sensory deprivation situation produces the anticipated state of low arousal. Indeed, sensory deprivation is sometimes associated with increased secretion of arousal-related hormones such as adrenaline and noradrenaline (Frankenhaeuser, Nordheden, Myrsten, & Post, 1971). Until such time as there is better experimental control over subjects' cognitive and motor activities while undergoing sensory deprivation, it is wise not to attach too much weight to the reported findings.

Matters appear to be more straightforward with respect to research on the hypothesis that the optimal or preferred level of stimulation is higher in extraverts than in introverts. The expectations that introverts are more likely than extraverts to want to decrease the level of stimulation whereas extraverts are more likely than introverts to seek an increase in stimulation were tested by Weisen (1965). He made use of an operant conditioning procedure in which the reinforcement for button pressing was either a 3-second period of sound and light stimulation (the onset condition) or a 3-second period of removal of sound and light stimulation (the offset condition). In line with theoretical predictions, extraverts produced many more reinforced button presses than introverts in the onset condition, whereas the opposite happened when absence of stimulation was the reward.

A somewhat similar study was reported by Davies, Hockey, and Taylor (1969). They found that extraverts requested 30-second periods of varied auditory stimulation more often than introverts during the performance of a visual vigilance task. In contrast, when continuous varied auditory stimulation was provided during a vigilance task, introverts made significantly more requests for periods of silence.

Less direct evidence that extraverts like greater stimulation than introverts was provided by Gale (1969) and by Hill (1975). Gale exposed his subjects to mild sensory deprivation and allowed them to obtain four different sound reinforcements by pressing buttons. Extraverts engaged in more button pressing than did introverts and also shifted more frequently among the four sounds. Hill made use of a rather monotonous task and discovered that extraverts produced a more varied pattern of responses than introverts. While the proper interpretation of this finding is obscure, Hill suggested that the extraverts were attempting to increase the amount of stimulation they received.

In view of all these positive findings, it is somewhat surprising that the most direct experimental assessment of the preferred level of stimulation failed to produce a significant difference between introverts and extraverts. Ludvigh and Happ (1974) asked people to adjust a light and a sound to the most comfortable or optimal level. Contrary to expectation, there was no effect of extraversion on the amount of sensory stimulation needed to produce optimal hedonic tone.

More promising findings were obtained by Geen. Subjects were asked to choose the optimal intensity level of noise bursts that were to be presented during a learning task. In the first of two experiments, the mean sound intensity selected by introverts was 55 dB against 72 dB for extraverts. Rather similar findings were obtained in the second experiment (54 dB and 70 dB, respectively). Thus, there was convincing evidence that the preferred level of stimulation was greater for extraverts than for introverts.

Geen collected further data that provide a plausible explanation of this personality-linked difference in the optimal level of stimulation. He ran a further condition in which each introvert received noise bursts at the preferred intensity level of an extraverted subject, and each extravert received noise bursts at the preferred level of an introverted subject. Physiological measures of arousal (pulse rate and skin resistance responses) then indicated that introverts were more aroused than extraverts when exposed to identical noise bursts, irrespective of whether these were at a level preferred by extraverts or by introverts. This suggests that the tendency of introverts to amplify stimulation accounts for their low optimal level of stimulation.

It will be recalled that a further prediction from the hypothesis under consideration is that introverts should have less tolerance than extraverts for intense or painful stimuli. This prediction was tested in the study by Ludvigh and Happ (1974), which has just been discussed. When they asked their subjects to adjust a light and a sound to a level slightly too high for comfort, extraverts selected a significantly higher level of stimulation than introverts. Elliott (1971) investigated tolerance of white noise and found that extraverts had a much greater tolerance of noise than introverts; indeed, the former subjects had a mean tolerance level which was more than 30 dB higher than that of the latter subjects.

Negative reactions to intense stimulation have often been assessed by measuring either the pain threshold or the ability to tolerate pain. The pain threshold often refers to the length of time an aversive stimulus is presented before pain is experienced, whereas pain tolerance is the length of time that an aversive stimulus can be endured. Most of the studies relating the pain threshold and pain tolerance to the personality dimension of extraversion were reviewed by Barnes (1975). In spite of the fact that several different kinds of aversive stimuli have been used (e.g., radiant heat, electric current, pressure), the results are fairly consistent. Of the five studies including a test of the relationship between extraversion and pain threshold, two reported that extraverts had significantly higher pain thresholds than introverts.

The findings are rather more impressive with respect to pain tolerance. Of the ten relevant studies discussed by Barnes (1975), five reported that pain tolerance was significantly greater in extraverts than in introverts, and there were nonsignificant tendencies in the same direction in at least two of the remaining studies. A few studies have appeared since the Barnes review. Bartol and Costello (1976) ascertained the number of seconds for which subjects could endure shock levels of 25, 50, 100, and 130 volts. Extraverts tolerated a greater duration of shock than introverts, especially at the higher intensities of shock. Shiomi (1978) obtained a significant correlation of approximately +.5 between extraversion and pain tolerance for hand immersion in extremely cold water.

Finally, Shiomi (1980) found that extraverts were able to pedal on an ergometer at a speed of 30 km/hour for longer than introverts.

These findings pose three questions. First, why does extraversion apparently affect pain tolerance more than pain threshold? It is possible that subjects find it difficult to decide precisely when a stimulus becomes painful but know when it becomes intolerable. Alternatively, it may be relevant that the pain-tolerance threshold is associated with a higher level of arousal than the pain threshold. It may simply be that differences between introverts and extraverts become more marked as the intensity of aversive stimulation increases.

Second, it may be wondered whether the greater sensitivity of introverts to painful stimuli is really attributable to their high level of arousal. Some support for this interpretation was obtained by Haslam (1967). She found that introverts had a significantly lower pain threshold than extraverts. Of greatest interest here was her discovery that caffeine citrate (a stimulant drug that increases arousal) also produced a marked reduction in the pain threshold. Although these data are by no means conclusive, they can most readily be interpreted by assuming that high arousal accentuates the negative effects of aversive stimulation.

Third, although the findings discussed above are entirely consistent with the notion that introverts augment incoming stimulation, there are other findings indicating that introverts often show a *reduced* reaction to intense stimulation. This paradoxical reaction is known as protective or transmarginal inhibition and was defined by Gray (1967) as "a response decrement when stimulus intensity is raised to a very high level." Some evidence that introverts are more likely to show transmarginal inhibition than are extraverts was obtained by S. B. G. Eysenck and H. J. Eysenck (1967). They found that introverts salivated much more than extraverts when lemon juice was applied to the tongue, a finding that is consistent with the putatively greater sensitivity to stimulation of introverts. However, when the stimulation was made more intense by requiring the subjects to swallow the lemon juice, extraverts salivated more than introverts. This latter finding may reflect transmarginal inhibition in introverts.

Rather stronger evidence that intense stimulation can actually reduce rather than augment the responsiveness of introverts was obtained by Fowles *et al.* (1977). They looked at the effects of tone loudness on arousal (measured by skin conductance level) in introverts and extraverts. When a stressful task preceded the presentation of the tones, introverts had greater skin conductance levels than extraverts with moderately intense tones, but the opposite was the case with the loudest tone

(103 dB). Indeed, introverts actually had lower skin conductance levels to the loudest tone than to less intense ones.

We have seen that the putatively greater sensitivity to stimulation of introverts than extraverts usually leads to predictable individual differences in hedonic tone to very weak or strong stimuli. The same theoretical notion is also applicable when we are endeavoring to predict the effects of extraversion on sensory thresholds. The argument is that introverts should find it easier than extraverts to detect the presence of, say, a very faint light or tone because of their tendency to augment incoming stimulation. In other words, introverts are expected to have lower sensory thresholds than extraverts.

The evidence is broadly in line with expectation, irrespective of the sense modality which is investigated. Fischer, Griffin, and Rockey (1966) studied the gustatory chemoreceptors and found that introverts had lower taste thresholds for quinine than extraverts. Dunstone, Dzendolet, and Henckeruth (1964) related the threshold of electrical vestibular stimulation to personality and discovered that the absolute threshold of response to low-frequency sinusoidal electrical stimulation of the human vestibular apparatus was much lower in introverts than in extraverts. Smith (1968) reported that introverts had lower absolute thresholds than extraverts to low-frequency tones. Stelmack and Campbell (1974) used the measures of signal-detection theory and discovered that introverts had greater sensitivity (d') with low-frequency tones but that the auditory threshold to high-frequency tones was unaffected by extraversion. In the visual modality, introverts were found to have much lower sensory thresholds than extraverts when subjects with high neuroticism scores were excluded, but the effect of extraversion became nonsignificant when no subjects were excluded on the grounds of high neuroticism (Siddle, Morrish, White, & Mangan, 1969). Finally, although the main effect of extraversion on detection thresholds for electrocutaneous stimulation was nonsignificant, neurotic introverts had lowest thresholds and neurotic extraverts the highest thresholds (Edman, Schalling, & Rissler, 1979).

It is clear that introverts consistently have lower sensory thresholds than extraverts, but is it appropriate to interpret this difference in terms of the higher level of cortical arousal in introverts? The effects of stimulant and depressant drugs on sensory thresholds are relatively uninformative about the possible role of arousal in determining thresholds. Nebylitsyn (1961) found that a stimulant drug (caffeine) greatly reduced the visual absolute threshold but had a much smaller and erratic effect on the auditory threshold. Smith (1968) found that neither nicotine nor Seconal had a significant effect on the auditory threshold.

Much stronger support for the notion that differences between introverts and extraverts in sensory thresholds are mediated by differences in the level of cortical arousal was obtained by Shigehisa and his colleagues (Shigehisa, Shigehisa, & Symons, 1973; Shigehisa & Symons, 1973). In the latter study there was no effect of extraversion on the auditory threshold when testing occurred in darkness. However, very different results were obtained when an arousing stimulus (visual stimulation) was presented during measurement of the auditory threshold. As can be seen in Figure 32, auditory sensitivity increased among both introverts and extraverts when the visual stimulation was relatively weak. With strong visual stimulation, however, auditory sensitivity decreased for introverts, whereas it continued to increase for extraverts. As a consequence, extraversion correlated −.7 or more with auditory threshold when the concurrent visual stimulation was intense.

What do these findings mean? First, the fact that the auditory threshold was systematically affected by the intensity of visual stimulation suggests that arousal influences sensitivity to weak stimuli. Second, the finding that intense visual stimulation had an adverse effect on introverts but not on extraverts can be explained in terms of the law of transmarginal inhibition: The presentation of increasingly intense stim-

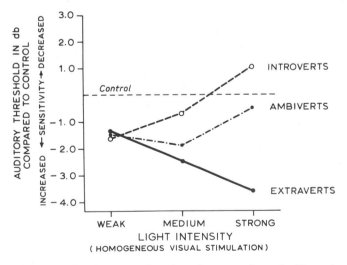

FIGURE 32. Auditory thresholds as a function of extraversion and of intensity of visual stimulation. (From "Effects of Intensity of Visual Stimulation on Auditory Sensitivity in Relation to Personality" by T. Shigehisa and J. R. Symons, *British Journal of Psychology*, 1973, *64*, 205–213. Copyright 1973 by the British Psychological Society. Reprinted by permission.)

ulation eventually leads the system to protect itself by producing a reduced reaction to the stimulation. The tendency of introverts to augment stimulation means that they reach the threshold of transmarginal inhibition at lower levels of stimulation than extraverts. As a consequence, the point at which sensory stimulation in one sensory modality ceases to be effective in lowering the threshold in another sensory modality is lower for introverts than for extraverts.

Shigehisa *et al.* (1973) carried out a study complementary to that of Shigehisa and Symons (1973). They investigated the effects of auditory stimulation on visual thresholds. Visual sensitivity in introverts decreased at medium and strong tone intensities, whereas it increased in extraverts, especially at the strongest intensities. With the loudest tone (85 dB), extraversion correlated −.74 with visual threshold. Once again, the results are in line with the notion that introverts have a lower threshold of transmarginal inhibition.

VIGILANCE

The vigilance task was originally introduced into psychology by Mackworth (1950). He was interested in the practical problem of simulating the task of maintaining radar watch for submarines under carefully controlled laboratory conditions. The solution he devised involved using a clock pointer which moved in a series of steps. The pointer occasionally gave a double jump, and it was the job of the subjects to detect and report these irregular movements. While later researchers have tended to use somewhat different tasks, the common thread running through vigilance tasks is that the observer (or listener) is required to detect inconspicuous visual or auditory signals over longish periods of one hour or more.

In view of the rather tedious nature of the task, it is perhaps not surprising that the probability of target detection tends to become lower over time. This is known as the vigilance decrement. In order to make theoretical sense of the vigilance decrement, it is necessary to take account of the number of false alarms (i.e., the reported detection of signals when none is actually presented). The usual finding is that both the number of detections and the number of false alarms decrease during the course of the experimental session, suggesting that the subjects simply become more cautious about indicating that they have detected a target. At a more formal level, information about detections and false alarms has been used by signal-detection theorists (e.g., Swets, 1977) to produce separate measures of observer sensitivity and of his decision criterion (i.e., his bias toward one or other of the response alternatives). Although the vigilance decrement is occasionally caused by a decline in sensitivity, especially if there is a rapid rate of stimulus presentation, the modal finding is that the vigilance decrement is largely attributable to a progressive increase in the strictness or cautiousness of the response criterion (Broadbent, 1971).

The effects of several different arousal-related factors on vigilance performance have been investigated (M. W. Eysenck, 1982). The findings have been particularly clear in the case of sleep deprivation, which usually produces reduced arousal, at least when the environment is relatively unstimulating or monotonous. Wilkinson (1960) found that sleep loss produced a marked increase in the vigilance decrement on a visual vigilance task, and similar results were obtained by Wilkinson (1964).

Further studies seem to have established that the effects of sleeplessness on vigilance-task performance are attributable to lowered arousal rather than to some other aspect of the sleep-deprived state. Corcoran (1962) discovered that white noise reduced the vigilance decrement associated with sleep loss and concluded, "This result is in accordance with a simple arousal theory, which would maintain that noise is arousing and loss of sleep de-arousing, so that their combination results in levels of arousal between these extremes" (p. 181). In similar vein, Bergström, Gillberg, and Arnberg (1973) found that the adverse effects of two nights' sleep loss on a radar-watching task could be reduced by the administration of painful electric shocks. An explanation of this result along arousal lines is supported by their additional finding that sleep loss reduced heart rate, whereas the electric shocks increased it.

The findings with other arousal manipulations are broadly consistent with the notion that vigilance performance is better in states of high arousal. For example, it has usually been assumed that arousal is relatively low early in the morning and increases thereafter, so that it would be expected that vigilance performance would be better in the evening than in the morning. This result was obtained by Mullin and Corcoran (1977), who discovered that auditory vigilance was much better at 20:30 than at 8:30. Furthermore, the introduction of an arousing stimulus (i.e., intense white noise) eliminated the time-of-day effect, largely by improving performance in the morning.

In sum, it is tempting to conclude that at least part of the vigilance decrement is due to the fact that the rather monotonous conditions characteristic of vigilance tasks produce a suboptimal level of arousal. This conclusion is strengthened by the results of various studies in which physiological recordings were taken during the course of vigilance-task performance (see Davies & Tune, 1970, for details). In essence, a progressive decrease in the level of arousal has been obtained when arousal has been measured by EEG activity, heart rate, skin resistance, and skin conductance. Especially striking results were obtained by O'Hanlon (1965), who discovered that the concentration of adrenalin (a substance associated with arousal) decreased during a very long visual vigilance task among subjects who showed a vigilance decrement, but not among those whose performance remained stable over time. Furthermore, the correlation between target-detection rate and adrenalin concentration was +.84 among those manifesting a decrement in performance.

Although it seems undeniable that arousal plays a part in determining vigilance performance, an arousal-based interpretation of the findings is disappointingly vague in various ways. For example, there are at least two possible reasons why those subjects showing no vigilance decrement in O'Hanlon's (1965) study maintained their concentration of adrenalin: (1) they did not habituate to the monotonous conditions and so found it easy to maintain their initial alert, aroused state, or (2) they resisted the arousal-lowering conditions by increased effort. The first explanation posits a *direct* effect of the environment on arousal level, whereas the second explanation points to an *indirect* effect.

There has been a fair amount of interest in the effects of extraversion on vigilance performance. The obvious prediction that introverts will perform better than extraverts because they are more aroused has been confirmed several times (Bakan, 1959; Bakan *et al.* 1963; Davies & Hockey, 1966; Gange *et al.* 1979; Gill, 1979; Hogan, 1966; Keister & McLaughlin, 1972; Mohan & Gill, 1979; Paramesh, 1963). It has also been discovered that extraverts show a greater vigilance decrement than introverts, that is, the superiority of introverts is more marked toward the end of the experimental session.

It is unfortunate that only a single task condition was used in most of the above studies, because it is probable that the precise effects of extraversion on vigilance performance depend on various characteristics of the vigilance task. Davies and Hockey (1966) discovered that the effects of extraversion were altered in complex ways by signal frequency, stage of practice, and noise. Kishimoto (1977) also found that signal frequency was important: Extraverts outperformed introverts when there was high signal frequency, but there was no effect of extraversion when there was low signal frequency.

Are these effects of extraversion on vigilance due to different levels of arousal between introverts and extraverts? It would appear so. Intense white noise virtually eliminates the inferiority of extraverts on a visual vigilance task (Davies & Hockey, 1966). Caffeine, a drug known to increase arousal, leaves the performance of introverts unaffected but eliminates and even reverses the normally inferior auditory vigilance performance of extraverts (Keister & McLaughlin, 1972). On the basis of their data, Keister and McLaughlin drew the following conclusion: "The effect of caffeine was to increase the cortical arousal of the extraverts and make them equal to the introverts who characteristically function at a high arousal level" (p. 10). Gange *et al.* (1979) took recordings of two physiological measures of arousal (heart rate and skin resistance) during a visual vigilance task. Introverts detected 81% of the signals, extraverts only 44%, and introverts appeared to be more aroused as indexed by both heart rate and skin resistance responses.

So far, it appears that an arousal theory account of differences in vigilance performance between introverts and extraverts is in perfect accord with the data. However, there is at least one puzzling problem. It will be remembered that the standard experimental result is that the vigilance decrement (theoretically ascribed largely to lowered arousal) is due primarily to increased cautiousness of responding. The analogous finding in the personality literature would be that the inferior performance of extraverted subjects (who are assumed to be of low arousal) is also due to the adoption of a stringent criterion for response. In general terms, this leads to the expectation that extraverts should have less of a tendency than introverts to say that a signal has occurred when it has not, that is, they should have a lower false alarm rate. In fact, the typical finding is the exact opposite. Carr (1971) and Krupski, Raskin, and Bakan (1971) discovered that extraverts made more false alarms than introverts.

A more precise assessment of the effects of extraversion on vigilance performance can be obtained by using the approach of signal-detection theory. This permits measurement of the two parameters, d' and β. D' is a measure of the subject's sensitivity to signals and β is a measure of the cautiousness of responding. Tune (1966) used these measures and found that extraverts had a lower β than introverts. Harkins and Geen (1975) also found that extraverts had a lower criterion for reporting signals than introverts; in addition, introverts had much greater sensitivity (d') than extraverts.

If the error proneness of extraverts on vigilance tasks cannot readily be handled by a simple arousal theory, how is it to be explained? One possible answer was supplied by M. W. Eysenck (1981, 1982). He pointed out that the setting of the response criterion is influenced by the subjective gains associated with correct responding and by the subjective costs associated with incorrect responding (i.e., false alarms). Now Gray (1973) argued that introverts are more susceptible to punishment than extraverts but are less susceptible to reward. If we combine these ideas, it seems plausible that introverts attach more importance than extraverts to the costs of false alarms and that extraverts are more affected than introverts by the potential gains of correct responding. As a result, introverts have a higher response threshold than extraverts.

In sum, the evidence provides partial support for an arousal interpretation of the effects on vigilance performance of extraversion. The basic findings are as predicted, and the interactions between extraversion and white noise and extraversion and caffeine administration strongly suggest that arousal is involved. However, more fine-grained analyses indicate the existence of various complexities. In addition to the curious effects of extraversion on the response criterion, the interpretation of some of the interaction effects is less obvious than might appear at first glance. There is evidence (discussed by H. J. Eysenck, 1967a) that introverts are more affected physiologically than extraverts by a stimulus of standard intensity, and yet intense white noise had no effect on the performance of introverts but produced substantial changes in that of extraverts (Davies & Hockey, 1966). Similarly, caffeine affected the performance of extraverts only (Keister & McLaughlin, 1972), and it is not clear why introverts showed no behavioral effects of caffeine administration.

VERBAL LEARNING AND MEMORY

The literature on the effects of extraversion on learning and memory has been summarized elsewhere (M. W. Eysenck, 1976b, 1977, 1981); the focus here will be on the major findings only, especially those relevant to arousal theory. An influential theory relating arousal to memory was proposed by Walker (1958). According to his action decrement theory, high arousal produces an active memory trace of longer duration; this in turn leads to enhanced consolidation and long-term memory. However, during the time that the process of consolidation is continuing, there is a transient inhibition of retrieval (referred to as "action decrement") which protects the active memory trace from disruption. As a consequence, although high arousal is beneficial for long-term retention, it impairs short-term retention for periods of time up to several minutes after learning.

Numerous studies have confirmed the predicted interaction between arousal and length of the retention interval (M. W. Eysenck, 1976a, provides a review). In a classic study by Kleinsmith and Kaplan (1963), the paired associates presented to each subject were assigned to a high-arousal or a low-arousal category on the basic of the galvanic skin response (GSR) produced by each paired associate at presentation. Low-arousal items were much better recalled than high-arousal items at the shortest retention interval (2 minutes), but the opposite was the case at retention intervals of 45 minutes, 1 day, and 1 week.

One of the problematical aspects of the experimental approach adopted by Kleinsmith and Kaplan is that it is not entirely clear why items differ in the physiological responses they evoke. It may be that the high-arousal items were better attended to than the low-arousal items, or they may have been more interesting, more meaningful, and so on. In other studies, experimental control over the assignment of items to high- and low-arousal categories has been achieved by pairing some of the items with white noise, which is known to increase the level of physio-logical arousal (e.g., Magoun, 1963). Most of the paired-associate learning studies using this approach have produced findings in line with Walker's (1958) action decrement hypothesis (e.g., Berlyne, Borsa, Craw, Gelman, & Mandell, 1965; McLean, 1969).

What happens when the memory performance of introverts and extraverts at different retention intervals is compared? The modal find-ing is that extraverts have better short-term recall than introverts but that this is reversed at longer retention intervals. This kind of interac-tion between extraversion and retention interval was most strikingly demonstrated by Howarth and H. J. Eysenck (1968); their results are shown in Figure 33. Similar results were obtained by McLean (1968), Opollot (1970), and Skanthakumari (1965), but there are at least four

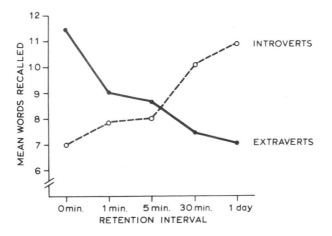

FIGURE 33. Memory as a function of extraversion and retention interval. (From "Extra-version, Arousal, and Paired-Associate Recall" by E. Howarth and H. J. Eysenck, *Journal of Experimental Research in Personality*, 1968, *3*, 114–116. Copyright 1968 by Academic Press, Inc. Adapted by permission.)

studies in which the anticipated interaction between extraversion and retention interval failed to materialize.

The modal finding can apparently be accounted for on the assumption that introverts are more aroused than extraverts (H. J. Eysenck, 1973c), and so the personality data provide further confirmation of Walker's hypothesis. However, this hypothesis cannot be accepted as it stands. Although high arousal usually reduces immediate retention in paired-associate learning, it often enhances immediate free recall and recognition (M. W. Eysenck, 1976a). Even within the confines of paired-associate learning, there are some findings that are embarrassing to the action decrement theory. Hamilton, Hockey, and Quinn (1972) considered the effects of white noise on paired-associate learning. Noise had no effect on initial learning when the order of the pairs was changed from one learning trial to the next, but it improved learning greatly when the order of presentation of the paired associates remained the same. Since there were short retention intervals between successive learning trials, the latter finding is the opposite of that predicted on the action decrement theory. As a result of these various problems with Walker's hypothesis, it must be concluded that we do not have a satisfactory explanation of the interaction between extraversion and length of the retention interval.

It is possible to make a definite prediction from arousal theory on the effects of varying task difficulty on the learning and memory performance of introverts and extraverts. If we combine the notion that introverts are more aroused cortically than extraverts with the theory (Yerkes & Dodson, 1908) that the optimal level of arousal is inversely related to task difficulty, then it follows that extraversion should interact with task difficulty and also that extraverts should perform relatively better than introverts on complex learning tasks.

The notion that high arousal is more likely to have adverse effects on highly demanding tasks than on simple ones has received reasonable support, whether arousal has been produced by incentive, intense noise, anxiety, or the use of introverted subjects (M. W. Eysenck, 1982). With respect to work on extraversion, the most thorough study was carried out by Jensen (1964). He considered performance on a wide range of learning tasks and then factor-analyzed the resultant data. One of his main conclusions was that extraversion correlated especially highly with resistance to response competition. This is pertinent to the Yerkes–Dodson theory, because learning tasks involving response competition are more difficult than those that do not.

A subsequent study by Howarth (1969b) broadly supported Jensen's conclusions. After an initial paired-associate list had been learned to criterion, a second list was learned, consisting of a re-pairing of the stimuli and responses from the first list. There was then another re-pairing of the same stimuli and responses to produce the third list. It was assumed that response competition increased from list to list. There was no effect of extraversion on speed of learning the first two lists, but extraverts acquired the final list significantly more quickly than introverts.

At the theoretical level, the major difficulty is that the Yerkes–Dodson formulation merely describes the predicted relationship between arousal and task difficulty and signally fails to provide an adequate explanation of precisely *why* arousal interacts with task difficulty. A partial explanation of the personality data was offered by M. W. Eysenck (1975). He found that the introvert inferiority on paired-associate lists involving response competition was due in large measure to the fact that introverts took longer than extraverts to retrieve the relevant information. He suggested that introverts are more likely than extraverts to retrieve readily accessible information, which harms their performance when accessible but incorrect responses are available (e.g., in conditions of response competition).

Some of the clearest findings in the literature relating extraversion and memory have emerged from studies on distraction. The usual finding is that introverts are more susceptible than extraverts to the deleterious effects of distraction on learning and memory. For example, Morgenstern *et al.* (1974) presented words auditorily for subsequent recall, and distraction consisted of further auditorily presented words, a German prose passage, or an English prose passage. Recall was unaffected by extraversion when there was no distraction, but introverts recalled significantly fewer words than extraverts under distracting conditions. Introverts were also found to be more distractible than extraverts in studies by Bergius (1939), Shanmugan and Santhanam (1964), and Howarth (1969a).

How is the differential susceptibility to distraction of introverts and extraverts to be explained? At the most general level, there are indications that various arousers or stressors (e.g., incentive, anxiety) increase distractibility, so that an explanation in terms of arousal may be warranted. However, the precise mechanisms or processes involved remain obscure. The simplest explanation is that the introduction of distracting stimuli increases task complexity, and so the findings provide more support for the notion (Yerkes & Dodson, 1908) that the optimal level of arousal is inversely related to task difficulty.

From the standpoint of contemporary cognitive psychology, most of the research on personality and memory is disappointingly vague. In particular, there have been relatively few attempts to identify the processing mechanisms affected by extraversion. An example of the kind of research that is needed was provided by Riding (1979), who investigated the hypothesis that extraverts tend to be verbalizers and introverts tend to be imagers. He discovered in a study of prose recall that extraverts showed superior recall of abstract details, whereas introverts did well in recall of spatial and directional details.

Somewhat stronger support for Riding's hypothesis was obtained by Riding and Dyer (1980). They made use of the verbal-imagery code test, in which each of 10 paragraphs was followed by questions that could be answered on the basis of images generated from information in the passage or by using verbal associations of concepts in the passage. The times taken to answer the various questions were measured, and the resultant verbal-imagery code ratio correlated $-.71$ with extraversion, that is, extraverts tended to be verbalizers rather than imagers. However, Riding and Dyer were less successful with a further experiment in which introverts and extraverts recalled visually descriptive and abstract passages. They predicted that extraversion would interact with passage type, but this interaction failed to materialize.

The findings from other studies are somewhat inconsistent. Gale, Morris, Lucas, and Richardson (1972) obtained a significant positive correlation between extraversion and the Betts Vividness of Imagery Scale, and Morris and Gale (1974) also found that vivid imagery was more prevalent among extraverts. Results that were more in line with Riding's hypothesis were obtained by Huckabee (1974). He asked introverts and extraverts to rate the ease with which concrete and abstract nouns evoked images. Higher imagery scores were obtained by extraverts on both classes of nouns.

What do these various findings add up to? A distinction that may be useful in providing an answer is that between coding ability and coding preference (Richardson, 1980). It is clear, for example, that a failure to use imaginal processing may be due either to poor imaginal processing ability or to a preference for alternative processing strategies. Most of the available evidence can be accounted for by hypothesizing that introverts and extraverts differ primarily with respect to their preference for using imaginal processing rather than with their ability to use it.

A further complication is that it is often unclear whether the observed differences in memory between extraverts and introverts are attributable to processes operating at the time of initial perception of the stimulus material or to subsequent processes of attention, rehearsal, consolidation, retrieval, or response production. However, it is possible to make use of paradigms that permit consideration of a single aspect of information processing. For example, suppose we ask people to retrieve well-learned information from semantic or permanent memory (e.g., names of four-footed animals), and we attempt to ensure that introverts and extraverts possess equivalent amounts of information or knowledge by making use of vocabularly tests. Under such circumstances it should be possible to assess the impact of extraversion on the retrieval process itself.

The paradigm described above (often known as verbal fluency) has been used several times. Generally speaking, extraverts are able to think of more words in a given period of time than introverts that adhere to some specified criterion (e.g., colors, words starting with the letter *T*). M. W. Eysenck (1974a) obtained the typical finding, with extraverts writing down significantly more words belonging to five semantic categories than introverts. Additional information was obtained when account was taken of each subject's level of self-reported arousal or activation immediately preceding task performance. As can be seen in Figure 34, although there was the usual advantage of extraverts over introverts when self-reported arousal was high, there was no effect of extraversion on the number of words recalled among those subjects who reported themselves to be low in arousal. Thus it appears that the retrieval deficit shown by introverts on verbal fluency tasks is due to a supraoptimal level of arousal.

In a long program of research, M. W. Eysenck explored in more detail the effects of extraversion on the speed of retrieval of information from semantic or permanent storage. Some limitations on earlier conclusions were discovered (M. W. Eysenck, 1974b). Recall required the production of a word belonging to a designated category and starting with a particular letter (e.g., article of furniture—*C*), whereas recognition required a decision as to whether or not a word belonged to a given category (e.g., article of furniture—chair). Task difficulty was manipulated

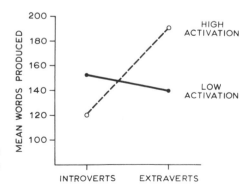

FIGURE 34. Retrieval from semantic memory as a function of extraversion and activation.

by varying item dominance on the basis of the normative data obtained by Battig and Montague (1969). They asked several people to think of items belonging to categories (e.g., articles of furniture) and listed the responses in decreasing order of frequency or dominance.

Extraverts responded faster than introverts on recall trials, but there was no effect of extraversion on recognition trials, where the retrieval demands were presumably much lower. Dominance was an important factor, since the superiority of low-arousal subjects (i.e., extraverts reporting low activation) over high-arousal subjects (i.e., introverts with high self-reported activation) was greater on low-dominance recall trials than on high-dominance recall trials. Similiar results were obtained in other studies, leading M. W. Eysenck (1976a) to propose the following hypothesis:

> High arousal has the effect of biasing the subject's search process towards readily accessible, or functionally dominant, stored information more than is the case with lower levels of arousal. (p. 401)

So far, we have considered only studies in which the information to be retrieved was defined by semantic criteria, and it is not clear whether similar results would be obtained if nonsemantic criteria were used. M. W. Eysenck and M. C. Eysenck (1979) presented a short list of words followed by a probe word. Subjects had to decide as rapidly as possible either whether there was a physical (i.e., identity) match between the probe and one of the list words or whether there was a semantic match (i.e., the probe was a member of one of the categories included in the memorized list). Extraversion interacted with the nature of the retrieval task: Extraverts scanned the memorized list faster than introverts under semantic matching conditions, but extraverts and introverts did not differ under physical matching conditions.

Related findings were reported by Schwartz (1979). Subjects were presented with two words at the same time. Sometimes they had to decide whether the two words were identical (physical identity condition), and at other times they decided whether the words were pronounced the same (homophone identity condition) or whether they belonged to the same semantic category (taxonomic category identity condition). Introverts were nonsignificantly slower than extraverts on physical identity trials, but they were much slower on homophone and taxonomic category identity trials.

In general, it appears that introverts are inferior to extraverts at retrieving most forms of semantic information but that there is no effect of extraversion on the retrieval of nonsemantic information. However, Schwartz's (1979) finding that introverts were slower than extraverts at making phonemic decisions in the homophone identity condition is inconsistent with this generalization. An alternative conclusion was suggested by M. W. Eysenck (1982): "It may be preferable to suppose that introverts take longer than extraverts to retrieve information from permanent storage, whether that information is semantic or phonemic in nature" (p. 135).

A final issue relating to the effects of extraversion on retrieval from memory concerns the criteria for response. It will be remembered from the earlier discussion of vigilance performance that introverts typically adopt a more stringent response criterion than extraverts. From the limited data that are available, the same appears to be true of memory. McLaughlin and Kary (1972) found on a recognition test that extraverts made more errors or false alarms than introverts, which suggests that they were responding less cautiously. When performance on a different recognition task was analyzed by means of the measures of signal detection theory, sensitivity or d' was unaffected by extraversion, but introverts set a higher criterion for responding than extraverts.

How well do the retrieval data fit the predictions from arousal theory? This is a difficult question to answer, because there is no well-established pattern of effects of high arousal on retrieval. The rather confusing state of affairs was summarized by M. W. Eysenck (1982):

> While there are some exceptions, it has usually been found that introversion and anxiety both impair the efficiency of retrieval. In contrast, noise seems to improve retrieval efficiency rather than to impair it, and incentive nearly always has no effect on retrieval. On the other hand, attempts to reduce arousal by means of relaxation instructions have usually facilitated retrieval. On the basis of such evidence it would be extremely difficult to claim that there is any consistent effect of arousal on retrieval. (p. 186)

The notion that differences between introverts and extraverts in the efficiency of retrieval can plausibly be attributed to the chronically higher arousal level of introverts also receives little support when arousal manipulations have been considered. M. W. Eysenck and M. C. Eysenck (1979) considered the effects of both extraversion and white noise on memory scanning. Extraversion affected the speed of scanning in a number of interesting ways, but white noise had no effect whatever. Of most theoretical relevance, extraversion did not interact with the presence or absence of white noise, so that there was no evidence that introversion and white noise were affecting the same arousal mechanism.

In sum, there are a number of reasonably consistent effects of extraversion on learning and memory. Introverts are affected more adversely than extraverts by increased task difficulty in the form of response competition, and the same is true of the effects of distracting stimulation. In addition, extraverts typically exhibit better memory performance than introverts at short retention intervals, but inferior memory at long retention intervals. Introverts are generally less efficient than extraverts at the retrieval of information from semantic or permanent storage, and introverts adopt a more cautious response criterion than extraverts.

This pattern of findings can be interpreted only loosely within arousal theory. Relatively little is known of the effects of arousal manipulations on learning tasks varying in the degree of response competition and amount of distraction. The interaction between extraversion and the length of the retention interval is predictable from arousal theory. However, the frequent failures of arousal manipulations to impair short-term retention and the fact that different arousers enhance long-term retention in a variety of ways point to unresolved issues. The effects of extraversion on the efficiency of retrieval are clearly at variance with those of other arousers. Practically nothing is known of the effects of other arousers on the response criterion. All in all, arousal theory provides a useful rule of thumb when attempting to account for differences in memorial functioning between introverts and extraverts but manifestly fails to offer a complete explanation.

PSYCHOMOTOR PERFORMANCE

The effects of extraversion on psychomotor performance have been explored on a variety of tasks, but the most popular task has been the pursuit rotor. The pursuit-rotor apparatus typically consists of a small metal disc embedded in a turntable that rotates fairly quickly. The subject is provided with a metal stylus and endeavors to keep it in contact with the metal disc for as much of the time as possible. In a conventional study, subjects are given a practice trial approximately 5 minutes in length, followed by a rest period of, say, 10 minutes, followed by an assessment of postrest performance.

The most important aspect of performance on the pursuit-rotor task for present purposes is the fact that performance immediately after the rest pause is nearly always considerably better than at any time prior to the rest pause. This phenomenon is known as reminiscence. It is perhaps natural to assume that reminiscence occurs because continued practice on the pursuit rotor leads to the accumulation of some inhibitory factor such as fatigue; this fatigue disappears during the rest interval and thus enables the subject to perform at a higher level than before the rest. A theory along these lines was proposed by Kimble (1949). He argued that massed practice on the pursuit rotor produces a buildup of reactive inhibition. This eventually induces involuntary rest pauses, which lower performance and also lead the subject to acquire a habit of nonresponding in addition to the habit of correct responding. Reactive inhibition dissipates during the rest period, with the result that reminiscence occurs during postrest performance. The marked improvement that is often found over the initial period of postrest practice (i.e., postrest upswing) was attributed to the extinction of the habit of nonresponding.

It is very straightforward to predict the effects of extraversion on reminiscence on the basis of H. J. Eysenck's (1957) theory, according to which extraverts develop inhibition more readily than introverts. Since Kimble's (1949) theory implies that the size of the reminiscence effect is a measure of the amount of inhibition accumulated prior to the rest period (provided that the rest pause is long enough to permit complete dissipation of inhibition), it follows that extraverts should show more reminiscence than introverts. The good news is that this prediction is very strongly supported by the data (H. J. Eysenck, 1962). However, the bad news is that the detailed pattern of the data is not as expected. Theoretically, extraverts should have lower prerest scores than introverts, because they accumulate more inhibition and inhibition impairs performance. In fact, the usual finding is that extraverts differ from introverts in having higher postrest scores rather than lower prerest scores (H. J. Eysenck, 1962).

Of course, the inhibition-theory account of pursuit-rotor performance is not necessarily accurate. The theory seems unduly complicated. For example, the theory can account for poor performance after a rest pause in several different ways: a failure of reactive inhibition to dissipate; the acquisition of a strong habit not to respond; or a failure to acquire the habit of correct responding. In addition to this theoretical looseness, it seems to be impossible to identify rest pauses in pursuit-rotor performance on the basis of the measure usually taken (i.e., time on target). This performance index does not enable us to distinguish between no attempt at performance (i.e., an involuntary rest pause) and an unsuccessful attempt.

H. J. Eysenck and Frith (1977) proposed an alternative theory of reminiscence based on consolidation. They argued that what has been learned on the pursuit rotor remains vulnerable to destruction until a rather lengthy consolidation process has run its course. The learning and consolidation processes pass through a total of three stages: (1) the learning is neither available for improving performance nor protected against destruction; (2) the learning can improve performance, but it is still not protected against destruction; and (3) the learning is available for improving performance and is also protected against destruction. After a short rest pause, some of the performance passes into the second stage of the consolidation process. Since some of what has been learned is available for improved learning, reminiscence occurs. However, since the learning is destroyed by the postrest practice, performance in the postrest period falls back after a time (this is known as postrest downswing).

This theory can be brought into contact with the effects of extraversion on reminiscence if we assume that consolidation of learning should be better at high levels of arousal (Walker, 1958). This leads to the prediction that introverts should show more reminiscence than extraverts, which suffers from the disadvantage that the actual finding is the exact opposite! However, it is perhaps reasonable to assume that complete consolidation takes longer to occur at high levels of arousal (Walker, 1958), in which cases the superior reminiscence of extraverts to introverts after short rest intervals of approximately 10 minutes might be reversed at longer intervals. Unfortunately, the relevant data are extremely inconsistent. Farley (1969) used rest periods of 10 minutes and 24 hours and found that extraverts showed more reminiscence than introverts at the shorter interval but showed less reminiscence than introverts at the longer interval. In contrast, Gray (1968) discovered that extraverts showed more reminiscence than introverts even one week after learning had taken place, and Seunath (1973) failed to detect any effects of extraversion on reminiscence after either 10 minutes or one week of rest.

In spite of this setback for the consolidation theory, it is becoming increasingly clear that arousal and consolidation both affect reminiscence. Some of the strongest evidence in favor of the consolidation theory was obtained by Rachman and Grassi (1965). They gave their subjects two 5-minute sessions on the pursuit rotor separated by a 4-hour rest. Subjects who practiced on a mirror-reversed pursuit rotor for 3 minutes during the first 10 minutes of the rest interval showed reduced reminiscence on the normal pursuit rotor. Of highest interest, the reduction in reminiscence was greatest for those subjects who practiced on the reversed pursuit motor during the first three minutes of the rest interval, a time at which the protection afforded by consolidation would be minimal.

There are various indications that arousal affects reminiscence. For example, it has been found that flashing a bright light or immersing the nontracking hand in cold water during rest intervals enhances reminiscence (Hammond, 1972). In other studies, arousal has been manipulated by leading some of the subjects to believe that their performance on the pursuit rotor would play a part in determining whether or not they were allowed to do a desired apprenticeship. The typical finding is that this motivating (and presumably arousing) state of affairs leads to greater reminiscence.

In view of the evidence that high arousal usually increases reminiscence, it is rather puzzling that it is the low-arousal extraverts rather than the high-arousal introverts who show more reminiscence. A step in the direction of accounting for the effects of extraversion on pursuit-rotor performance was taken by Frith (1971), who used a version of the pursuit rotor in which subjects attempted to keep the end of a stylus on top of a moving light that described the shape of an equilateral triangle. There was no difference between introverts and extraverts in terms of the total time on target, but there was a strong association between personality and the performance strategy that was adopted. Introverts tended to keep the stylus on the track but often misjudged the speed or velocity of the target, whereas extraverts matched the velocity of the target but often cut the corners of the triangle. Perhaps, on this task at least, extraverts tend to construct motor programs incorporating velocity information, whereas introverts attend more to the exact location of the target and depend on visual feedback for successful performance.

H. J. Eysenck and Frith (1977) speculated that it was primarily the motor programs that needed rest in order for consolidation to occur. This led them to the following explanation of the effects of extraversion on reminiscence:

> It is the laying down of motor programs that underlies reminiscence and indeed improvement in pursuit-rotor performance in general. Hence those subjects who choose to rely more on motor programs, the extraverts, will show the greater reminiscence and the better post-rest performance. (p. 391)

H. J. Eysenck's (1957) theory attached considerable importance to the notion that extraverts should produce more involuntary rest pauses than introverts in most circumstances. The reason for this is that involuntary rest pauses allegedly occur when inhibition has built up, and a crucial difference between extraverts and introverts is that inhibition develops more rapidly in extraverts. Since the pursuit-rotor task is not a suitable one for detecting these involuntary rest pauses, researchers have considered other psychomotor tasks.

Promising initial results were obtained with a tapping task, in which subjects tap a brass plate with a metal stylus as rapidly as possible (Spielman, 1963). While there was no effect of extraversion on the number of taps produced, unusually long intertap intervals or involuntary rest pauses were much more common in the performance of the extraverts than that of the introverts. Indeed, during a total of 25 minutes on the tapping task (5 minutes on each of five successive days), five extraverted subjects produced a total of 370 involuntary rest pauses, whereas five introverted subjects produced only 25 rest pauses. H. J. Eysenck (1964c) subsequently reported somewhat similar findings.

Although these data seem to provide striking support for inhibition theory, an alternative explanation was suggested by Frith (1973). He reanalyzed Spielman's data and discovered that most of the rest pauses were almost exactly double the length of the normal intertap interval. In his own study, Frith took recordings of muscle action potentials during tapping performance; he discovered that the rest pauses usually occurred when the subject attempted to make a tap, but the strength of the response was insufficient to produce proper contact. It would be possible to reformulate the inhibition hypothesis so that it stated that inhibition produces a reduction in effort rather than a complete cessation of responding. However, it is perhaps simpler to argue that introverts are more cautious than extraverts and so apply more force in order to make sure that they do not miss any taps.

It is possible that differences in tapping performance should be interpreted in terms of the higher level of cortical arousal in introverts. First, H. J. Eysenck (1964c) found that highly aroused subjects (candidates for a desired apprenticeship) had fewer involuntary rest pauses than normally motivated subjects, just as introverts had fewer involuntary rest pauses than extraverts. Second, Amelang and Breit (1983) instructed their subjects to tap a morse key as rapidly as possible for 2 minutes and found that low impulsives (introverts) had shorter intertap times than high impulsives (extraverts). However, the incidence of extremely delayed responses was not affected by personality. This general slowing of performance is in line with the notion that extraverts are underaroused but inconsistent with the prediction from inhibition theory that extraverts should only occasionally be slower than introverts.

A further interesting aspect of the study reported by Amelang and Breit (1983) is that tapping performance was affected by the impulsivity component of extraversion rather than the sociability component. Barratt, Patton, Olsson, and Zuker (1981) also found that speed of tapping was affected by impulsivity, but they discovered that impulsivity was positively related to tapping rate. These findings bring in question whether tapping performance is affected more by extraversion or by the impulsivity component of extraversion.

One of the more promising avenues of research relates to possible strategic differences between introverts and extraverts in psychomotor performance. For example, introverts seem to favor relatively small and precise movements on the pursuit rotor, whereas extraverts prefer more sweeping and less precise movements that literally involve corner-cutting (Frith, 1971). Findings from other paradigms point in the same direction. Rachman (1961) measured the motor reactions of people who were presented with conflict situations in which incompatible signals were provided. His key finding was that extreme extraverts responded more extensively than extreme introverts. Wallach and Gahm (1960) attempted to measure graphic constriction and expansiveness as a function of personality by considering the area filled with doodling. Extraverts were more expansive than introverts in their doodling among those low in neuroticism, whereas neurotic introverts were more expansive than neurotic extraverts. Similar results with a variety of writing tasks were obtained by Taft (1967).

In sum, it may be the case that extraverts produce larger movements than introverts provided that there is minimal stress. If the subjects are susceptible to stress (e.g., high on neuroticism), the results may be rather different. In line with this argument are the data obtained by Davis (1948) in a study in which performance under stressful conditions in a simulated cockpit was assessed. Dysthymics (who can be regarded as neurotic introverts) tended to overcorrect errors, whereas hysterics (who are neurotic ambiverts) undercorrected errors.

There is another aspect of performance style that is of some theoretical interest. It is possible to identify two extremes, one in which speed of performance is emphasized at the expense of making errors and another in which performance is accurate but slow. Most of the available evidence suggests that extraverts tend toward fast but inaccurate performance, whereas introverts are slow and accurate. The greater error proneness of extraverts was shown by Brebner and Flavel (1978). They carried out a reaction-time study in which there were numerous catch trials (i.e., the warning signal was not followed by the stimulus requiring response). Extraverts were four times as likely as introverts to respond erroneously on catch trials. More dramatically, extraverts responded prematurely a total of 97 times, whereas introverts did not produce any anticipatory responses at all.

Further evidence that extraverts are less accurate than introverts in their performance comes from some of the vigilance and memory studies discussed earlier in this chapter. This is true whether error proneness is measured in terms of the number of false alarms or the cautiousness of the response criterion (Carr, 1971; Gillespie & M. W. Eysenck, 1980; Harkins & Geen, 1975; Krupski et al., 1971; Tune, 1966).

Information about the effects of extraversion on performance speed has emerged from research on intelligence. Jensen (1966) found that extraversion correlated +.44 with the speed of solution of the Matrices intelligence test and that extraverts made nonsignificantly more errors than introverts. Similar results were obtained by Farley (1966) with the

Nufferno speed test, which requires the subject to identify a recurring pattern of letters and then to write down the next letter in the series. Extraverts performed this task much faster than introverts. This finding was replicated by Goh and Farley (1977), at least under nonstressed conditions.

In spite of the empirical support for the notion that introverts are slow and methodical and extraverts are fast and inaccurate, there are some complications in the data. There are a number of studies in which extraverts responded more rapidly than introverts at the start of the task, with introverts subsequently catching up or even surpassing the performance speed of extraverts later in the task. This has been found with tapping tasks (Amelang & Breit, 1983; Wilson, Tunstall, & Eysenck, 1971), a reaction-time task (Brebner & Cooper, 1974), and a continuous serial reaction task (Thackray, Jones, & Touchstone, 1974).

In sum, it appears that extraverts perform faster than introverts under relatively arousing conditions (e.g., intrinsically interesting or short task), whereas introverts respond faster than extraverts when long and monotonous tasks are used. It is of considerable interest that the typical finding that the low-arousal extraverts perform rapidly and inaccurately, whereas the high-arousal introverts act slowly and accurately, is precisely the opposite of what happens with most other arousal manipulations. For example, sleep deprivation reduces the level of arousal in most circumstances and characteristically leads to slow but accurate performance (see Kjellberg, 1977, for a review). In contrast, there is reasonable evidence that increasing arousal by incentive manipulations often increases the speed of responding and reduces the accuracy of performance (Feldman, 1964; Maller & Zubin, 1932), and the same is true of white noise (Hartley & Carpenter, 1974; Wilkinson, 1963).

What is going on here? M. W. Eysenck (1982) suggested that the natural effect of arousal is to produce fast and inaccurate performance, and so it is the reversal of this tendency in introverts that requires explanation. Speculatively, it may be the greater cognitive control of performance shown by introverts that prevents their high level of arousal from producing impulsive behavior. Alternatively, introverts' greater fear of punishment (Gray, 1973) may make them behave in a cautious manner.

PERCEPTUAL PHENOMENA

The basic Eysenckian approach to perception has been to attempt to identify perceptual phenomena that depend in an important way on inhibition or arousal level. The effects of extraversion on many such phenomena have been investigated over the years; we will here consider only some of the more important ones, including figural aftereffects, the spiral aftereffect, visual masking, and critical flicker fusion.

Figural aftereffects involve distortion of sensory input as a consequence of preceding stimulation in the same area. Research into the effects of personality on figural aftereffects has concentrated on the kinesthetic and visual figural aftereffects. The usual paradigm for investigating kinesthetic figural aftereffects involves judgments of the width of a wooden block on the basis of touch alone both before and after repeated stroking of a block of a different size. The typical finding is that the apparent width of the first block is decreased by experience of a wider interpolated block. This is the kinesthetic figural aftereffect.

Some of the most extensive work on the kinesthetic figural aftereffect has been carried out by Petrie (1978). She noted that some subjects react to stimulation of the fingers by underestimating the size of a subsequently touched stimulus object; such individuals were termed *reducers*. In contrast, other individuals (*augmenters*) produced overestimates under the same conditions. She proposed that these differences in the kinesthetic figural aftereffect reflect crucial individual differences in stimulus intensity modulation across several sensory modalities.

The likely relationship between augmentation-reduction and the personality dimension of extraversion can be worked out on the basis of Figure 31. Since introverts tend to perceive any given stimulus as more intense than extraverts, they are clearly amplifying sensory stimulation and thus should tend to be augmenters. On the other hand, extraverts should tend to be reducers. Much of the research discussed by Petrie (1978) is consistent with these expectations, but there are some negative findings (Barnes, 1976).

An apparent problem with the kinesthetic figural aftereffect is its rather poor test–retest reliability, which has led several researchers to regard research on this phenomenon as of little value. However, what appears to be happening is that long-term effects of the first experimental session affect performance on later sessions in various ways. In fact, first-session performance is reasonably valid (Baker, Mishara, Kostin, & Parker, 1976). Since Petrie relied largely on scores obtained from the first (or only) session, her work cannot be dismissed on the grounds of unreliable and invalid measures.

Visual figural aftereffects are observed when the presentation of an inspection figure produces distortions in the perception of a subsequently presented test figure. For example, Köhler and Wallach (1944) found that a previously fixated black square caused two white squares that were presented above and below the area that had contained the black square to appear further apart.

The exact mechanisms producing these figural aftereffects are still unclear. However, it has nearly always been assumed that they have a purely physiological basis and are caused by adaptation or satiation in particular neural structures. In the case of visual figural aftereffects, it may be appropriate to consider the theory that there are many separate channels in the visual system, each of which is tuned to a narrow range of spatial frequencies (Blakemore & Campbell, 1969). Channels that are specialized for low spatial frequencies are activated only when there are large objects in the visual field, whereas channels that are specifically for high spatial frequencies respond to fine details. If the inspection figure is small, this would produce adaptation of the higher frequency channels. As a consequence, only the low-frequency channels are available, and the subjective size of the test figure (or the space between the test figures in the study of Köhler and Wallach, 1944) is increased.

H. J. Eysenck (1955a, 1967a) argued that figural aftereffects were caused by satiation or adaptation. Since he also assumed that extraverts develop inhibition faster than introverts, he predicted tentatively that extraverts would show greater satiation, and thus larger figural aftereffects. Although he discovered (1955) that hysterics (neurotic ambiverts) had greater kinesthetic figural aftereffects than dysthymics (neurotic introverts), several studies failed to substantiate the predicted effect.

How to resolve this? According to H. J. Eysenck (1960a), extraverts are more prone than introverts to involuntary rest pauses and so they receive less total sensory input from the original stimulation (i.e., fixation of the inspection figure or prolonged stroking of the wooden block). This reduces figural aftereffects in extraverts relative to introverts and thus attenuates or even eliminates the extravert superiority predicted on the basis of their stronger buildup of satiation.

According to this revised theory, extraverts should show greater figural aftereffects than introverts when the original sensory stimulation is relatively brief so that extraverts have no difficulty in maintaining attention, but the opposite should be the case with long duration of the original sensory stimulation because of the large reduction in effective sensory input suffered by the extraverts under those conditions.

The available data are rather inconsistent. Many studies of visual figural aftereffects have failed to obtain significant relationships between extraversion and measures of figural aftereffects (e.g., Gardner, 1961; Spitz & Lipman, 1960; Wertheimer, 1955). However, Rechtschaffen (1958) reported a suggestive correlation of +.14 between extraversion and the size of the visual aftereffect with an inspection time of 40 seconds, and Holland (1965) obtained a similar result with a 90-second inspection time. Very few studies have made use of short inspection times, but an exception was that of Holland and Gomez (1963). They used very short inspection times that varied between ⅙ second and 5 seconds and considered the effects of a depressant drug Amytal and a stimulant drug Dexedrine on visual figural aftereffects. The theoretical expectation is that depressant drugs increase the amount of satiation that is created, whereas stimulant drugs decrease it. It follows that visual figural aftereffects should be enhanced by Amytal and reduced by Dexedrine. In fact, the depressant Amytal did increase the magnitude of the visual figural aftereffect, but the stimulant Dexedrine produced results resembling those produced by the placebo. Byth (1972) used the very short inspection time of 500 milliseconds. He obtained the typical visual figural aftereffect, but there was no effect of extraversion on performance.

More promising results were obtained by Singh and Gupta (1981) in a study of the visual figural aftereffect. Extraverts had a greater aftereffect than introverts, and the depressant drug phenobarbital tended to increase the size of the aftereffect. In another drug study, Gupta and Kaur (1978) found that extraverts had larger kinesthetic figural aftereffects than introverts under placebo conditions. The stimulant drug dextroamphetamine reduced the aftereffect in extraverts but increased it in introverts.

It is obviously difficult to draw any firm conclusions about the effects of personality on figural aftereffects. The position is equally equivocal with respect to performance on the spiral aftereffect. Various ways of producing the spiral aftereffect have been used, but one popular method is to rotate a disc on which an Archimedes spiral is painted. The spiral appears to expand if the disc is rotated clockwise and to contract if it is rotated counterclockwise. If the spiral is stopped after the rotation has been observed for a number of seconds, it appears to move in the opposite direction (i.e., expanding if it had been contracting or contracting if it had been expanding). This is the spiral aftereffect. The same kind of effect can be achieved if you look at a waterfall and then at a stationary object; the object will appear to move in the opposite direction to that previously experienced (i.e., it will move upwards).

How is the spiral aftereffect produced? We know from microelectrode evidence that there are visual brain cells that are responsive to movement in particular directions. When the appropriate moving stimulation is received, cells sensitive to motion in that direction increase their firing rate considerably above the spontaneous firing level. When that stimulation ceases, however, the firing rate of those cells drops to well below that of the resting level for some time. Evidence for this contention was obtained by Sekuler and Ganz (1963). They discovered that exposure to a stimulus moving in a particular direction produced a higher than normal threshold for subsequent motion in the same direction, indicating that stimulation of specific cells can cause them to develop inhibition. With respect to the spiral aftereffect, it seems plausible that such an effect would occur if the cells responsive to motion in the direction of the Archimedes spiral became inhibited, whereas those attuned to motion in the opposite direction continue to fire at the spontaneous level.

If we can assume that the spiral aftereffect depends on the build up of inhibition, then the natural prediction on H. J. Eysenck's (1957) inhibition theory would seem to be that extraverts (who develop inhibition rapidly and strongly) should show a longer duration of aftereffect than introverts. In fact, Eysenck argued that the duration of the spiral aftereffect should be *shorter* in extraverts than in introverts for two reasons: (1) the greater susceptibility of extraverts to inhibition means that they receive less effective stimulation than introverts because of the occurrence of involuntary rest pauses, and (2) the physiological processes responsible for maintaining the aftereffect are themselves subject to inhibition and will thus cease more rapidly in extraverts.

A related but simpler line of argument is possible from arousal theory. Since introverts tend to augment stimulation more than extraverts, any effects of the rotating spiral stimulus should be correspondingly augmented.

There is certainly some experimental support for the predicted effect of extraversion on the spiral aftereffect. Lynn (1960) found that introverts had a longer-lasting aftereffect than extraverts, and Levy and Lang (1966) reported that subjects low on impulsivity (i.e., introverts) had a mean duration of aftereffect that was almost twice as long as that of subjects high on impulsivity (i.e., extraverts). Positive results were also reported by Holland in some of his early studies (e.g., Holland, 1959) and by Paramesh (1963). In addition, Knowles and Krasner (1965) obtained significantly longer duration of the spiral aftereffect in introverts than extraverts, but only among subjects who were high in neuroticism. However, there are several unpublished studies reporting no relationship between extraversion and the spiral aftereffect (see Holland, 1965, for details).

One of the obvious difficulties with evaluating the evidence is the likelihood that there are individual differences in the criteria used for reporting on the subjective duration of the aftereffect. It is of interest in this connection that Holland (1961) was able to alter the reported aftereffect by manipulating the instructional emphasis on confidence or certainty. Why is the apparent duration of the spiral aftereffect affected by instructions? According to Kristjansson and Brown (1973), part of the answer lies in the fact that the aftereffect dies away in two phases, the first of which is a short phase of rapid decrease in the speed of the after-movement and the second of which is longer-lasting and less distinct. It is possible that the spiral aftereffect seems to last longer for introverts than for extraverts because extraverts are more likely to jump to the conclusion that the spiral aftereffect has ended at the conclusion of the first phase. However, Kristjansson and Brown did not obtain support for this idea. There was no relationship between extraversion and the duration of the spiral aftereffect with conventional instructions (to indicate when the aftereffect appears to stop), but extraverts had a significantly shorter aftereffect than introverts when the instructions were unambiguous (to report the end of the first phase of decay).

It would clearly be desirable to make use of some measure of the spiral aftereffect that is less susceptible to distortion than is verbal report. Precisely this was done by Claridge and Herrington (1963a). They ran a condition in which the subjects were asked to close their eyes when the rotating spiral stopped. They took as their measure of the spiral aftereffect the duration of alpha blocking in the EEG record, that is, the length of time until alpha activity returned to its resting level. The duration of alpha blocking was greater in dysthymics (neurotic introverts) than in hysterics (neurotic ambiverts), thus suggesting that there is a greater spiral aftereffect in introverts. This conclusion was supported by their data based on the conventional assessment of the spiral aftereffect: The mean subjective judgment of the duration of the spiral aftereffect was 17.2 seconds in dysthymics and 10.9 seconds in hysterics.

If the effects of personality on the spiral aftereffect are mediated by individual differences in inhibition, then fairly clear predictions can be made about the ways in which the spiral aftereffect should be affected by stimulant and depressant drugs. Since the theoretical expectation is that inhibition shortens the duration of the aftereffect, it follows that stimulant drugs should increase the spiral aftereffect, whereas depressant drugs should reduce it. H. J. Eysenck, Holland, and Trouton (1957) obtained partial support for these predictions. The depressant drug Sodium Amytal significantly shortened the duration of the illusion, but the stimulant drug Dexedrine had no effect. In other studies, Costello (1963) discovered that a depressant drug reduced the duration of the spiral aftereffect, and H. J. Eysenck and Easterbrook (1960) found that amphetamine increased the duration.

As we have already noted, it is possible to propose an arousal-theory account according to which the spiral aftereffect is greater in introverts or in individuals given stimulant drugs than in extraverts or individuals given depressant drugs because high arousal leads to an augmentation of the stimulus input provided by the rotating spiral and thus enhances its aftereffects. However, such a theory must predict that highly motivated (and so aroused) subjects should have a longer-lasting spiral aftereffect than relatively unmotivated subjects. In fact, precisely the opposite results have been obtained when high motivation was created by leading subjects to believe that the perceptual task formed part of the selection procedure for a desired apprenticeship (H. J. Eysenck & Holland, 1960; H. J. Eysenck, Willett, & Slater, 1962).

All in all, work on the spiral aftereffect leads to the conclusion that we are dealing with some interesting findings that are still lacking an adequate interpretation, in spite of the fact that various explanatory notions have been invoked (inhibition produced by the rotating spiral, reduced effective stimulation due to involuntary rest pauses, inhibition of the processes maintaining the spiral aftereffect, augmentation of the stimulation received from the rotating spiral). Although there is evidence from other tasks that introverts and extraverts differ with respect to the last three factors at least, there is little or no compelling evidence that these factors are actually affecting performance on the spiral aftereffect in the ways specified theoretically.

Another perceptual phenomenon that has been examined in light of individual differences in personality is that of metacontrast or the masking effect. Metacontrast can be demonstrated by presenting a disc briefly and following it almost immediately with the presentation of an annulus. If matters are arranged so that the inner contour of the annulus is presented in exactly the same location as that previously occupied by the outer contour of the disc, then subjects are sometimes unaware that a disc has been presented at all. Not surprisingly, this effect is obtained only when the interval between the onset of the first stimulus and the masking stimulus is very short.

It has been suggested that this paradigm can be used to measure the speed of perceptual processing, with the time taken to encode the first stimulus being indexed by the smallest interval beween the two stimuli at which no metacontrast is obtained. One interpretation of metacontrast is that the contour defining the first stimulus does not have sufficient time to be established by the time that the second or masking stimulus is presented and destroys it (Werner, 1935). A more contemporary view was expressed by Weisstein (1968). He proposed that metacontrast can be regarded as a form of lateral inhibition in which the inhibitory effect of the mask develops more rapidly than the excitatory effect of the initial stimulus. The evidence was reviewed by Lefton (1973), who concluded that there is much support for an inhibition explanation of metacontrast.

At the theoretical level, there are at least two potential ways in which extraversion might affect metacontrast. First, since extraverts are thought to generate inhibition more rapidly and strongly than introverts, the inhibitory effect of the masking stimulus should be greater in extraverts. Second, in view of the higher arousal level of introverts than of extraverts, it might be expected that the excitatory effect of the first stimulus would be greater for introverts. Either way, the natural prediction is that introverts should be less susceptible than extraverts to metacontrast.

McLaughlin and H. J. Eysenck (1966) reported empirical support for this prediction. They obtained a significant correlation of +.36 between extraversion and the minimum interstimulus interval at which there was no metacontrast.

Holland (1963) investigated the effects of various stimulant and depressant drugs on metacontrast. The stimulant drug Dexedrine reduced the interstimulus interval required to produce masking, which is consistent with an arousal-theory account. The depressant drug Sodium Amytal failed to increase the critical interstimulus interval as had been predicted, but the expected results were obtained with the depressant drug meprobamate. Holland also made use of nitrous oxide and oxygen, which are depressant and stimulant gases respectively. Both produced the expected effects on metacontrast, especially nitrous oxide.

A related visual masking effect was investigated by Aiba (1963). He discovered that there was a reduced masking effect with the stimulant drug Dexedrine and a smaller effect in the opposite direction with the depressant drug amobarbital. Those results were obtained when the luminance of the first stimulus was manipulated. In a further study, stimulus duration was varied and there was increased masking with Sodium Amytal (a depressant) and decreased masking with Dexedrine.

In sum, the data on visual masking effects are fairly consistent. High arousal typically reduces masking effects, whereas low arousal has the opposite effect. While no detailed interpretation of these findings is available, it seems reasonable to assume that cortical arousal serves to facilitate discriminations such as those between two successive stimuli presented close together in time and place.

The theoretical account that has been offered to explain metacontrast may also be applicable to an apparently related perceptual phenomenon known as critical flicker fusion. Critical flicker fusion is defined as the rate of successive light flashes from an unmoving source of light at which the light appears to stop flickering and becomes steady or continuous. Thus metacontrast and critical flicker fusion do at least resemble each other in that they both demonstrate limitations on temporal resolution within the visual system. The precise mechanisms underlying critical flicker fusion remain unclear, but some interesting theoretical speculations were put forward by Kelly (1971, 1972). He argued that the critical fusion frequency reflects the rate at which photoreceptors are able to change from one level of excitation to another. If this is correct, then research into critical flicker fusion permits an examination of one of the fundamental temporal limitations of the visual system.

H. J. Eysenck (1967a) related the personality dimension of extraversion to critical flicker fusion. He suggested that critical flicker frequency provides a measure of the efficiency of cortical resolution of stimuli. Since introverts have a chronically higher level of cortical arousal than extraverts, they should be more efficient at resolving stimuli and thus have a higher flicker fusion frequency than extraverts. The prediction can be looked at from a slightly different perspective. Critical flicker fusion thresholds increase in line with higher stimulus intensity (Gray, 1964), and introverts augment stimulation. The expected finding was reported by Washburn, Hughes, Steward, and Sligh (1930), but the difference between the extraverted and introverted subjects was not significant. More pronounced effects in the same direction were reported by Madlung (1936) and Simonson and Brozek (1952).

Is it reasonable to attribute the superior performance of introverts on the critical flicker fusion task to their higher level of cortical arousal? The most direct evidence for this contention was obtained by Frith (1967). He compared the effects of 50 dB noise on the performance of introverts and extraverts, discovering that this arousing stimulus improved the critical flicker fusion performance of extraverts but had no effect at all on introverts. The clear implication of this result is that the relatively poor level of performance usually achieved by extraverts on the critical flicker fusion task is due, at least in part, to their low level of arousal.

There is other evidence implicating arousal as one of the determinants of critical flicker performance. It has been found, for example, that the discrimination of two flashes in the visual cortex of cats was improved when the reticular formation (which plays a major role in cortical arousal) was stimulated electrically (Lindsley, 1957). Predictable effects of drugs on critical flicker fusion performance have also been reported. Granger (1960) discovered that performance was improved by stimulant drugs and impaired by depressant drugs.

In sum, the degree of success of attempts to relate personality to perceptual phenomena has varied considerably from phenomenon to phenomenon. We may provisionally divide the phenomena into two categories, the first of which includes figural aftereffects and the spiral aftereffect, and the second of which includes metacontrast and critical flicker fusion. In the first category, the data are rather inconsistent, and somewhat cumbersome explanations have been put forward. In the second category, matters appear to be discernibly clearer both empirically and theoretically. One of the differences between the two categories is that perceptual phenomena in the first category were considered from the perspective of H. J. Eysenck's (1957) inhibition theory, whereas those in the second category lend themselves to interpretation by the newer arousal theory (H. J. Eysenck, 1967a). Thus, findings in this area bear testimony to the superiority of the arousal approach over that based on inhibition.

Another relevant factor may be the complexity of the perceptual tasks. There are reasons for supposing that metacontrast and critical flicker fusion are simpler phenomena than the spiral aftereffect and figural aftereffects. In particular, the crucial events involved in producing the former phenomena extend over a much shorter period of time than those associated with the latter phenomena; as a result, there may very well be fewer mechanisms and processes involved in metacontrast and critical flicker fusion than in the spiral aftereffect and figural aftereffects. Thus, the lesson to be learned from this body of research is that consistent and predictable effects of personality on perceptual performance are likely to be obtained only when relatively simple and well-understood tasks are used.

SUMMARY AND CONCLUSIONS

Our review of the experimental literature on the effects of extraversion on performance has revealed the existence of numerous consistent behavioral differences between introverts and extraverts. These differences extend from classical conditioning to verbal learning and from vigilance to critical flicker fusion. Far and away the most successful attempt to provide a unified account of this gallimaufry of findings is the arousal theory of H. J. Eysenck (1967a). This theory has generated a multitude of predictions, most of which have found experimental support. Of particular importance is the notion that the higher level of cortical arousal in introverts than in extraverts causes them to augment stimulation relative to extraverts. This notion is most directly relevant to research on sensory deprivation, the optimal level of stimulation, pain tolerance, and sensory thresholds, but it may also help to account for the effects of extraversion on classical conditioning and on perceptual phenomena such as critical flicker fusion and metacontrast.

This theoretical approach manifestly has the advantage of parsimony. However, although the arousal theory captures some of the crucial differences between introverts and extraverts, there are increasing doubts as to its ability to provide a completely adequate interpretation of the behavioral data. Some of these doubts and other problems are discussed at length below:

1. It is not altogether clear that arousal theory makes clearcut predictions about behavior in most situations. As we saw in Chapter 7, although there are certain similarities in the behavioral effects of different arousal manipulations, there are also discernible differences. If we evaluate H. J. Eysenck's (1967a) arousal theory of extraversion by comparing the effects of extraversion with those of various arousers, then certain discrepancies become apparent. In particular, although high arousal typically affects speed–accuracy tradeoff by increasing speed at the expense of accuracy, introverts characteristically show the opposite pattern of reduced speed accompanied by increased accuracy. The implication of these findings is that introverts are more cautious than extraverts, and this is supported by other observations. It makes intuitive sense that this should be so, since introverts are low in impulsivity. However, it is not clear that it makes theoretical sense. At a more general level, the difficulty is that the imprecision of arousal theory often means that we are unsure whether a particular finding is consistent or inconsistent with H. J. Eysenck's (1967a) theory of extraversion.

2. It has been assumed (H. J. Eysenck, 1967a) that introverts have a chronically higher level of cortical arousal than extraverts. This assumption provides a plausible underpinning for the stable personality differences between introverts and extraverts but is difficult to reconcile with recent evidence that introverts are more aroused than extraverts only at certain times of day. The initial evidence that introverts are not necessarily more aroused than extraverts throughout the day was obtained by Blake (1967). He discovered that introverts had a higher body temperature than extraverts in the morning (which is consistent with the notion that they are more aroused), but during the late afternoon and evening it was the extraverts who had the higher temperature.

Of course, it would be easy to dismiss Blake's evidence by arguing that body temperature is an unsatisfactory and inadequate measure of arousal level. However, stronger evidence that extraverts may be more aroused than introverts during the evening has been reported by Revelle et al. (1980). They carried out a series of experiments in which subjects were given an academic-type test similar to the American Graduate

Record Examination either in the morning or in the evening. In addition, the subjects who did the test had previously taken either the stimulant caffeine or a placebo. On the basis of the Yerkes–Dodson law, according to which an intermediate level of arousal is optimal for performance, caffeine should have enhanced the test scores of those subjects who were underaroused and should have reduced the scores of those who were already optimally aroused.

The results obtained by Revelle *et al.* (1980) across a total of seven experiments are shown in Figure 35. The morning data are very much in line with what would be expected on H. J. Eysenck's (1967) arousal theory, with caffeine helping the underaroused extraverts and harming the more aroused introverts. However, the evening data do pose problems and are most readily explained by assuming that extraverts were more aroused than introverts.

What weight should be placed on these data? According to Gray (1981), the findings of Revelle *et al.* (1980) and of Blake (1967) together form "a dagger that goes to the heart of the Eysenckian theory" (p. 258). This seems to be overstating the case. Of course, it is difficult to relate stable differences between introverts and extraverts to the apparently complex effects of time of day on their levels of arousal. Furthermore, there is the nasty suspicion that the effects of extraversion on performance discussed in this chapter owe much of their consistency to the fact

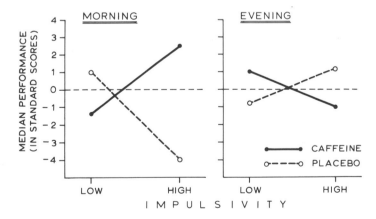

FIGURE 35. Performance of demanding verbal tasks as a function of time of day, impulsivity, and caffeine. (From "The Interactive Effect of Personality, Time of Day and Caffeine: A Test of the Arousal Model" by W. Revelle, M. S. Humphreys, L. Simon, and K. Gilliland, *Journal of Experimental Psychology: General*, 1980, *109*, 1–31. Copyright 1980 by the American Psychological Association. Adapted by permission of author.)

that testing rarely occurred in the evening. If the results obtained by Revelle *et al.* can be extrapolated to other tasks and situations, then the implication is that many of the findings derived from experiments carried out during the normal working day would be reversed if the same experiments were carried out in the evening. However, it is not clear that such extrapolation is appropriate. As M. W. Eysenck and Folkard (1980) pointed out, Revelle *et al.*'s choice of task was unfortunate in at least two respects: (1) the psychological processes involved are largely unknown, and (2) H. J. Eysenck's (1967a) theory makes no direct predictions about the effects of extraversion on it. According, although it is possible that interactions between extraversion and time of day may eventually necessitate some modification of Eysenck's arousal theory, the currently available evidence is insufficient to justify any change.

3. The findings obtained by Revelle *et al.* (1980) bring a further issue into focus. We have so far reported their data in terms of interactions involving extraversion, but they also conducted further analyses based on its two major components, impulsivity and sociability. The consistent finding was that interactions between extraversion and time of day were almost entirely attributable to impulsivity rather than to sociability. This led them to the challenging conclusion that impulsivity rather than extraversion *per se* is the primary causal determinant of many of the major findings reported in the literature. They argued that their time-of-day data reflected a phase difference of several hours between introverts and extraverts in the circadian rhythm of arousal or, more precisely, between low and high impulsives. Some support for Revelle *et al.*'s position was reported by M. W. Eysenck and Folkard (1980). They discovered that the phase difference in body temperature between introverts and extraverts was due more to the impulsivity component of extraversion than to the sociability component.

It is rather difficult to decide whether extraversion or impulsivity is the more important factor in determining performance. First, the substantial overlap between the two personality dimensions makes it complicated to disentangle their effects at an empirical level. Second, there appear to be a number of different impulsivity factors, and some of the impulsivity items on the Eysenck Personality Inventory used by Revelle *et al.* (1980) correlate more highly with psychoticism than with extraversion. Once that is said, it is interesting to note that most of the relevant research suggests that impulsivity is the component of extraversion mediating the effects of extraversion on performance. This has been found in work on classical conditioning (H. J. Eysenck & Levey, 1972), vigilance decrement (Thackray *et al.*, 1974), driver safety (Loo, 1979b), and tapping performance (Amelang & Breit, 1983).

The implication of the various findings seems to be that researchers should in the future compare systematically the effects on performance of extraversion and its two main components. Although there is some support for impulsivity as the crucial factor, it is probable that this is a function of the experimental situations that have been explored. It seems plausible that sociability would play a larger role in determining performance if behavioral measures related to sociability (e.g., talkativeness, amount of eye contact) were taken in explicitly social settings.

4. It is unfortunate that such limited data have been collected in most of the studies reported in this chapter. Far too often researchers have asked only the simplest question, namely, "Does extraversion improve or impair performance on this task?" Such an approach by its very nature cannot provide strong support for the arousal theory of extraversion. This is because there are any number of reasons other than the greater cortical arousal of introverts why there might be a main effect of extraversion on performance. In contrast, the discovery of predictable interactions between extraversion and the extent to which the experimental situation is arousing (e.g., H. J. Eysenck & Levey, 1967; Shigehisa & Symons, 1973) is much more difficult to account for without reference to arousal.

This plea for more informative data is also relevant to the research strategy of comparing the performance effects of extraversion and some arousal manipulation in two different experiments. If significant main effects are obtained in both experiments, there is a minimum probability of .5 that the findings will be in accord with arousal theory! For example, some of the early studies on classical conditioning showed that introverts conditioned better than extraverts and that stimulant drugs enhanced conditioning. This is, of course, consistent with H. J. Eysenck's (1967a) arousal theory. However, it is now known that the precise effects of extraversion on conditioning depend critically on task parameters and on the performance measures that are recorded (Jones et al., 1981). It would be a tremendous coup for Eysenck's arousal theory if the detailed effects of a stimulant drug resembled those of introversion, and it would be informative if they did not.

5. Someone who has read this chapter carefully may feel that although arousal theory has proved its usefulness when it comes to predicting certain limited kinds of performance, it has not been demonstrated convincingly that individual differences in cortical arousal are of relevance to most human behavior. More specifically, the level of cortical arousal may have a direct effect on the sensitivity to stimulation or on critical flicker fusion frequency, but its effects on, for example, thought processes are likely to be no more than indirect. It is certainly true that

the impact of cortical arousal on performance varies as a function of the task in question, and it is basic physiological and biological processes that are likely to be most affected. However, as is shown in Chapter 11, introverts and extraverts do differ in many everyday situations in ways that are very much in line with arousal theory.

In sum, the simple notion that introverts have a chronically higher level of cortical arousal than extraverts has proved extraordinarily successful in accounting for an enormous variety of findings. Not surprisingly, there are cases in which the theory's fit to the data is imperfect, and a more complex formulation may be needed at some point. However, the general rule in science is that a theory is usually discarded only when a superior theory replaces it, and this has not happened so far.

CHAPTER TEN

Neuroticism, Anxiety, and Performance

This book is concerned primarily with the three personality dimensions of extraversion, neuroticism, and psychoticism. Laboratory research looking at the behavioral correlates of extraversion was discussed in the previous chapter, and rather little research in the laboratory has examined the psychoticism dimension. That leaves the neuroticism dimension. However, most researchers have not investigated the effects of neuroticism on performance but have instead opted to study the related personality dimension of anxiety. The anxiety dimension of personality as measured by a test such as the Manifest Anxiety Scale (Taylor, 1953) falls within the two-dimensional space defined by the extraversion and neuroticism dimensions, correlating approximately $+.6$ to $+.7$ with neuroticism and $-.3$ to $-.4$ with extraversion.

As we saw in Chapter 7, there has been some theoretical controversy as to whether it is preferable to regard anxiety as a unified personality dimension (Gray, 1973) or to proceed on the basis that there are two separable components of anxiety, one relating to neuroticism and the other to introversion (H. J. Eysenck, 1967a). However, in terms of predicting behavioral data, the distinction between these two theoretical approaches is sometimes terminological rather than substantive. For example, H. J. Eysenck (1973c) identified two major effects of anxiety, one attributable to the neuroticism component of anxiety and the other produced by the introversion component, but theorists of otherwise very different persuasions have also argued that anxiety has two kinds of effects (e.g., Morris, Davis, & Hutchings, 1981). The details of these and other theoretical formulations will be discussed later in the context of the available evidence.

THE STATE–TRAIT APPROACH

Several theorists (e.g., M. W. Eysenck, 1979; Spielberger, 1966) have favored a conceptual distinction between trait anxiety (defined by Spielberger, Gorsuch, & Lushene, 1970, p. 3, as "relatively stable individual differences in anxiety proneness") and state anxiety (defined by Spielberger *et al.*, 1970, p. 3, as "characterized by subjective, consciously perceived feelings of tension and apprehension, and heightened autonomic nervous system activity"). These two kinds of anxiety are often measured by means of the State–Trait Anxiety Inventory (Spielberger *et al.*, 1970), which contains some items relating to how people generally feel (trait anxiety) and others dealing with how they feel "right now" (state anxiety).

According to the state–trait approach to anxiety, the amount of anxiety actually experienced by an individual at any given time (state anxiety) is determined interactively by the degree of stress present in the situation and by that individual's susceptibility to anxiety (trait anxiety). This assumption is incorporated into Figure 36, as is the further assumption that performance is affected to a greater extent by state anxiety than by trait anxiety.

An obvious prediction of this theoretical approach stems from the notion that state anxiety is affected by the degree of situational stress, whereas trait anxiety is not. As expected, trait anxiety is more stable than state anxiety across situations varying in their stressfulness (Allen, 1970; Martuza & Kallstrom, 1974). The further prediction that the difference in state anxiety between groups high and low in trait anxiety should be enhanced as the degree of situational stress increases has also been confirmed a number of times (see Shedletsky & Endler, 1974). However, an important qualification on that statement was discovered by Hodges (1968). As can be seen in Figure 37, the predicted interaction between trait anxiety and situational stress occurred when stress consisted of a threat to self-esteem (failure threat) but not when it consisted of phys-

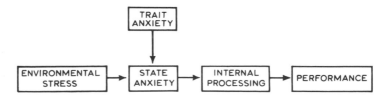

FIGURE 36. The basic state–trait theory of anxiety.

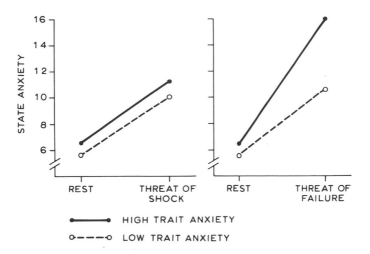

FIGURE 37. State anxiety as a function of trait anxiety, failure stress, and shock stress. (From "Effects of Ego Threat and Threat of Pain on State Anxiety" by W. F. Hodges, *Journal of Personality and Social Psychology*, 1968, *8*, 364–372. Copyright 1968 by the American Psychological Association. Adapted by permission.)

ical danger (shock threat). It appears to be the case that measures of trait anxiety focus on reactions to ego threat rather than to physical danger (Endler, Magnusson, Ekehammar, & Okada, 1976).

A crucial implication of the state–trait approach to anxiety is the notion that the behavioral consequences of state anxiety produced by external stress are equivalent to those stemming from high levels of trait anxiety. There are several confirmations of this equivalence, but a single example will make the point. Rosenbaum (1953) discovered that environmental stress increased generalization of a voluntary response, and found later (1956) that clinically anxious subjects showed greater generalization than control subjects. Such findings both support the state–trait theory of anxiety and help to forge a link between experimental and personality research.

THEORIES OF ANXIETY AND PERFORMANCE

Having dealt with the issue of the factors determining the amount of anxiety experienced by an individual in a particular situation, we shall turn to the mechanisms theoretically mediating the effects of state anxiety on performance. Although there are various differences from one theory to another, there is some agreement on the idea that anxiety has two main effects, one motivational and beneficial and the other disruptive and detrimental. For example, H. J. Eysenck (1973c) proposed the following theory of anxiety:

> It is introversion (characterized by high cortical arousal) that is responsible for the drive-properties of the MAS+ (high Manifest Anxiety Scale scoring) subjects, rather than neuroticism (which when aroused through ego-involving instructions, or some other manipulation of the situation, produces the drive stimuli that interfere with performance). (p. 401)

A similar theory was proposed by Wine (1971). She attributed the adverse effects of anxiety on performance to task-irrelevant thoughts, whereas the positive effects were accounted for in motivational terms:

> The low-test-anxious person is focused on task-relevant variables while performing tasks. The highly test-anxious subject is internally focused on self-evaluative, self-deprecatory thinking, and perception of his autonomic responses.... Non-evaluative situational conditions do not elicit the highly test-anxious subject's self-directed interfering responses, thus it is possible for him to direct his full attention to the task. Conversely, non-evaluative testing conditions presumably do not excite the nonanxious subject's motivation and interest. (pp. 92–96)

A third example of an anxiety theory incorporating separate cognitive and noncognitive components is the formulation proposed by Liebert and Morris (1967) and subsequently modified and expanded (e.g., Morris, Brown, & Halbert, 1977; Morris et al., 1981). Their major assertion was that test anxiety comprises two conceptually distinct components, worry and emotionality. Worry is the cognitive aspect of anxiety and consists of concern about one's performance and its consequences, negative self-evaluations, and negative task expectations. In contrast, emotionality refers to physiological changes and the accompanying unpleasant feelings of uneasiness, nervousness, and tension.

Worry and emotionality scores have been obtained through questionnaire measurement in several studies, and the evidence suggests that these two kinds of anxiety often follow a different time course. The worry scores of graduate students were elevated five days before an important examination, but their emotionality scores were not (Spiegler, Morris, & Liebert, 1968). The worry scores did not change from immediately before to immediately after the examination, whereas the emotionality scores decreased significantly.

A further important difference between worry and emotionality was discovered by Morris and Liebert (1973). Worry scores increased following failure threat but not shock threat, whereas emotionality scores increased after shock threat but not failure threat. These results suggest quite strongly that there are at least two separate components of anxiety, each of which is affected by rather different environmental stressors.

What are the expected effects of worry and emotionality on performance? According to Morris et al. (1981), worry should impair performance because it involves a misdirection of attention away from the task at hand. On the other hand, emotionality should not have this adverse effect except on those rare occasions when one becomes preoccupied with one's own physiological processes.

It is possible to relate some of these theoretical notions to personality theory. Since introverts generally condition better than extraverts, it might be expected that neurotic introverts would be more susceptible than neurotic extraverts to conditioned anxiety in the form of worry or psychic anxiety. On the other hand, neurotic extraverts with their low level of arousal and high level of autonomic activation may be liable to the more somatic components of anxiety resembling emotionality.

Most of the available evidence indicates that the relative importance of psychic and somatic anxiety in determining the overall level of anxiety is affected by an individual's position on the extraversion dimension. For example, Schalling et al. (1975) obtained a highly significant negative correlation between psychic anxiety and extraversion in a group of psychiatric patients. They also found that extraversion (or at least its impulsiveness component) correlated positively with somatic anxiety.

WORRY AND PERFORMANCE

There is plenty of indirect support for the notion that any adverse effects of anxiety on performance are attributable to task-irrelevant processing activities such as worry. For example, Morris and Liebert (1970) reported that correlations between worry and final examination scores, with emotionality partialled out, were significantly negative. In contrast, the correlations between emotionality scores and examination performance, with worry partialled out, were statistically nonsignificant.

Similar findings were reported by Spielberger, Gonzalez, Taylor, Algaze, and Anton (1978). Worry scores on the Test Anxiety Inventory correlated $-.47$ with grade point average (a measure of academic achievement) among male students, whereas emotionality correlated only $-.13$ with grade point average. The correlations for female students were $-.35$ and $.00$, respectively.

The literature was reviewed by Morris et al. (1981), who concluded that "the inverse relationship between anxiety and various performance variables under appropriate conditions is attributable primarily to the worry-performance relationship" (p. 541). However, it must not be forgotten that we are dealing with correlational evidence, that is, those people who report worried and self-evaluative thoughts tend to be the same people who exhibit poor levels of performance. Since correlations cannot prove causes, the evidence is equivocal. Even if it turns out that there is a causal relationship between worry and performance, it may well be that the direction of causation is, in fact, the opposite of that usually envisaged. That is to say, poor performance or merely the anticipation of poor performance may produce greater concern about competence and thus to increased worry. However, irrespective of the direction of causation it is still the case that emotionality or the physiological aspect of anxiety is of less relevance to an understanding of its detrimental effects than is the cognitive or worry aspect, at least with the modest levels of anxiety experienced in laboratory conditions.

EFFICIENCY AND EFFECTIVENESS

M. W. Eysenck (1979, 1981, 1982, 1983) has argued that the most important (but often ignored) result to have emerged from the experimental research in this area is the typical smallness or even absence of any effects of anxiety on performance. We have for too long reacted in the same way as Inspector Gregory did in "The Silver Blaze" when he was asked by Sherlock Holmes to consider "the curious incident of the dog in the night-time." The slow-witted Inspector Gregory replied, "The dog did nothing in the night-time," to which Holmes retorted in triumph, "That *was* the curious incident" (Doyle, 1974, p. 33). Holmes could see that it would have meant little if the dog had barked at the intruder, but the dog's silence proved that the intruder was known to the animal. In similar vein, the fact that anxiety often has much smaller effects on performance than might have been anticipated may be of considerable importance.

M. W. Eysenck (1979) attempted to clarify matters theoretically by distinguishing between efficiency and effectiveness. Since the meanings he attached to those terms were the opposite of those common in the literature, they will here be defined in conventional terms. *Performance effectiveness* is simply the quality of performance, and the behavioral measures taken in most studies provide a straightforward assessment of effectiveness. On the other hand, *processing efficiency* refers to the relationship between the effectiveness of performance and the amount of effort or processing resources invested in it. As a first approximation, processing efficiency can be measured by making use of the formula

$$\text{processing efficiency} = \frac{\text{performance effectiveness}}{\text{effort}}$$

with *effort* referring to the allocation of processing resources.

M. W. Eysenck (1979) claimed that the distinction between performance effectiveness and processing efficiency was an important one, in part because the effects of anxiety on effectiveness and efficiency are rather different. More specifically, his basic hypothesis was that anxiety is more likely to have an adverse effect on efficiency than on effectiveness. The most obvious way of testing this hypothesis is to consider those situations in which there are no effects of anxiety on performance effectiveness. The conventional interpretation of this finding (or nonfinding) is that anxiety has had no effects whatever, whereas the hypothesis just proposed leads to the prediction that nonsignificant effects of anxiety on performance are often camouflaging a detrimental effect of anxiety on processing efficiency.

There are various ways of assessing processing efficiency. In interesting research on motor skills by Weinberg and Hunt (1976), continuous physiological recordings were obtained. Electromyography (EMG) was used to provide several measures of muscle activity during a throwing task, including the length of time from initial muscle activity to the major muscle contraction involved in actually throwing the ball (anticipation), the length of time of major muscle contraction (duration), and the length of time for which there was continued contraction of decreasing amplitude following the major muscle action (perseveration). An additional measure of efficiency was based on the pattern of muscle activity. Sequential contraction, in which the antagonists are used before the agonists, is an efficient pattern, whereas simultaneous contraction of the antagonists and agonists (i.e., cocontraction) is not.

In terms of throwing performance, there was no effect of trait anxiety prior to the introduction of feedback. However, there were striking effects of anxiety on muscle activity. The high-anxiety subjects showed much greater perseveration than the low-anxiety subjects and were also much more likely to produce cocontraction rather than sequential contraction. Weinberg and Hunt (1976) provided the following summary of their findings:

> High-anxious subjects anticipated significantly longer with the agonists and shorter with the antagonists than did the low-anxious group. Therefore, they were preparing for the throw in all of the muscles while low-anxious

subjects were preparing mostly with the antagonist muscles. This implies that high-anxious subjects were using more energy than necessary, and expending it over a greater period of time, than were low-anxious subjects. (p. 223)

A similar set of findings with the same throwing task was obtained by Weinberg (1978). He discovered that there was no effect of trait anxiety on throwing performance prior to feedback but that anxiety had highly significant effects on cocontraction, perseveration, and duration. Indeed, it was possible purely on the basis of EMG data to classify correctly 83% of the subjects as high or low in anxiety.

The research of Weinberg and Hunt (1976) and of Weinberg (1978) is important because it illustrates so clearly that there can be a great discrepancy between the effects of anxiety on the effectiveness of performance and on the efficiency of performance. It is possible to demonstrate this, of course, only when the research design provides separate measures of effectiveness and of efficiency, and this has rarely been done.

A rather different approach was adopted by Dornic (1977). He compared the performance of stable extraverts (who are low in trait anxiety) with that of neurotic introverts (who are high in trait anxiety) across versions of a task varying in task load (i.e., the number of information sources) and extra-task load (i.e., amount of extra-task stimulation). As can be seen in Figure 38, there was practically no effect of anxiety on performance effectiveness. However, the two groups did differ markedly in terms of self-reports of perceived effort. There was a significant triple

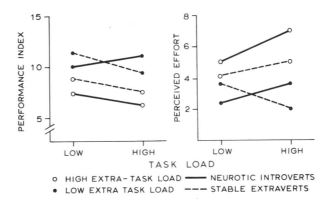

FIGURE 38. Effects of anxiety (stable extraversion versus neurotic introversion) on performance and on perceived effort as a function of task load and extra-task load. (From "Mental Load, Effort, and Individual Differences" by S. Dornic, *Reports from the Department of Psychology*, University of Stockholm, No. 509, 1977. Copyright 1977 by the University of Stockholm. Adapted by permission.)

interaction involving task load, extra-task load, and personality. The neurotic introverts expended more effort than the stable extraverts, especially in the more demanding conditions. Rather similar findings were reported by Dornic (1980) with two different tasks (visual search and counting backwards). If these self-reports of effort expenditure can be accepted as valid, then the implication is that anxiety had adverse effects on processing efficiency that were not apparent in the performance data.

An alternative method of assessing the efficiency of processing is suggested by the theoretical analysis provided by Kahneman (1973). He claimed that spare processing capacity is inversely related to the amount of resources or effort invested in a primary task and that this spare processing capacity can be measured by looking at performance of a secondary task that is performed at the same time as the primary task.

The results from such dual-task studies were reviewed by M. W. Eysenck (1982). The typical finding is that anxiety has no effect on the effectiveness with which the primary task is performed. However, in the 16 experiments in which anxiety failed to alter performance of the main task there was a significantly detrimental effect of anxiety on the performance of the secondary task in 11 cases. In all of the remaining experiments there was no effect of anxiety on secondary task performance.

The implication of these findings is that anxious subjects invested more resources than nonanxious subjects in the primary task and as a consequence had less spare processing capacity for coping with the secondary task. However, many of the studies are open to an alternative interpretation. The usual secondary task was incidental learning, and so it is possible that anxious subjects were simply more likely than nonanxious subjects to decide not to process the apparently irrelevant stimulus information from the secondary task, even though they could have done so if they had chosen to.

Some resolution of this interpretative ambiguity was provided by Hamilton (1978). His primary task was remembering random digits in the correct order, and a secondary reaction-time task was interpolated between presentation of the digit string and its subsequent recall. In contrast to most of the earlier dual-task studies, the subjects were instructed to invest all of their spare processing capacity in the secondary task. High-anxiety subjects had significantly longer reaction times than low-anxiety subjects when the digit-string task was especially demanding, suggesting quite strongly that anxiety reduced spare processing capacity.

A final way of attempting to demonstrate that nonsignificant effects of anxiety on performance are often camouflaging adverse effects of anxiety on processing efficiency involves a *loading* design. Subjects perform a task either on its own or in conjunction with a second task or *load*. This method was first applied to anxiety by Anna Eliatamby (1984). Subjects high and low in anxiety solved anagrams on their own or while counting backwards by threes or while simply repeating overlearned material (i.e., articulatory supression). The results on five-letter anagrams for those subjects who were told that their intellectual ability was being tested on the anagram task are shown in Figure 39. Not surprisingly, subjects who had the additional load of counting backwards took longest to solve the anagrams. However, the key findings relate to the effects of anxiety on anagram-solution times. Anxiety had no effect on anagram performance in the control condition but produced a considerable impairment when counting backwards was required. It appears that anxiety reduced the efficiency of processing on the anagram task but that this inefficiency only became manifest in performance when the total processing demands were increased by combining the anagram task with counting backwards.

In sum, we have discussed four different kinds of research strategy that have all indicated that anxiety can reduce processing efficiency without impairing performance effectiveness. It is noteworthy that these strategies are very diverse, with processing efficiency being asssessed physiologically, by self-reported effort, by dual-task performance, or by observing the effects of loading. Although no single finding may be persuasive, the totality of the evidence is quite convincing. The quintessential conclusion was expressed by M. W. Eysenck (1983) in terms of a metaphor. He compared anxious subjects to a car with a heavy trailer and

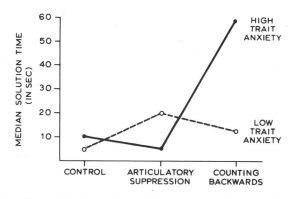

FIGURE 39. The effects of trait anxiety and concurrent task on solution speed for five-letter anagrams (Eliatamby, 1984).

nonanxious subjects to an identical car without the trailer. Effort or the amount of resources invested in the task corresponds to use of the accelerator. The car with the trailer moves less efficiently than the other one, in that it proceeds more slowly with identical use of the accelerator. However, the effects of the trailer (or anxiety) on the car's speed (or task performance) are more conjectural:

> The question as to whether the car pulling the trailer will travel faster or slower than the other car has no definitive answer: while its progress is slowed by the trailer, sufficient use of the accelerator will compensate for this. A crucial point is that even if the two cars travel at the same speed, this does not mean that they are functioning in the same way. Applied to people, anxious subjects may perform as efficiently as non-anxious subjects, but only at greater 'subjective cost' to the system. (M. W. Eysenck, 1983, pp. 286-287)

ANXIETY × TASK INTERACTIONS

One of the most informative approaches to an understanding of the workings of anxiety is to compare the effects of anxiety on different tasks. Although anxiety typically impairs performance, it would be quite untrue to say that anxiety affects every task in the same way or to the same extent. Indeed, a perusal of the literature (M. W. Eysenck, 1982) indicates that there are more than 20 studies in which anxiety was found to interact with task difficulty. What happened in nearly all of these studies was that anxiety had a more adverse effect on difficult or complex tasks than on simple ones. For example, Mayer (1977) compared the effects of trait anxiety on a number of rote or simple problems (e.g., visual search, simple mathematical operations) and various much more complex cognitive tasks such as anagrams and water-jar problems. Anxiety had no effect on performance of the rote problems but reduced the percentage of cognitive problems solved correctly from approximately 80% to just over 40%.

Such findings are very much in line with the Yerkes–Dodson law (Yerkes & Dodson, 1908), according to which the optimal level of arousal varies inversely with task difficulty. However, the notion of *task difficulty* is amorphous, and describing the relationship between anxiety and task difficulty is by no means the same as explaining it. Spence and Spence (1966) argued that the crucial ingredient in task difficulty is response competition. In essence, they suggested that anxiety acts as a drive which increases the probability of the strongest response in any given situation being produced. Thus, anxiety should facilitate performance when the strongest response happens to be the correct one but should impair performance when one or more incorrect or competing responses are stronger than the correct one.

A classic study based on this approach was carried out by Spence, Taylor, and Ketchel (1956). High- and low-anxiety subjects learned a list of noncompetitive paired associates and a list of competitive paired associates. The noncompetitive list consisted of highly associated pairs of adjectives, whereas the competitive list contained stimulus words that were unrelated to their paired responses but strongly associated with the response words in other pairs. Anxiety interacted significantly with list competitiveness, with the high-anxiety subjects learning faster than the low-anxiety subjects on the noncompetitive list but with the opposite happening on the competitive list.

On the face of it, Spence *et al.* (1956) were right to interpret these findings in terms of the way in which response competition mediates the effects of anxiety on learning. However, it is clear that there is a confounding of response competition and list difficulty in this and other studies, because the competitive list was much harder to learn than the noncompetitive list. Saltz and Hoehn (1957) managed to unconfound these two factors and discovered that anxiety interacted with task difficulty rather than with response competition *per se.*

An intriguing interpretation of the differential effects of anxiety on easy and difficult tasks was offered by Weiner and Schneider (1971). They argued that difficult tasks provide more failure experiences and feelings of anxiety than easy tasks because they take longer to perform and lead to a greater incidence of error. Tennyson and Wooley (1971) and Spielberger, O'Neil, and Hansen (1972) found that state anxiety was higher when learning tasks were difficult than when they were easy. What may be happening is simply that high-anxiety individuals do not learn efficiently when they experience failure (e.g., on difficult tasks) but do learn rapidly when they experience success (e.g., on simple tasks). Weiner and Schneider obtained empirical support for this notion by discovering that the effects of anxiety on learning are determined by the kind of feedback presented during acquisition (i.e., success or failure) rather than by task difficulty itself.

Although there is undoubtedly some validity in the notion that the effects of anxiety on task performance are determined in part by the extent to which success and failure are experienced, it seems unlikely that this is the whole story. The theoretical position espoused by Weiner and Schneider (1971) seems to imply that the failure experienced during the performance of complex or difficult tasks prevents high-anxiety subjects from performing adequately. In other words, anxiety should have a very general adverse effect in such circumstances. However, the actual effects of anxiety on the performance of a difficult task are often more complicated than that. M. W. Eysenck (in preparation) used a letter-transformation task in which four had to be added to each of four presented letters (e.g., the answer to $FLUR + 4 = ?$ is $JPYV$). The processing of each letter required three separate stages: (1) access of long-term memory in order to locate the appropriate part of the alphabet, (2) the transformation itself, and (3) the storage and rehearsal of the accumulating answer. High-anxiety subjects took much longer than low-anxiety subjects to arrive at solutions to these problems, but the effects of anxiety were by no means general. When the 12 component times were calculated (i.e., the three separate stages for each of the four letters), it turned out that the detrimental effects of anxiety were limited to the stage of storage and rehearsal and did not include access to long-term memory or transformation time. Furthermore, even these limited adverse effects of anxiety occurred only toward the end of each problem.

What do these findings suggest? The most obvious implication is that anxiety affects some kinds of internal processes more than others. The fact that anxiety did not disrupt either access to long-term memory or the speed of transformation suggests that simple and well-practiced skills are not affected by anxiety. On the other hand, the finding that anxiety slowed down storage and rehearsal toward the end of the problem suggests that anxiety interferes with the transient storage of task-relevant information. It is, of course, toward the end of the problem that short-term storage demands are at their greatest.

There have been several attempts to assess the effects of anxiety on short-term storage capacity, usually based on measures of digit span. The results have been rather inconsistent when trait anxiety has been considered, but a much clearer picture emerges when the effects of state anxiety or situational stress on short-term capacity have been measured. M. W. Eysenck (1979) reviewed the relevant digit-span studies and found in 11 out of the 12 studies reporting significant effects of anxiety that high anxiety or stress reduced short-term storage capacity.

It may well be that a major reason why some tasks are much more difficult than others is simply that they involve greater reliance on the transient storage of information. Thus, a problem in mental arithmetic such as $49 \times 26 = ?$ is a lot harder than one such as $9 \times 6 = ?$ primarily because of the extra storage demands imposed by the former problem. If anxiety impairs the efficiency of short-term storage, and if a crucial aspect of task difficulty is the short-term storage requirement, then we may begin to understand why it is that performance is interactively determined by anxiety and task difficulty.

There have been some recent important theoretical developments with respect to the theoretical construct of the short-term store, and it would be desirable to incorporate them into theories of anxiety. The views of Baddeley and Hitch (1974) and Hitch and Baddeley (1976) are of especial relevance here. They proposed that the construct of the short-term store should be replaced with that of working memory. As they conceptualized it, the working memory system comprised three major components: a modality-free central executive, an articulatory loop, and a visuospatial scratch pad. The central executive has limited capacity and closely resembles an attentional system, whereas the articulatory loop and the visuospatial scratch pad are both slave systems that are used for various specific purposes by the central executive. Verbal information can be rehearsed in a relatively mindless way in the articulatory loop, whereas the visuospatial scratch pad is specialized for spatial information.

Which component of working memory is most affected by anxiety? The question is an important one in view of the involvement of the working memory system across a wide range of tasks, but no definitive answer is available as yet. The initial study in this area was carried out by a student at London University (Anna Eliatamby), who followed up research by Hitch and Baddeley (1976) in which they discovered that some verbal reasoning problems placed greater demands on working memory than others. She used the same reasoning task as Hitch and Baddeley and obtained a significant interaction between anxiety and problem complexity in which anxiety had a much greater adverse effect on the more complex problems involving the greatest demands on working memory.

More detailed evidence was obtained by Eliatamby in a study on anagram solving (see Figure 39). Anxiety had little effect on anagram-solution time when subjects performed a second task at the same time that involved primarily the articulatory loop (i.e., the articulatory suppression task), but it lengthened solution times considerably when the concurrent task made use of the central executive (i.e., counting backwards). The most natural interpretation of these findings is that anxiety impairs the functioning of the central executive component of working memory.

In sum, the fact that anxiety frequently interacts with task difficulty probably has important implications for theory. However, there are several different ways in which one task can be more difficult than another, and there is still no consensus as to which aspect of task difficulty is crucial in producing the interaction with anxiety. The degree of response competition no longer seems of major importance, but the amount of involvement of some transient storage system or working memory remains a strong contender. However, the details of the effects of anxiety on working memory are still not clear.

ATTENTIONAL MECHANISMS

One of the earliest attempts to explain the effects of anxiety on performance in attentional terms was made by Easterbrook (1959). He theorized that states of high anxiety, emotionality, and arousal produced a reduction in the range of cue utilization. In essence, the idea was that anxiety and arousal produce narrowed attention or increased attentional selectivity. Perhaps the strongest evidence to support this theory comes from studies in which there are concurrent main and subsidiary tasks. The modal finding is that anxiety impairs performance on the subsidiary task to a greater extent than on the main task (M. W. Eysenck, 1982), which is certainly consistent with the hypothesis that anxiety increases attentional selectivity.

Easterbrook's hypothesis can readily be extended to accommodate the finding that difficult tasks are more adversely affected by anxiety than are easy ones. This follows from the basic assumption that anxiety reduces the number of cues used, provided that we are willing to assume that difficult tasks involve more cues than easy tasks.

A somewhat different aspect of the effects of anxiety on attentional mechanisms concerns the question of attentional control and susceptibility to distraction. Easterbrook (1959) argued that high anxiety leads to great concentration on some of the task stimuli, and this suggests that anxiety reduces the susceptibility to distraction. In contrast, Wachtel (1967) claimed that anxiety often increases distractibility. He compared attention to a beam of light and argued that it was important to distinguish between the width of the beam and movements of the beam. Wachtel agreed with Easterbrook that anxiety narrows the attentional beam, but he disagreed in suggesting that anxiety causes the beam to roam all over the perceptual field, thus producing an inability to concentrate and great distractibility.

Does anxiety lead to good concentration and resistance to distraction or to poor concentration and increased distractibility? No unequivocal answer has emerged from the relevant empirical research, perhaps because it is usual for only two levels of anxiety to be compared. If several different levels of anxiety were compared, then it is possible that what is happening is that modest levels of anxiety are associated with low distractibility, whereas high anxiety levels produce high distractibility.

It is certainly true that rather low levels of anxiety were apparently involved in those studies in which support was obtained for Easterbrook's (1959) hypothesis. In all of these studies (Bruning, Capage, Kozuh, Young, & Young, 1968; Geen, 1976; Zaffy & Bruning, 1966), acquisition of the to-be-learned stimulus material occurred in the presence of no cues, task-relevant cues, or task-irrelevant cues. High-anxiety subjects appeared to be less affected than low-anxiety subjects by these additional, potentially distracting stimuli, since the beneficial effects of relevant cues on learning and the detrimental effects of irrelevant cues were smaller among those high in anxiety.

A rather different picture emerges from other investigations. Korchin's (1964) clinical observations of patients suffering from anxiety led him to the view that there is a general breakdown of organized behavior at extreme levels of anxiety. He concluded, "The anxious patient is unable to concentrate, hyper-responsive, and hyper-distractible." Supporting evidence of a more experimental nature was obtained by Dornic (1977). He compared performance of a complex counting-backward task that was carried out either in the absence or the presence of interference in the form of visual stimuli (words and pictures) that the subjects were told to ignore completely. Subjects high in anxiety (neurotic introverts) suffered more than those low in anxiety (stable extraverts) from interference.

Even if there is some dispute about the effects of anxiety on susceptibility to distraction by *external* stimuli, it has nearly always been assumed that anxiety increases the susceptibility to distraction from *internal* stimuli. The tendency of high-anxiety subjects to experience worry, self-concern, and so on clearly reflects an inability to maintain concentration on task stimuli. Indirect evidence that anxiety reduces the ability to concentrate on the task in hand was obtained by Deffenbacher (1978), who gave high- and low-anxiety subjects an anagram task under evaluative or nonevaluative conditions. In the condition of greatest anxiety (i.e., high-anxiety subjects exposed to evaluation) subjects reported that they spent only 60% of the time attending to the anagram task, against an average of over 80% in each of the other three conditions. In addition, high-anxiety subjects reported much greater distraction of attention to increased arousal, more task-produced interfering responses, and more worry than low-anxiety subjects.

In sum, there appears to be mileage in the general notion that many of the effects of anxiety on performance are mediated by attentional mechanisms. There is some evidence that anxiety produces a narrowing of attention, and anxiety also seems in many circumstances to reduce attentional control and to increase distractibility. In view of these various effects of anxiety on attention, it is not surprising that anxiety predominantly impairs performance.

LEARNING AND MEMORY

Much of the laboratory research concerned with the effects of anxiety on psychological processes has focused on learning and memory. There are various reasons for this, but perhaps the most important one is that experimental psychology consisted largely of work on learning and memory until comparatively recently. One obvious research strategy is to make use of the dichotomy between short-term and long-term memory stores (e.g., Atkinson & Shiffrin, 1968) and to investigate the effects of anxiety on each store. We have already seen that anxiety usually reduces short-term storage capacity, but the evidence relating to long-term memory has not yet been considered.

A lengthy series of studies on anxiety and long-term memory has been carried out by Mueller and his associates (e.g., Mueller, 1976, 1977, 1978; Mueller, Carlomusto, & Marler, 1977). M. W. Eysenck (1979) reviewed these studies and discovered that trait anxiety produced a significant impairment of long-term memory in 6 out of a total of 11 experimental comparisons, with the effects of anxiety being nonsignificant in the other 5 cases. Free recall was the retention test used in most of the studies, but Mueller *et al.* (1977) used picture recognition memory. Mueller typically used a relatively short retention interval, but similar effects of anxiety on memory have been obtained at much longer retention intervals (e.g., Pagano & Katahn, 1967). If anything, high-anxiety subjects show a greater retention loss than low-anxiety subjects over time (e.g., Ray, Katahn, & Snyder, 1971).

Why does anxiety reduce long-term memory? One possibility is that anxious subjects tend to process the stimulus material less deeply or semantically than nonanxious subjects. Since semantic processing usually enhances long-term memory (Craik & Lockhart, 1972), this could account for the data. The strongest support for this hypothesis was obtained by Schwartz (1975). Paired associates were the learning materials in his first experiment, and the response members of each pair were either phonemically or semantically similar. Since it is known that response similarity slows learning because of the interference it generates, it seems reasonable to assume that semantic response similarity would impair learning only among those subjects processing at a deep or semantic level and that phonemic response similarity would have a detrimental effect on learning only when the stimulus material was processed at a shallow or phonemic level. Neurotic introverts (who are high in anxiety) were less affected by semantic similarity than stable extraverts (who are low in anxiety), and so it can be concluded on the basis of the assumptions that anxiety reduced semantic processing.

The same conclusion is applicable to the second experiment reported by Schwartz (1975), in which there was free recall of a categorized word list presented in a random order. There was a tendency for neurotic introverts to recall the list in a less semantically organized fashion than subjects with lower levels of anxiety.

In spite of these apparent successes for a depth-of-processing explanation of anxiety effects, more recent research has rather altered the picture. Craig, Humphreys, Rocklin, and Revelle (1979) attempted a modified replication of Schwartz's (1975) first experiment, but their results were quite different. Mueller has also explored the depth-of-processing hypothesis by presenting lists of words that can be organized along either shallow or semantic lines. While the depth hypothesis predicts that anxiety should reduce semantic organization more than shallow organization, this pattern of results was not found in any of nine experimental comparisons. Instead, there were comparable effects of anxiety on retention of semantic and shallow word features in eight cases, and the opposite of the predicted pattern was found in the remaining case.

Mueller (1979) argued that the depth hypothesis should be replaced by an elaboration hypothesis. The basic idea is that anxiety leads to reduced processing, with the consequence that fewer stimulus attributes or features are processed. In other words, anxiety reduces the elaboration or extensiveness of processing rather than the depth of processing itself. More direct evidence in favor of an elaboration hypothesis was obtained by M. W. Eysenck and M. C. Eysenck (unpublished). A list of words was presented and followed by cued recall. Thus, for example, memory for the list word "ball" might be tested by means of a cue such as "BAT associated with _____." Of particular relevance here, some of the cues were strongly related to list words whereas others were only weakly related.

The key finding was that there was little difference in recall between neurotic introverts (high in anxiety) and stable extraverts (low in anxiety) when strong recall cues were presented but that high-anxiety subjects had much lower levels of recall than low-anxiety subjects with weak recall cues. On the assumption that elaborate encoding of the list words is necessary in order for recall to weak cues to occur but is not needed for recall to strong cues, these results indicate that anxiety reduces the elaboration of processing.

We have discussed so far the effects of individual differences in anxiety on learning and memory. What about the effects of anxiety-inducing manipulations such as failure feedback and electric shock on learning and memory? Spence and Spence (1966) made the reasonable assumption that the learning of high-anxiety subjects is more affected than that of low-anxiety subjects by both kinds of manipulation. However, an examination of the research literature by Saltz (1970) indicated that matters were actually more complex. It is true that failure or threat of failure disrupts learning to a greater extent for high-anxiety individuals than for low-anxiety individuals, but the effects of shock or the threat of shock are quite different. M. W. Eysenck (1979) summarized the findings from 14 experimental comparisons of the effects of shock on learning. The learning performance of high-anxiety subjects was improved in 9 cases and reduced in 5 cases by shock, whereas shock impaired learning among low-anxiety subjects on 13 occasions. In 13 out of 14 experimental comparisons, the performance of high-anxiety subjects was either less impaired or more improved by shock.

What do these findings mean? According to Saltz (1970), failure or the threat of it produces more anxiety among high-anxiety subjects than among low-anxiety subjects, whereas shock generates greater anxiety in low-anxiety than in high-anxiety subjects. The evidence (e.g., Hodges, 1968, Figure 37) supports Saltz's contention with respect to failure but not to shock. However, there is practically no evidence that high- and low-anxiety subjects differ in their sensitivity to shock. There is thus the rather inexplicable situation that shock has comparable effects on the state anxiety of individuals high and low in trait anxiety but impairs the learning of low-anxiety subjects much more than that of high-anxiety subjects.

It is still not clear exactly what is happening. However, an intriguing lead was provided by Morris and Liebert (1973) in the study discussed earlier in the chapter. They investigated the effects of threat of shock and threat of failure on the worry and emotionality components of state anxiety. The results were very clearcut: Threat of failure only increased worry, and threat of shock only increased emotionality. As a first approximation, the performance of high-anxiety subjects may be especially liable to disruption by worry, whereas that of low-anxiety subjects is more affected by emotionality.

There is a very different way of considering the relationship between anxiety and memory. It has been established (see Eich, 1980) that information can more readily be recalled when the mood or state at the time of retrieval resembles that at the time of learning than when it does not. Thus, for example, happy events are more likely to be recalled when an individual is in a contented mood than when he is unhappy. This phenomenon of recall depending on the congruence (or lack of it) between internal states at learning and retrieval is usually referred to as "state-dependent retrieval."

There have been a few attempts to demonstrate that state-dependent retrieval can be affected by anxiety. Macht, Spear, and Levis (1977) presented a list of nouns for learning in the presence or absence of mild electric shocks and then gave a recall test in the presence or absence of shocks. If we may assume that the shocks increased the level of anxiety, then the results indicated that congruence of state anxiety produced better retention than a lack of congruence (55% versus 38% recall, respectively). Mayo (1983) discovered that people high in neuroticism were better able to recall unhappy events than those low in neuroticism, presumably because neurotic individuals tend to have rather negative emotional states.

It should be noted in passing that the phenomenon of state-dependent retrieval may have considerable theoretical significance. It suggests that part of the misery experienced by chronically anxious individuals is caused by a kind of vicious circle: an anxious mood facilitates retrieval of anxiety-laden memories which in turn enhance anxiety. In addition, the phenomenon of state-dependent retrieval helps to explain why it is that the performance of anxious individuals is often disrupted by a variety of negative thoughts (e.g., worry and self-criticism).

In sum, although it has been customary to emphasize only one or two of the effects of anxiety on learning and memory, there are actually several effects that require interpretation. The circumstances in which trait anxiety is most likely to impair learning and memory are those involving a difficult learning task and negative evaluation in the form of failure feedback; conversely, trait anxiety is least likely to reduce learning and memory when the learning task is simple and neutral or success feedback is provided.

The differential effect of failure feedback on high-anxiety and low-anxiety subjects presumably stems from the fact that failure increases state anxiety (and especially the worry component) much more among those high in trait anxiety. The fact that anxiety has a greater adverse effect on difficult than on easy learning tasks can be accounted for it we assume that the working memory system is more heavily involved in difficult learning tasks than in easy ones and that anxiety impairs the efficiency of the working memory system.

The same explanation may also be applicable to the findings relating to the short-term and long-term storage systems. If, for example, anxiety reduces the efficiency of the central executive component of working memory, then it would be expected that short-term and long-term memory would both be adversely affected by anxiety. This is, of course, the standard result. Why does anxiety reduce the efficiency of the working memory system? In spite of the absence of any direct evidence, the most plausible explanation is that state-dependent retrieval causes the anxious subject to retrieve anxiety-related memories that are irrelevant to the task in hand but which preempt some of the resources of working memory.

CONCLUSIONS

The conclusions that we will draw are based on laboratory research which, for ethical or other reasons, has involved relatively modest levels of anxiety. At extreme levels of anxiety, complete disorganization of performance is often found. For example, more than 200 of the muzzle-loading rifles used in the battle of Gettysburg during the American Civil War were found to have been loaded five or more times without being fired (Walker & Burkhardt, 1965). In similar fashion, Patrick (1934a,b) gave subjects the task of discovering which of four doors was unlocked. The same door was never unlocked on two successive trials, so that the optimal strategy was to ignore the door that had been unlocked on the previous trial and to try each of the three remaining doors once. Under normal conditions, approximately 60% of the solutions were optimal and only 2% involved attempting to open the same door several times. However, when the subjects had cold water streams directed at them, or had their ears blasted by a car horn, or were given continuous electric shock until they located the unlocked door, only 20% of the solutions were optimal and 45% involved repeated attempts to open the same door.

It seems probable that the anxiety experienced in everyday life resembles the less dramatic kind usually investigated in the laboratory. The general opinion appears to be that the adverse effects of such anxiety on task performance are due to the cognitive or worry component of anxiety rather than to its physiological or emotionality component. Striking support for this viewpoint was obtained by Weiner and Samuel (1975). They told their subjects that the tests that they would perform were measures of intelligence. The subjects were given a placebo pill and told either that it would produce small increases in the strength of their heartbeat, some moistness of the palms, and a very slight feeling of queasiness in the stomach or that it would lead to symptoms characteristic of a state of calm. High-test-anxiety subjects did much worse than low-test-anxiety subjects on an anagram task when they were led to believe that the pill would make them calm, but when the subjects could attribute their physiological unease to the pill there was no effect of test anxiety on anagram performance. This suggests that the cognitive interpretation placed on an individual's physiological state importantly determines the effectiveness of performance.

Accurate predictions of the effects of anxiety on performance necessitate some consideration of the pattern of anxiety effects on component processes. It seems reasonably well established that anxiety reduces the efficiency of short-term storage and long-term memory, that it increases attentional selectivity, that it decreases accuracy without affecting performance speed, and that it increases distractibility. In view of these various effects, anxiety is unlikely to impair performance on tasks when the primary requirement is speed but will usually have a disruptive effect when the task imposes great demands on short-term storage capacity.

It might be argued that this is simply reexpressing the old notion that anxiety disrupts performance of difficult tasks to a greater extent than that of easy tasks, but that would not be true. Task difficulty has usually been conceptualized as a unidimensional construct, whereas the emphasis here is on the idea that tasks must be thought of as varying along several different dimensions. It is unfortunate that little systematic attention has been paid to the issue of identifying the major component processes involved in task performance, since as a consequence we may well have ignored some of them. Future research should be concerned both with the task of discovering the main internal processing mechanisms and with that of establishing the effects of anxiety on each one.

Of course, as we have argued forcefully in an earlier section of this chapter, a misleading impression of the effects of anxiety on internal processes can be created if we focus exclusively on the effectiveness of performance. Anxiety may have no apparent effect whatever on the quality of performance, and yet other evidence may indicate strongly that anxiety has reduced processing efficiency. If we accept (as we must) that it is a hazardous procedure to infer the characteristics of internal processes on the basis of external, behavioral evidence, then it becomes clear that we should strive to obtain independent evidence (e.g., physiological, self-report) concerning processing efficiency. When this has been done, it has usually been discovered that the effects of anxiety on processing efficiency and on performance effectiveness are by no means comparable. It is to be hoped that the precise reasons for these discrepancies will soon be discovered.

CHAPTER ELEVEN

Social Behavior

If extraversion, neuroticism, and psychoticism are the three major dimensions of personality, then it seems entirely reasonable to assume that individual differences along these dimensions should be important in everyday life. The relationships between personality and real-life behavior can be explored in two separate but related ways: (1) we can use our theoretical understanding of personality to predict the impact of various personality types on social behavior; or (2) we can proceed in a less theoretical way to assess the relevance of personality to individual differences in behavior in social, educational, occupational, and other settings. In practice, of course, researchers typically attempt to combine both approaches, proceeding from personality to behavior and from behavior to personality.

One of the potential advantages of this kind of extension of personality research is that it can provide a measure of validation for the personality dimensions postulated by H. J. Eysenck (1957, 1967a). In addition, of course, it can help to allay fears that the theoretical formulations derived from laboratory research are relevant only to that setting. On the other hand, there are some fairly obvious difficulties associated with such real-life research. Even if personality is one of the determinants of, for example, psychiatric disorder or criminality, it is clear that there are numerous other factors (e.g., parental influences, peer-group pressures, traumatic life events) that are also relevant and important. It is thus a relatively difficult achievement to demonstrate clear relationships between personality variables and major real-life activities.

Within the original Eysenckian two-dimensional space formed by the extraversion and neuroticism dimensions, it was natural to wonder about the social impact of individuals having extreme scores on these two personality dimensions. It would be entirely appropriate to consider

311

whether there are any interesting peculiarities about the real-life behavior of stable introverts and stable extraverts, but little or no work has been addressed to this issue. In contrast, there has been considerable interest in the potential consequences of neurotic introversion and neurotic extraversion. According to H. J. Eysenck (1957), neurotic introverts were the people most likely to suffer from dysthymia, a term which covers phobias, obsessional-compulsive rituals, anxiety states, and neurotic depression. Neurotic extraverts, on the other hand, were allegedly the people most susceptible to hysteria, that is, the classic paralytic or anaesthetic symptoms of conversion hysteria.

This ambitious attempt to relate personality theory to psychiatric classification was only a partial successs. Dysthymics do indeed tend to be neurotic introverts, but hysterics are usually neurotic ambiverts rather than neurotic extraverts (e.g., Sigal, Star, & Franks, 1958). In the revised version of the theory (H. J. Eysenck, 1967a), the link between neurotic introversion and dysthymia was retained, but the real-life manifestation of neurotic extraversion became criminality or psychopathy (i.e., antisocial and illegal activity). The evidence relating to this theoretical formulation will be discussed later in the chapter.

The whole issue of attempting to make theoretical predictions about the real-life consequences of different kinds of personality became much more complex when a third personality dimension (psychoticism) entered the picture. We now have eight different kinds of extreme personality, ranging from individuals high on psychoticism, neuroticism, and extraversion to individuals low on all of these three dimensions. We also have to consider the possibility that important aspects of social behavior may be affected by one personality dimension on its own, or by some combination of two or even all three personality dimensions. The progress that has been made in elucidating these matters is the subject matter of the remainder of this chapter.

SOCIAL INTERACTION

Many, if not most, social psychologists believe that the ways in which people behave in any particular social situation are primarily determined by the social norms and conventions appropriate to that situation. As Barker and Wright (1955) pithily remarked, "When we are in church we behave church, when we are in school we behave school." Within such a theoretical framework, individual differences in personality are regarded as making no more than a modest contribution to the determination of behavior. The issue of the relative importance of the situation and of individual personality in social and other situations has generated fierce controversy (e.g., Bowers, 1973; M. W. Eysenck & H. J. Eysenck, 1980; Mischel, 1977), and we do not want to enter into an extended discussion of it at this point. However, it is worth considering the factual evidence. Sarason *et al.* (1975) reviewed 138 analyses of variance reported in the social research literature in which both situational and personality factors were included. The situation accounted for 10% of the variance on average, personality accounted for 9% of the variance, and the interaction between personality and situation for nearly 5% of the variance. Thus it would be completely inappropriate to regard situational determinants of social behavior as of far greater consequence than personality factors.

Furthermore, the distinction between personality and situation is in some ways misleading, especially when applied to social behavior. The reason is that the social situations in which individuals find themselves are determined, at least in part, by their personality. Which aspect of personality should we consider first? Since one of the major components of extraversion is sociability, there are good reasons for focusing on the differences in social behavior between introverts and extraverts. Furnham (1981) investigated the effects of extraversion on situational selection and preference in a study with female students who completed the Eysenck Personality Questionnaire, the Leisure Scale, the Free-Time Activity Scale, the Social Situation Scale, and the Stressful Situation Scale. In essence, extraverts reported taking part in social interaction and physical pursuits to a greater extent than introverts, and they also indicated a greater preference for stimulating, active, and unusual situations and less tendency to avoid stressful situations. The implication of these findings is that a person's life-style or pattern of social interaction reflects his or her personality.

How can these findings be interpreted? Social situations often produce powerful arousing effects, and introverts tend to be more aroused than extraverts. As a consequence, extraverts may seek out personal contacts in order to prevent the level of arousal from becoming too low, whereas introverts may tend to avoid such contacts because of the danger of overarousal. Such an arousal model is clearly consistent with the data obtained by Furnham (1981), but it would be desirable to have more direct evidence. For example, alcohol reduces the level of arousal and generally seems to increase sociability; if alcohol were found to increase the liking for stimulating, active, and unusual social situations, then we could be more confident that arousal is playing an important part.

Much of the early evidence concerning the effects of social contact on arousal was reviewed by Zajonc (1965). He concluded that an individual's level of arousal was typically increased by social contact. Increased arousal was also the norm even when other people were merely present as observers while the subject performed a task.

Arousal theory has also been applied to interpersonal intimacy (Patterson, 1976). The basic contention is that various kinds of nonverbal behavior in social settings (e.g., gazing at the other person, moving close to the other person) increase the level of arousal. The increased arousal produced by enhanced interpersonal intimacy may be experienced negatively as anxiety or discomfort or positively as liking or love, depending on the setting and other factors. Positive emotional states lead to still greater interpersonal intimacy, whereas negative emotional states precipitate compensatory behavior designed to produce a more comfortable level of interpersonal intimacy. Somewhat surprisingly, Patterson assumed that the intensity of arousal produced has no effect on whether a positive or a negative emotional state is experienced.

If we assume that interpersonal intimacy produces arousal (an assumption for which there is some evidence—see Patterson, 1976), and that the highly aroused introvert is more likely than the less aroused extravert to interpret this arousal in negative terms, then it follows that introverts will show a greater tendency than extraverts to reduce interpersonal intimacy. Contrariwise, underaroused extraverts will be more likely than introverts to increase interpersonal intimacy. The evidence is broadly consistent with these predictions. Patterson and Holmes (1966) reported that extraverts approached an interviewer more closely and talked longer while answering questions than did introverts. Campbell and Rushton (1978) videotaped female occupational therapy students while they were discussing their holiday plans with a female experimenter, and found that the extraverts talked much more than the introverts. Carment, Miles, and Cervin (1965) asked pairs of subjects to discuss a topic on which they initially held opposite opinions. When one subject was introverted and the other was extraverted, the extravert spoke first in 28 out of 33 pairs and spoke more in 23 out of 33 cases.

With respect to interpersonal distance, Leipold (1963) and Williams (1963) both reported a slight tendency for extraverts to come closer than introverts. However, the effect was only statistically significant for very close contact such as the preference for dancing cheek to cheek (a very dated notion nowadays) and the minimum distance chosen for comfortable conversation.

A few studies have dealt with the effects of extraversion on gazing behavior. Mobbs (1968) stared continuously at his subjects while engaging them in conversation. The extraverts were slightly more likely than introverts to gaze back, and their glances were on average almost twice as long (3.1 seconds versus 1.7 seconds). Kendon and Cook (1969) also found that extraversion correlated modestly with the amount of gazing.

While all of these findings are in line with arousal theory, it has to be pointed out that they are equally consistent with other theoretical notions. For example, it would presumably be expected that sociable or extraverted individuals would display greater interpersonal intimacy than unsociable or introverted individuals, whether or not the former individuals were less aroused than the latter. More convincing evidence that arousal is directly implicated in social behavior was obtained by Laverty (1958), who administered the depressant drug Sodium Amytal to normal and neurotic groups. The drug produced an increase in spontaneous speech during a card-sorting task, especially for the neurotic subjects. More striking evidence of the effects of Amytal on sociability was reported in an informal way by Laverty (1958):

> A second dysthymic subject went to a dance on the evening after injection of amytal and surprised those who knew him by his sociable attitude; formerly in hospital he had been persistently shy. A third dysthymic patient went out with her friends the day after injection of amytal; one reported, "she was brighter than she ever was before, she read our fortunes out of cups—I never knew her to do that before." (p. 52)

The effects of neuroticism on social behavior seem less clear. Since neurotic people possess a labile and overactive autonomic nervous system, they are susceptible to fear and anxiety. It might thus be assumed that they would attempt to avoid stressful kinds of social situations, but Furnham (1981) was unable to obtain any support for this prediction. However, he did find that neurotic individuals avoided stimulating, active, and unusual situations more than stable individuals.

There has been some research concerned with the relationship between neuroticism and measures of interpersonal intimacy. It would seem likely that interpersonal intimacy would tend to be experienced negatively by those high in neuroticism, and this might then lead to attempts to reduce the level of intimacy (cf. Patterson, 1976). This expectation was confirmed by Campbell and Rushton (1978) in a study to which reference has already been made. The behavioral measure that was most consistently related to neuroticism was gaze aversion, that is, turning the eyes away from contact with the experimenter. It is of some interest that similar findings have also been reported by Kendon and Cook (1969), Rutter and Stephenson (1972), and Williams (1974). A more detailed analysis of the effects of neuroticism or anxiety on gazing was offered by Daly (1978). She assessed neuroticism or anxiety on the basis of the Watson-Friend Scale of Social Anxiety and discovered that highly anxious people looked less than others while they were speaking to the experimenter. However, when they were listening there was a tendency for the anxious subjects to demonstrate an extreme pattern of gazing behavior, either showing long-term fixation or glancing away very quickly.

There has been some theoretical controversy about the usual effects of neuroticism or anxiety on interpersonal intimacy. Schachter (1959) argued that since isolation increases anxiety there would be an increase in affiliative behavior among those experiencing anxiety. In support of this contention, he found that female students waiting to participate in an experiment involving electric shock preferred the company of others to remaining alone. On the other hand, Teichman (1973) argued that an individual's attention is directed toward himself when he is personally threatened. This in turn arouses a tendency toward isolation and thus the need for affiliation diminishes.

Some clarification was achieved by Brady and Walker (1978). They measured how far away from the experimenter each subject placed his chair prior to a two-person conversation. Anxiety was aroused in some of the subjects by telling them that their social competence would be assessed by someone behind a one-way screen. Brady and Walker discovered that anxiety led to a significant increase in interpersonal distance. It may well be that anxiety in the form of a threat to self-esteem leads to reduced interpersonal intimacy, whereas the threat of an external, fearful situation (as used by Schacter, 1959) has the opposite effect.

A rather underresearched aspect of social interaction is shyness and its personality correlates. According to H. J. Eysenck and S. B. G. Eysenck (1969), there are two conceptually distinct forms of shyness which they referred to as introverted social shyness and neurotic social shyness. The basic idea is that the introvert appears to be shy because he prefers to be alone even though he does have the ability to function effectively in company, whereas the neurotic individual may desire the company of others but is rather fearful of it at the same time. It is certainly true that shyness appears to involve some features of introversion (keeping in the background, preferring one's own company) and of neuroticism (feelings of inadequacy and worry, emotional arousal), and it is also true that shyness typically correlates negatively with extraversion and positively with neuroticism (Crozier, 1979).

Why is it the neurotic and introverted individuals who are most susceptible to shyness? A plausible answer presents itself if we consider those situations that are usually reported as being most likely to elicit shyness. In general terms, such situations tend to be those that make great demands on competence (meeting new people) or those that may involve criticism or disapproval (interacting with authority figures or being the focus of attention). In other words, shyness is most likely to be experienced in situations that produce heightened arousal and/or stress and anxiety, and it is introverted people who are most likely to become overaroused and neurotic people who are most susceptible to stress.

SEXUAL BEHAVIOR

At least since the time of Freud, it has been assumed that an individual's sexual behavior is a reflection of his or her personality and psychological well-being. It is a somewhat speculative matter to predict the precise effects of extraversion, neuroticism, and psychoticism on sexual attitudes and behavior, but certain plausible assumptions can be made. Extraverts are characteristically less aroused than introverts and so might be more likely to seek social and physical stimulation in order to attain the optimal level of arousal.

The same prediction can readily be made on the basis of Gray's (1970, 1973) theoretical assumption that extraverts are more susceptible than introverts to reward, whereas introverts are more susceptible than extraverts to punishment. Sexual advances can be seen as involving an approach–avoidance conflict in which there are both potential rewards (e.g., sexual satisfaction) and potential punishments (e.g., rejection). If extraverts tend to focus on the potential rewards and introverts on the potential punishments, then it makes sense that extraverts should engage in more sexual activity than introverts.

Neurotic individuals are more susceptible than stable individuals to fear and anxiety, and this may reduce the availability of direct sexual outlets. However, since there is no reason to assume that sexual drive is much influenced by neuroticism, it seems likely that neurotic individuals might seek substitute sexual outlets such as pornography, prostitution, and masturbation.

It is perhaps most difficult to make specific predictions with respect to the effects of psychoticism on sexual behavior. However, it is known that high scorers on psychoticism typically have a lack of personal involvement and human feeling. As a result, it seems likely that high levels of psychoticism would be associated with abnormal sexual activities.

Considerable evidence relating to these and other predictions has been obtained by Giese and Schmidt (1968) and by H. J. Eysenck (1976b). In the former study, approximately 6,000 students completed questionnaires dealing with sexual behavior, as well as a personality inventory measuring extraversion and neuroticism. Sexual behavior was affected considerably more by extraversion than by neuroticism. Fifteen percent of introverted male students experienced sexual intercourse by the age of 19, against 45% of extraverted male students. The respective figures for female students were 12% and 29%. In addition, extraverted students who were sexually experienced were more adventurous than introverted students. Of the extraverted male nonvirgins, 25% had had intercourse with four or more different partners during the previous 12 months, against only 7% of the introverted male nonvirgin students. For female students, the figures were 17% and 4%. Extraverted males indulged in longer precoital sex play than introverted males and were more likely to have used more than three different coital positions. These trends were not confirmed among the female students, presumably because men tend to determine sexual practices. Finally, cunnilingus and fellatio were more common among extraverted than introverted students.

Giese and Schmidt (1968) also discovered a few significant effects of neuroticism on reported sexual activity. Male students who were high on neuroticism masturbated more often, expressed greater desire for sexual intercourse, and claimed to have spontaneous erections more often than those low on neuroticism. Females with high neuroticism scores experi-

TABLE 16. Fourteen Primary Factors in Sexual Attitudes

Factor	Sample item
Satisfaction	I have not been deprived sexually.
Excitement	It doesn't take much to get me excited sexually.
Experimentation	Young people should be allowed out at night without being too closely checked.
Curiosity	I would agree to see a "blue" film.
Premarital sex	One should not experiment with sex before marriage.
Promiscuity	I have been involved in more than one sex affair at the same time.
Homosexuality	I understand homosexuals.
Hostility	I have felt hostile to my sex partner.
Prudishness	I don't enjoy petting.
Censorship	Prostitution should not be legally permitted.
Repression	I think only rarely about sex.
Inhibition	My parents' influence has inhibited me sexually.
Nervousness	I feel nervous with the opposite sex.
Guilt	My conscience bothers me too much.

Note. From *Sex and Personality* by H. J. Eysenck, London: Open Books, 1976. Copyright 1976 by Open Books Publishing Ltd. Reprinted by permission.

TABLE 17. Sexual Attitudes and Personality[a]

Factor	E	N	P
Satisfaction	+	– – –	–
Excitement	+	+ +	+
Nervousness	– – –	+ +	0
Curiosity	0	+	+ +
Premarital sex	+	0	+ +
Repression	0	0	–
Prudishness	–	+	+
Experimentation	+	0	0
Homosexuality	0	+	+
Censorship	–	0	–
Promiscuity	+ +	0	+ + +
Hostility	0	+ + +	+ + +
Guilt	0	+ + +	0
Inhibition	0	+ + +	+

Superfactor			
Sexual pathology (versus satisfaction)	–	+ + +	+ +
Libido	+	+	+ + +

Note. From Sex and Personality by H. J. Eysenck, London: Open Books, 1976. Copyright 1976 by Open Books Publishing Ltd. Reprinted by permission.
[a] + = positive relationship; – = negative relationship; 0 = no relationship; the number of signs indicates the strength of the relationship.

enced orgasm during sexual intercourse less frequently than those with low scores.

Most of these findings were replicated by H. J. Eysenck (1976b), who studied the interrelationships among personality, sexual behavior, and sexual attitudes. He included the psychoticism dimension in his research and found that high psychoticism scorers were distinguished from low psychoticism scorers mainly by a liking for oral sex. They tended to be very experienced and reported that they had indulged in a wide variety of sexual practices.

H. J. Eysenck (1976b) also reported some interesting connections between personality and attitudes toward sex. A factor analysis of questionnaire items relating to sexual attitudes produced a total of 14 primary factors. These factors are listed in Table 16, and items representative of each factor are also included. The strength of the correlations between these attitude factors and the three personality dimensions of psychoticism, neuroticism, and extraversion are shown in Table 17.

Extraverts were high on promiscuity and low on nervousness and prudishness; high scorers on neuroticism were high on excitement, nervousness, hostility, guilt, and inhibition, and low on satisfaction; and high psychoticism scorers were high on curiosity, premarital sex, promiscuity, and hostility. In other words, there were marked differences in sexual attitudes as a function of personality.

These 14 primary factors were then reduced to two broader factors, which were called *libido* and *satisfaction*. Libido appears to reflect a general sex drive and includes different aspects of active sexuality and permissiveness, whereas satisfaction refers to sexual difficulties and deprivation. The factor of libido was most strongly associated with high psychoticism, but it was also modestly related to high neuroticism and extraversion. The factor of satisfaction showed a strong negative correlation with neuroticism, a moderate negative correlation with psychoticism, and a small positive correlation with extraversion.

In sum, there appear to be more powerful effects of personality on sexual attitudes and behavior than on most of the other real world activities that have been investigated. It is not entirely clear why this should be so, but a few speculations may be in order. We are dealing with an area of life in which strong emotional states, high physiological arousal, and powerful rewards and punishments are involved. Since all of these factors are theoretically of great relevance to the personality dimensions of extraversion, neuroticism, and psychoticism, it should perhaps come as no surprise to discover the close links that exist between personality and sexual activity. In addition, an individual's sexual life is usually of central importance to him or her, and we might expect personality to have more impact on the major aspects of life than on more trivial matters.

EDUCATIONAL ACHIEVEMENT

The most obvious issue about educational achievement from the perspective of personality theory concerns the identification of those personality characteristics conducive to academic success. As we will see shortly, this issue has attracted a fair amount of research interest. However, it would be rather simple-minded to expect it to be a straightforward issue to resolve, since the relevance of particular personality characteristics for academic success is likely to be a function of the subject studied, the teaching methods used, the age of the learner, and so on.

In spite of these complexities, a few reasonably robust generalizations have emerged from research in which academic attainment has been correlated with personality scores. For example, it is generally true at all ages from about 13 or 14 upwards that introverts show superior academic attainment to extraverts. This has been demonstrated several times with British university students (e.g., Furneaux, 1957; Lynn, 1959), and it has also been found in the very different cultural setting of the Cape Coast in Ghana (Kline, 1966).

Anthony (1977) was interested in the shift between a positive relationship between extraversion and attainment in younger children and a negative relationship between those two variables in older children and adults. One possibility is that the more able children become more introverted over time, whereas the less able children become more extraverted. Another possibility is that the extraverted children tend to fall behind as academic skills develop while the introverted children make more rapid progress. Anthony reanalyzed some data collected by Rushton (1969) in his longitudinal study of 266 children of above-average intelligence who were tested at 10–11 years and then retested at 15–16 years. In essence, it turned out that both of the possibilities explored by Anthony were partially correct.

One of the basic problems with attempting to interpret the superior academic achievement of introverts to extraverts in adolescence and adulthood is that the evidence is correlational in nature. As Anthony's (1977) findings suggest, it may be that, over and above any direct effect of introversion on academic achievement, there are also reciprocal effects of the efforts required for superior academic achievement on introversion. Even if introversion has a causal effect on academic attainment, it is still rather obscure how this actually happens. It may be that the low arousal level of extraverts makes it difficult for them to maintain concentration over the long periods of time required for successful academic study (cf. Chapter 9). Campbell and Hawley (1982) carried out a study among students in a university library. Extraverts reported taking more study breaks than introverts, and they were more concerned than introverts to select a study location that offered socializing opportunities.

Readers of an ingenious turn of mind should have little difficulty in generating alternative explanations. It may simply be that extraverts have a greater variety of interesting social engagements competing for their time, so that they spend less time studying than introverts. If we can assume that the greater sociability of extraverts only begins to disrupt their study habits from early adolescence onwards, this would account for the finding that introversion is not associated with superior academic attainment until the age of 13 or so. Goh and Moore (1978) did not discover any significant differences between introverts and extraverts in the number of hours spent studying, but Banks and Finlayson (1973) found that introverted boys had a greater commitment to homework than extraverted boys.

A theoretically interesting hypothesis is suggested by research carried out by Leigh and his colleagues. In a typical study (Leigh & Wisdom, 1970), subjects had to learn how to solve odd-man-out problems involving figures and spatial patterns. Introverts learned better when they followed a carefully sequenced, highly prompted learning structure, whereas extraverts were more successful when presented with a random arrangement. Other researchers (e.g., Shadbolt, 1978) have also discovered that introverts learn better in structured learning environments, whereas extraverts learn better in relatively unstructured learning environments. At least until quite recently, educational provision has tended to approximate more closely to a structured than an unstructured system. As a consequence, the academic superiority of introverts may be due in large measure to the fact that the educational system is more closely geared to their needs than to those of extraverts.

Of the remaining dimensions of personality, there has been very little research relating academic attainment to psychoticism. One of the few exceptions was a study of Goh and Moore (1978), in which they looked at attainment in the form of grade point average in three different educational settings (a university, a vocational technical institute, and a high school). Psychoticism correlated negatively with grade point average for students at the vocational institute and also for university students taking social studies. The uncaring and hostile nature of those high in psychoticism makes the direction of the relationship between psychoticism and academic attainment seem reasonable.

The relationship between academic attainment and neuroticism or anxiety appears to be rather conjectural. Neuroticism is sometimes positively related to attainment, sometimes negatively related, and sometimes there is a curvilinear relationship between neuroticism and attainment, with intermediate levels of neuroticism being associated with the greatest academic success. An attempt to introduce some order into an apparently chaotic situation was made by H. J. Eysenck (1971a). He pointed out that neuroticism tends to be negatively correlated with attainment in groups that have not been subjected to a selection process (e.g., school children). Adverse effects of neuroticism on academic performance in children were noted by Child (1964) on pupils within the 11–15 age range and by Entwistle and Cunningham (1968) on children aged about 13 years. In contrast, neuroticism tends to be positively correlated with attainment in groups that have survived a difficult selection process (e.g., university students). Lynn (1959) found that university students were significantly more neurotic than control subjects, and Furneaux (1957) reported that students who do well at university score higher on neuroticism than those who are less successful.

What is happening here? The evidence discussed in Chapter 10 suggested that neuroticism or anxiety has two major effects: (1) it increases motivation, and (2) it disrupts performance by producing worry and other task-irrelevant thoughts. Perhaps educational selection processes tend to reject those neurotic students who are especially inclined to worry, leaving those for whom anxiety has an energizing or motivating effect.

An alternative possibility was suggested by Spielberger (1966). He pointed out that there is much evidence to indicate that anxiety facilitates performance on easy tasks but impairs it on difficult tasks. Whether or not any particular task is easy or difficult depends upon the expertise and intelligence of the person performing it. Thus, anxiety may be more likely to enhance attainment in highly selected groups because such groups consist of very intelligent people for whom most academic tasks are relatively easy.

Spielberger (1966) tested the hypothesis that the effects of anxiety on academic performance at university level are influenced by the level of intelligence. He compared the academic attainment (based on grade point average) of high and low scorers on the Manifest Anxiety Scale as a function of their performance on the ACE Psychological Examination, which is a measure of scholastic aptitude usually regarded as a valid index of intelligence. Five levels of scholastic aptitude or intelligence were identified. As can be seen in Figure 40, academic attainment was mostly lower in high-anxiety subjects than in low-anxiety subjects. However, there was little effect of anxiety on attainment among the least able students, probably because of a "floor" effect. Of greatest interest, there was also rather little effect of anxiety on grade point average among the most able students. Indeed, if we consider only those students in the brightest group whose ACE scores were above the median for that group, then the high-anxiety students actually had a better grade point average than the low-anxiety students.

It is likely that the impact of neuroticism or anxiety on academic achievement depends in various subtle ways on the teaching methods that are used. For example, Trown and Leith (1975) exposed children of approximately 10 years of age to a learning task that was taught either by an inductive, learner-centered, exploratory teaching strategy or by a more deductive, teacher-centered, supportive strategy. The two teaching strategies were of equal overall effectiveness, but the key finding was that the relative effectiveness of the two strategies was affected by the

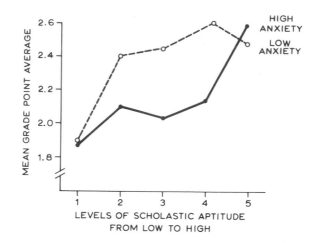

FIGURE 40. Academic achievement (grade point average) as a function of anxiety and intelligence. (From "The Effects of Anxiety on Complex Learning and Academic Achievement" by C. D. Spielberger. In C. D. Spielberger (Ed.), *Anxiety: Current Trends in Theory and Research*, Vol. 1, New York: Academic Press, 1966. Copyright 1966 by Academic Press, Inc. Adapted by permission.)

children's anxiety level. More specifically, children high in anxiety learned better with the supportive teaching strategy than the exploratory strategy, whereas the opposite was true of children of low anxiety. It may be that individuals high in anxiety or neuroticism learn most effectively when they are taught in ways that minimize situational stress.

Gray's (1973) theory can be used as the basis for predicting interactions between teaching strategy and personality. In general terms, the expectation is that extraverts should show superior learning to introverts when teachers emphasize rewards and praise, whereas introverts should outperform extraverts when teachers rely predominantly on threats of punishment. McCord and Wakefield (1981) tested these hypotheses among nine- and ten-year-olds in five different classes varying in the ratio of teacher-presented reward to teacher-presented punishment. There was a highly significant interaction between this ratio and the degree of extraversion of the children for arithmetic achievement, and the interaction was of the expected form. However, rather different findings might have been obtained if, for example, strong punishments had been administered.

In sum, there are two main research strategies that have been used in this area. One approach is simply to correlate personality scores with measures of academic attainment. This can provide an overview of what is happening, but it is usually extremely difficult to decide exactly how personality is having its effect. The alternative approach is more experimental in nature and involves comparing the effects of different teaching strategies on learners varying in personality. This approach has not been used extensively but appears to offer much greater promise of discovering the teaching methods that are optimal for each individual.

OCCUPATIONAL PERFORMANCE

It seems probable that individual differences in personality are of relevance in accounting for both occupational choice and job performance. With respect to occupational choice, it should be natural for extraverts to prefer jobs that involve dealing with other people, whereas introverts might well have a preference for jobs that do not require numerous social contacts. Some support for these ideas was obtained by Bendig (1963), who made use of the Strong Vocational Interest Blank. He discovered that introverts preferred scientific and theoretical jobs such as journalism, architecture, and the teaching of mathematics, whereas extraverts expressed more interest in occupations involving more social contact (e.g., selling life insurance and social work).

Further evidence along similar lines was obtained by Wankowski (1973), who investigated a random sample of students at Birmingham University. Extraverted students tended to choose practical or people-oriented courses, whereas introverted students preferred more theoretical subjects. When he looked at examination success, Wankowski discovered that introverts had greater examination success than extraverts in the physical sciences. Neuroticism was also implicated, since low neuroticism scorers opted for practically biased courses, whereas high neuroticism scorers preferred people-oriented courses. In terms of examination success, low neuroticism was associated with success in the applied sciences.

A rather more complex picture emerged in a study by Rim (1977), in which several job applicants completed the Eysenck Personality Inventory and rated statements according to how well they described their ideal job. Among the male subjects, the neurotic extraverts had the most distinctive ratings, valuing the social contact, economic, social position, patterning of time, and power functions of work more than neurotic introverts, stable extraverts, or stable introverts. There were only modest and uninterpretable effects of personality on the description of the ideal job among female subjects.

Since neurotic individuals in general and neurotic introverts in particular are especially susceptible to stress, it might be thought that such people would prefer jobs that involve minimal stress. However, there is scant evidence in favor of this hypothesis. Rim (1977) did not find any large differences in the ideal job as a function of either neuroticism or neurotic introversion, and Bendig (1963) reported only that high neuroticism was associated with a dislike of business-type occupations such as banking, office management, and accountancy.

We have already discussed the notion that extraverts are more likely than introverts to prefer occupations that involve plenty of social contact. In view of the greater arousal level of introverts than extraverts, there is the danger that introverted workers may become overaroused if their jobs involve considerable extraorganizational contact and a relative absence of routine. Blunt (1978) argued that introverted managers would thus tend to choose positions involving relatively routine duties (finance, production, or technical managers), whereas extraverted managers would be more likely to select jobs in sales, marketing, or transport. The results were broadly as hypothesized, except that transport managers were less extraverted and production managers more extraverted than predicted.

The notion that introverted workers are better able than extraverted ones to handle routine work activities was also investigated by Cooper and Payne (1967) in a study carried out in the packing department of a tobacco factory where the work was repetitive and light. Job adjustment as assessed by two supervisors was negatively related to extraversion, and those workers who left the job in the 12 months following testing were significantly more extraverted than those who remained. Neuroticism was also implicated, being related to poor job adjustment and to frequency of nonpermitted absence. It is unfortunate that Cooper and Payne did not extend their study to include other kinds of jobs.

It is possible that the personality differences between the employed and unemployed are even greater than those between holders of different jobs. MacLean (1977) administered Cattell's 16 Personality Factor Questionnaire to unemployed married men between the ages of 25 and 40 who had at least average intelligence and education and no physical disability. These men were more neurotic, psychotic, and introverted than matched controls, especially when they had been without work for a long time. There is some doubt about causality here, since their inadequate personalities may have rendered them unemployable, or the degradation of unemployment may have affected their personality, or both.

The research discussed so far may suggest that personality tests would prove useful in the process of personnel selection. However, while it is undeniable that it is to everyone's advantage to have a good fit between a worker and his work environment, there are problems of faking when desirable job opportunities are at stake. Not surprisingly, job applicants tend to obtain higher lie scores and lower neuroticism scores on personality questionnaires than those who complete questionnaires with nothing at stake. It has usually been assumed that the high motivation of job applicants leads them to answer questionnaire items in a socially desirable fashion. Elliott (1981) discovered that there was greater distortion when testing occurred under stressful conditions (being sent to an external assessment center and being pressed into testing). He argued that scores on the lie scale may reflect the perceived stress of the testee rather than the level of motivation, and the reduction in neuroticism scores may measure coping with that stress.

What happens if an individual finds himself in a job that is ill-suited to his personality? If he remains in that job, then the obvious answer is that his job performance will tend to be relatively poor. An alternative possibility that has rarely been considered is that his personality may alter as a result of being exposed to a particular job environment. Turnbull (1976) found that there was no tendency for success among male student salesmen to be related to extraversion. However, the experience of selling and making numerous contacts with strangers produced a highly significant increase in the average level of extraversion.

The relationship between personality and occupational success has been examined many times. Fairly striking findings were obtained among trainee pilots by Jessup and Jessup (1971). They tested would-be pilots with the Eysenck Personality Inventory early in their course and discovered that the subsequent failure rate varied considerably as a function of personality. Specifically, 60% of the neurotic introverts failed, against 37% of the neurotic extraverts, 32% of the stable extraverts, and only 14% of the stable introverts. In other words, high levels of neuroticism had a much greater adverse effect on introverts than on extraverts.

Similar findings were reported by Reinhardt (1970), who carried out a battery of personality tests on a sample of the United States Navy's best pilots. Their mean score on the neuroticism scale of the Maudsley Personality Inventory was only 11, compared with a mean of 20 among American college students. Okaue, Nakamura, and Niura (1977) divided the extraversion and neuroticism scores of military pilots into three categories (high, average, and low) on each dimension. Of the sample of 75 pilots, 38 fell into the stable extravert category, with the highest frequency in any of the other eight categories being only 8. In more recent research with military pilots in the United Kingdom, Bartram and Dale (1982) found a tendency for successful pilots to be more stable and more extraverted than those who failed flying training.

The consistent finding that neuroticism is negatively related to flying success makes intuitive sense. Flying can obviously be stressful, with a single mistake proving fatal. In such circumstances, pilots who are especially susceptible to stress are likely to perform less well than those who are more stable.

Personality has been shown to be predictive of success in other occupations. It has been found that successful businessmen tend to be stable introverts (H. J. Eysenck, 1967b), and the same personality profile characterizes entrepreneurs (Lynn, 1969). However, it should be noted that these studies were carried out in the United Kingdom, where it is likely that extraversion is less highly valued than in the United States. In contrast, creative painters and sculptors tend on average to be neurotic introverts (Götz & Götz, 1973).

There has been some interest in the relevance of personality to performance under rather monotonous conditions. By analogy with laboratory research on vigilance (see Chapter 9), it might be expected that underaroused extraverts would find it more difficult than introverts to maintain performance over time. Extraverts showed a greater deterioration than introverts in driving performance over a four-hour period (Fagerström & Lisper, 1977). In line with an arousal interpretation of the extraverts' performance decrement, their performance improved more than that of introverts when someone talked to them or the car radio was turned on.

One index of poor driving performance is obviously the number of accidents that a driver has. Shaw and Sichel (1970) compared the personality characteristics of accident-prone and safe South African bus drivers. Most of the accident-prone drivers were neurotic extraverts, whereas the safe drivers were predominantly stable introverts (see Figure 41). As might have been expected, it is the impulsiveness component of extraversion rather than the sociability component that is more closely related to poor driving and accident proneness (Loo, 1979b).

In sum, it appears that preferences for different kinds of occupation and occupational success are both determined to some extent by personality. The research to date mostly suffers from the disadvantage that job characteristics are discussed in an *ad hoc* fashion. A major dimension along which jobs can be ordered is the extent to which the behavior of an individual doing that job is constrained by external factors. For

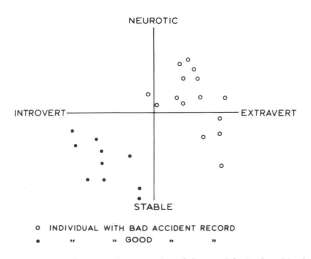

FIGURE 41. Personality differences between bus drivers with good and bad accident records. (From *Accident Proneness* by L. Shaw and H. Sichel, Oxford: Pergamon, 1970. Copyright 1970 by Pergamon Press, Inc. Reprinted by permission.)

example, a car worker on an assembly line has minimal control over his work activities, whereas a university lecturer has great control. It seems likely that personality will be a more consequential determinant of job satisfaction and success when severe constraints exist. It may be no coincidence that two of the occupations wherein personality has been found to be relevant (flying and driving) both involve considerable constraints. In other words, the fit of a worker to his job is especially important when the worker has little scope for tailoring the work environment to his needs.

ANTISOCIAL BEHAVIOR AND CRIME

A matter of great concern in modern society is the increasing incidence of various forms of antisocial behavior. The major focus has been on manifestly criminal activities such as mugging, rape, and murder, but it is perhaps fruitful to identify a continuum of antisocial behavior ranging from minor misdemeanors such as petty theft and riding a bicycle at night without a light at one end to major criminal offences at the other end.

It has usually been assumed that social and cultural factors (e.g., social deprivation, poverty, broken homes) play a major role in producing criminals, and it is undoubtedly true that many such factors are of importance. However, the fact that crime has increased markedly during the twentieth century while poverty has decreased considerably suggests that additional influences must be considered. Of particular relevance in the context of this book is the possibility that individual differences in personality may play a part in determining who acts in an antisocial and criminal fashion. Of course, it would be simplistic to assume that criminal activity is determined exclusively either by situational factors or by personality characteristics; more realistically, situation and personality combine interactively to determine the occurrence of criminality.

If there is any mileage in the notion of a "criminal personality," how might we expect the personalities of criminals to differ from those of noncriminals? According to H. J. Eysenck (1967a), the answer to that question is that criminals are more extraverted and more neurotic than the normal population. Extraverted individuals are more likely than introverts to turn to crime because their poor conditionability (discussed in Chapter 9) tends to prevent them from acquiring social rules as readily as introverts. As a consequence, they experience less inhibition with respect to behaving in an antisocial manner. Individuals who are high on neuroticism are relatively anxious, and anxiety acts as a drive that multiplies with habit. This means that someone who has acquired antisocial responses will be especially likely to engage in those responses if he or she is high on neuroticism.

More recently, H. J. Eysenck (1977a) has argued that criminals should also be relatively high on psychoticism. High scorers on psychoticism tend to be uncaring with respect to people and are unlikely to feel guilt, empathy, or sensitivity to the feelings of others. It seems reasonable to assume that individuals with these characteristics would experience relatively few qualms about behaving antisocially.

On the face of it, these hypotheses can be tested relatively easily. All we need to do is to compare the psychoticism, extraversion, and neuroticism scores of criminals and noncriminals. The prediction is that the criminals will, on average, have higher scores on all three personality dimensions. However, there are various complexities. First, it is not always possible to be sure that the control group is appropriate. Second, it may be doubted whether criminals who have been apprehended and incarcerated constitute a representative sample of the criminal fraternity. It may be mainly those criminals who fail to plan their crimes meticulously who are caught, and it is likely that the amount of planning involved is itself related to personality. Third, there is the problem of attempting to interpret what is basically correlational evidence. If, for example, prisoners are found to be high on neuroticism, this may be because their neurotic personalities predisposed them toward crime. However, it is also possible that the experience of being convicted and locked up in prison has made them neurotic. Fourth, it is probable that there is more than one "criminal personality." People who become muggers, counterfeiters, pubroom brawlers, flashers, and conmen presumably differ from each other in terms of their personality. In other words, when investigating the personality characteristics of criminals, it is clearly desirable to take account of the precise crimes that have been committed. In this connection, it may be useful to distinguish between *primary* psychopathy and *neurotic* or *secondary* psychopathy. Primary psychopaths characteristically have a low level of anxiety and are unresponsive to threats of punishment. In contrast, secondary psychopaths exhibit antisocial and aggressive behavior but suffer from severe emotional frustration and inner conflict. Primary psychopathy may be associated with the psychoticism dimension, whereas secondary psychopathy is associated with the neuroticism and extraversion dimensions.

Since the relevant literature has been reviewed a number of times (e.g., Feldman, 1977; Passingham, 1972), we will content ourselves here with a brief summary and a description of one or two recent studies. Of the three personality dimensions, extraversion is the one that is least reliably and strongly related to criminality (H. J. Eysenck, 1977a). When a detailed analysis was carried out in order to determine which extraversion items differentiated criminals from noncriminals, it transpired that it was the impulsivity component of extraversion rather than the sociability component that was primarily responsible for the modest link between extraversion and criminality (S. B. G. Eysenck & H. J. Eysenck, 1971). However, some of these impulsivity items correlate quite highly with psychoticism, which complicates interpretation. One of the reasons for the disappointingly small association between extraversion and criminality may be that the prison environment is hardly conducive to being sociable and leading an active social life. Heskin, Smith, Bannister, and Belton (1973) discovered that long-term prisoners were significantly more introverted than short-term prisoners.

With respect to neuroticism, the typical finding is that criminals are more neurotic than noncriminals. However, a somewhat more complex picture emerged in a study by Hare and Schalling (1978). They found that primary psychopaths tended to have low neuroticism scores, whereas secondary psychopaths had high neuroticism scores.

It is only in recent years that adequate measures of psychoticism have become available, and as a consequence there have been relatively few studies in which the relationship between psychoticism and criminality has been examined. However, the evidence to date indicates fairly consistently that criminals are high on psychoticism. Putnins (1982) found that juvenile delinquents differed from controls only with respect to their psychoticism scores. Hare (1982) found that primary psychopaths had significantly higher psychoticism scores than controls but that there was no difference between the two groups on extraversion or neuroticism. In contrast, Wilson and Maclean (1974) found that prisoners scored higher than controls on psychoticism, neuroticism, and extraversion.

We mentioned earlier that it is usually difficult to know whether it is the personality that is causing the criminality or the unpleasant consequences of unsuccessful crime that affect personality. An interesting exception is a study by Putnins (1982), in which boys between the ages of 13 and 15 were followed up through juvenile court and juvenile aid court records for one year after they had completed a personality test. The aim was to see whether the personality assessment had any predictive power. High psychoticism scorers were significantly more likely than low psychoticism scorers to commit offenses during the one-year period.

Virtually all of the studies discussed so far were carried out either in the United States or in the United Kingdom. However, a few relevant studies have been carried out in Eastern European countries (see H. J. Eysenck, 1977a, for details), and these studies enable us to make interesting cross-cultural comparisons. According to environmentalist theories that attribute crime to factors such as capitalism or inequality, there is no particular reason to suppose that the personality of criminals in very different societies would be similar. In contrast, Eysenck argued that similarity of criminal personality should occur cross-culturally because of the importance of hereditary factors.

The usual finding is that the personality of criminals in Eastern European countries corresponds to that of criminals in North America and Britain. In one study, male Hungarian criminals were found to be much higher than noncriminals on psychoticism and neuroticism and a little higher on extraversion. Similar findings were reported for female Hungarian criminals. In another study, comparable findings were obtained in Czechoslovakia.

We have concentrated so far on each of the three dimensions of personality (e.g., extraversion, neuroticism, and psychoticism) one at a time. However, since the relevant theories assume that there are particular combinations of personality characteristics that produce a predisposition toward criminality, it makes sense to consider the pattern of scores on the three dimensions. The value of this approach was demonstrated by Burgess (1972). He failed to discover significant differences between prisoners and controls on either extraversion or neuroticism; however, further analysis revealed that significantly more prisoners than controls fell within the neurotic extravert quadrant.

A related approach was adopted by McGurk and McDougall (1981). They made use of cluster analysis and found that two personality types were present in a delinquent sample but not in a comparison group. One of these personality types comprised high neuroticism and extraversion, and the other involved a combination of high psychoticism, extraversion, and neuroticism. These findings provide good support for some of Eysenck's theoretical ideas. In addition, they demonstrate the heterogeneity of personality that exists within the criminal population.

Rather similar findings were obtained by McEwan (1983). He performed a cluster analysis on the Eysenck Personality Questionnaire responses of delinquents between the ages of 14 and 16 in a closed custodial institution. Four clusters were identified, two of which (high psychoticism and extraversion, and high neuroticism and extraversion) resembled the clusters reported by McGurk and McDougall (1981). Delinquents belonging to the various clusters did not differ in terms of the nature of their offences, but those delinquents high on both psychoticism and extraversion had the greatest average number of previous convictions.

Before turning our attention to a theoretical discussion of the evidence, we should devote some attention to a related research strategy, in which the natural covariation of antisocial behavior and personality is examined in normal groups. This strategy differs from that dealt with so far in two major ways: (1) it deals with antisocial behavior in the broader sense and includes many forms of behavior that are less serious than actual crimes; and (2) no control groups are needed, thus removing the problems of matching and sampling biases.

There have been several studies in which this strategy has been adopted. Shapland, Rushton, and Campbell (1975) administered a self-report measure of delinquency and measures of extraversion and neuroticism to boys and girls. Those children with higher delinquency scores tended to be more extraverted and neurotic than those with lower delinquency scores. Allsopp and Feldman (1976) discovered that boys who had engaged in antisocial behavior were higher in psychoticism, neuroticism, and extraversion than well-behaved boys. However, the results of this study are rather difficult to interpret unequivocally because the delinquent children scored much higher than the nondelinquent ones on the lie scale. Saklofske (1977) considered the personalities of groups of 10- and 11-year-old boys whose behavior at school was regarded as either good or problematical by their teachers. The badly behaved boys were significantly higher on psychoticism and lower on the lie scale. Somewhat surprisingly, they were significantly lower on extraversion, and there was no relationship between antisocial behavior and neuroticism.

Powell (1976, 1977) investigated the personality characteristics of nonconforming children (i.e., those who admitted fighting and potentially delinquent acts such as trespass or minor theft; those who smoked early; and those who had antisocial, progressive, or nontraditional attitudes). Nonconformers were high on extraversion, neuroticism, and psychoticism and low on the lie scale. However, since nonconformity scores were based on self-report, it is possible that they are inaccurate or distorted in some way. Powell and Stewart (1983) tested this by comparing these self-report data with teachers' ratings of the antisocial behavior of the children. In general, the two sets of ratings were comparable. However, the teachers' antisocial ratings were strongly related to psychoticism in the children, but not to extraversion or neuroticism.

An unusually thorough study was reported by Rushton and Chrisjohn (1981). In the first part of their study, seven different samples completed measures of self-reported delinquency and of personality. Extraversion correlated positively and significantly with self-reported delinquency in five of the seven samples, and psychoticism did the same in five out of the six samples in which it was measured. In contrast, neuroticism showed practically no relationship with antisocial behavior. As in some of the earlier studies, interpretation of the data is complicated by the further finding that those who scored high on the lie scale tended to have low scores on self-reported delinquency, extraversion, and psychoticism.

Rushton and Chrisjohn (1981) used a further group of 128 psychology students. Extraversion correlated $+.35$ with self-reported delinquency, psychoticism correlated $+.20$, and neuroticism correlated only $+.02$. In other words, there was a very consistent general tendency for self-reported delinquency to be associated with high extraversion and psychoticism but not with neuroticism.

In sum, fairly reliable findings have emerged from studies of the natural covariation between personality and antisocial behavior. Those who engage in antisocial behavior are typically high in psychoticism and extraversion and sometimes high in neuroticism. However, while these findings are very much in line with theoretical expectation, there are some interpretative problems. Consider the fact that antisocial behavior and the socially undesirable "criminal personality" are frequently associated with low scores on the lie scale. It is at least possible that the apparent association between good behavior and low psychoticism, neuroticism, and extraversion is attributable in part to "faking good," that is, the attempt to present a socially desirable impression of oneself. In addition, it is obviously rather simplistic to equate all forms of antisocial behavior. Various subgroups of delinquents can be identified on the basis of the nature, seriousness, and motivation of their antisocial acts. Such subgroups would undoubtedly differ in terms of personality. A possible categorization was suggested by Jurkovic and Prentice (1977). They distinguished among psychopathic delinquents, who show little remorse; neurotic delinquents, who suffer from feelings of guilt, depression, and anxiety; and subcultural delinquents, who are socialized into delinquent behavior by their peers. Jurkovic and Prentice discovered that these three subgroups differed in cognitive development, social responsiveness, and moral reasoning.

We have seen that criminal and antisocial behavior tends to be committed by individuals who are high in psychoticism, extraversion, and neuroticism. However, the association between crime and extraversion is weak at best, and the same is true of the association between antisocial behavior and neuroticism. Furthermore, there are probably a number of discriminable "criminal types," and distinctions such as that between primary psychopathy and secondary psychopathy appear to be important.

Although the various associations between antisocial behavior and personality characteristics are broadly consistent with Eysenckian theory, it is still very difficult to explain these associations. H. J. Eysenck's (1977a) theory of criminality assumes that children learn to inhibit antisocial responses as a consequence of the development of conscience, which comprises a set of conditioned emotional responses to those stimuli associated with antisocial acts. In other words, conditioned fear acts to prevent antisocial behavior, and it is those individuals who do not form classically conditioned responses readily who are most likely to become criminals.

According to this theory, then, individual differences in conditionability are of paramount importance. The relevant evidence was discussed by Passingham (1972). He reviewed ten studies concerned with conditioning in criminals, seven of which supported the prediction that criminals condition poorly and three of which were inconclusive. Although these results are quite encouraging, it would be misleading to assume that some individuals always condition well and others poorly. The evidence reviewed in Chapter 9 showed beyond peradventure that whether a particular individual conditions relatively well or not depends in great measure on the precise parameters of the conditioning situation. It is thus likely that conscience development depends on an interaction between individual personality and social milieu.

Interesting findings consistent with this interactive perspective were obtained by Raine and Venables (1981). They exposed school children to a skin-conductance classical conditioning paradigm and assessed the degree of socialization by means of self-report measures and by teacher ratings of antisocial behavior. Conditioning performance was interactively determined by antisocial tendencies and social class: Good conditioning was associated with antisocial behavior among lower-class children, but good conditioning was linked with good behavior in upper-class children.

Why was the opposite of the expected result obtained with the lower-class children? Raine and Venables (1981) argued that a process which H. J. Eysenck (1977a) termed "antisocialization" might be involved. The basic notion is that children who are easily conditioned and whose parents are antisocial will tend to become socialized into antisocial habits.

Gray (1981) has applied his theory to the issue of the relationship between personality and criminality. Although he makes predictions rather similar to those of H. J. Eysenck (1977a), he does not rely on individual differences in conditionability as an explanatory principle. Instead, he argued that criminality should be greatest among impulsive individuals because these are the people who have the greatest sensitivity to reward. In the original formulation of Gray's theory, the prediction was that psychopaths or criminals should tend to be neurotic extraverts. The fact that psychopaths also tend to be high on psychoticism can be handled by rotating the plane in which the dimensions of anxiety and impulsivity lie into the dimension of psychoticism in order to produce a positive correlation between extraversion and psychoticism. As a consequence, psychopaths still tend to be impulsive, but high impulsivity involves high neuroticism, extraversion, and psychoticism.

To conclude, there are a number of interesting personality differences between those who display more and less antisocial behavior than most people. The fact that psychoticism, extraversion, and neuroticism are all implicated helps to strengthen the argument that these are the three most important dimensions of personality. However, matters are more conjectural at the explanatory level, and there is a suspicion that conditionability is merely one of several factors mediating between personality and antisocial behavior.

PSYCHIATRIC DISORDERS

Since the Eysenckian approach to personality identifies neuroticism and psychoticism as two of the major dimensions of personality, it is not surprising that H. J. Eysenck (e.g., 1970a) has considered the relationship between personality and the various forms of psychiatric disorder. Of particular interest is the fact that the personality and psychiatric approaches differ greatly in their initial assumptions. Psychiatrists often assume that there are qualitative differences both between the normal and the abnormal and between different kinds of abnormality. These alleged discontinuities find expression in their diagnostic categories (e.g., schizophrenia, manic-depressive psychosis). In contrast, H. J. Eysenck (1970a) argued that it was fruitless for those interested in mental illness to mimic the medical approach to physical illness; rather, neurotics and psychotics differ in quantitative fashion from normals. As a result, gradations of abnormality exist and are reflected in scores along the neuroticism and psychoticism dimensions.

A further point of controversy concerns the relationship between neurosis and psychosis. Freud apparently argued that neurotic and psychotic disorders lie along the same continuum, with psychosis simply representing a more serious disorder involving greater regression than neurosis. In contrast, it has usually been assumed in conventional psychiatric theory that neuroses and psychoses are essentially unrelated to each other. Eysenck's assumption that his neuroticism and psychoticism dimensions are independent of each other is clearly more in line with psychiatric orthodoxy than with Freudian theory.

The available evidence indicates that neurotics do not differ qualitatively from normals (H. J. Eysenck, 1950), and the same appears to be true of psychotics (H. J. Eysenck, 1952a). Additional support for the notion that there are gradations of psychosis was obtained by McPherson, Presley, Armstrong, and Curtis (1974). They divided schizophrenic patients into three groups on the basis of symptom severity (no delusions, integrated delusions only, and nonintegrated delusions). Patients having the more severe symptoms tended to have higher scores on psychoticism than those reporting less severe symptoms. Thus, it appears preferable to consider neurosis and psychosis within a dimensional rather than a categorical framework.

The further issue of whether neurotic and psychotic disorders lie along the same dimension or two independent dimensions has received rather more experimental attention. S. B. G. Eysenck (1956) administered a number of objective tests to normals, neurotics, and psychotics. When a discriminant function analysis was carried out, it turned out that two dimensions were required to account for the data. The same conclusion was suggested by Cowie (1961). She argued that the children of psychotic parents should not show any more neuroticism than the children of normal parents if it is actually true that psychotic and neurotic disorders are independent of each other. Perhaps surprisingly, the children of psychotic parents were, if anything, less neurotic than the children of normal parents. Finally, Trouton and Maxwell (1956) had a random sample of 819 patients rated on 45 neurotic and psychotic symptoms. Two independent factors of neuroticism and psychoticism were identified, and virtually all of the symptoms were much more closely related to one factor than to the other.

We have seen that it is reasonable to argue that normal and abnormal individuals differ quantitatively from each other along one or both of the independent dimensions of neuroticism and psychoticism. With respect to personality data, it is natural to assume that neurotic patients should have greatly elevated neuroticism scores, whereas psychotic patients should have inflated psychoticism scores. Some relevant data taken from the Manual of the Eysenck Personality Questionnaire are shown in Table 18. The pattern of scores is very much as expected, with psychotics having higher scores than neurotics or normals on psychoticism and neurotics having higher scores than psychotics or normals on neuroticism. The psychotics tended to have rather high lie scores. Since their psychoticism and neuroticism scores both correlated negatively with their lie scores, it is probable that the mean scores on those two dimensions underestimate the true values for psychotics.

The important notion that categorical terms within the conventional psychiatric system can be translated in a meaningful fashion into locations within the dimensional framework of personality theory was tested by H. J. Eysenck and S. B. G. Eysenck (1976). They selected individuals belonging to seven different psychiatric status groups (i.e., normals, criminals, personality disorders, reactive depressives, dysthymics or anxiety states, endogenous depressives, and schizophrenics). Scores

TABLE 18. Mean Neuroticism and Psychoticism Scores on the Eysenck Personality Questionnaire among Normals, Psychotics, and Neurotics

	Normals	Psychotics	Neurotics
Males			
Psychoticism	3.78	5.66	4.19
Neuroticism	9.83	13.39	16.56
Females			
Psychoticism	2.63	4.08	3.25
Neuroticism	12.74	14.56	17.88

Note. From *Psychoticism as a Dimension of Personality* by H. J. Eysenck and S. B. G. Eysenck, London: Hodder & Stroughton, 1976. Copyright 1976 by Hodder & Stoughton Ltd. Reprinted by permission.

from the Eysenck Personality Questionnaire were obtained from all of the subjects and the resultant data submitted to discriminant function analysis in order to provide the maximum discrimination among the groups.

The results of this analysis are shown in Figure 42. The first component had the normal controls at one extreme and the psychiatric groups at the other extreme; this is clearly a psychiatric abnormality component. The second component contrasted the neurotic disorders (reactive depression and dysthymia) with the psychotic ones (endogenous depression and schizophrenia). The implication is that it is possible to locate diagnostic categories from the psychiatric system within a dimensional framework.

A similar conclusion emerged from a study by Wakefield *et al.* (1974). They considered nine out of the ten scales of the Minnesota Multiphasic Personality Inventory, most of which correspond fairly closely to psychiatric diagnoses (e.g., paranoia, schizophrenia, psychopathic deviate, and hysteria). They were able to demonstrate that these scales could be represented as points within Eysenck's three-dimensional framework. The notion that there are meaningful relationships between personality and psychiatric symptomatology was further strengthened in a subsequent article by Wakefield *et al.* (1975).

Should the dimensional framework of personality replace the traditional psychiatric classification system altogether? Probably not. There are many individuals whose psychoticism, neuroticism, and extraversion scores resemble those of the average manic-depressive or schizophrenic and yet who manage to lead relatively contented, symptom-free lives. Perhaps an individual's location within Eysenckian three-dimensional space can appropriately be regarded as a measure of his or her vulnerability to different kinds of mental illness. If the psychiatric and personality systems fulfil rather different functions, then they should be regarded as complementary rather than antagonistic.

One of the ways in which personality assessment can prove useful in the clinical situation was demonstrated by Rachman and S. B. G. Eysenck (1978). They administered the Eysenck Personality Questionnaire to patients who had been diagnosed as neurotic. The length of time taken for these patients to respond to treatment was positively correlated with their psychoticism scores. Many of those with high psychoticism scores were eventually reclassified as having personality disorders. It is possible that knowledge of these high psychoticism scores at an early stage would have facilitated rapid correct diagnosis and thus would have enabled the appropriate form of treatment to have been chosen rather sooner.

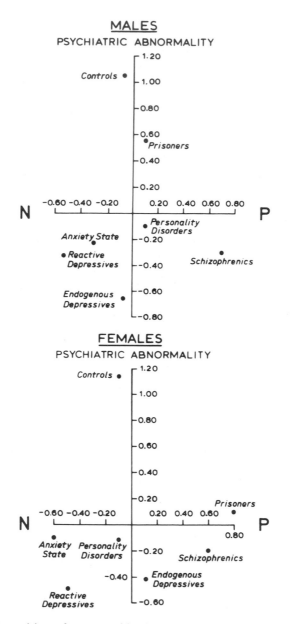

FIGURE 42. The positions of seven psychiatric groups on two main discriminant function variates. (From *Psychoticism as a Dimension of Personality* by H. J. Eysenck and S. B. G. Eysenck, London: Hodder & Stoughton, 1976. Copyright 1976 by Hodder & Stoughton Ltd. Adapted by permission.)

There is further intriguing evidence that treatment can be facilitated when account is taken of the patient's personality. Di Loreto (1971) compared the effectiveness of different kinds of therapy in the treatment of social and general anxiety and discovered that the personality of the patient played an important part. Rogers's client-centered therapy was effective only with extraverted patients, whereas Ellis's rational-emotive therapy was effective only with introverted patients. In contrast, systematic desensitization was effective with both introverted and extraverted patients. This pattern of findings is broadly in line with Gray's (1973) notion that extraverts are primarily responsive to reward and introverts to punishment, since Rogerian therapy is basically supportive and rewarding and rational-emotive therapy is more punitive.

At a more theoretical level, it is of much interest to attempt to elucidate the causal determinants of neurosis in general or dysthymia in particular. Eysenck has consistently argued that dysthymics tend to be individuals who are high on neuroticism and low on extraversion, and the main reason that they become dysthymic is because they readily develop fear responses. According to H. J. Eysenck (1976c):

> The main process leading to neurotic fear/anxiety responses is probably Pavlovian (classical) conditioning. This may be either of traumatic, single-trial kind, or may be sub-traumatic, repeated presentation of the CS-UCS combination. (p. 263).

This view of the factors producing dysthymia was disputed by Gray (1973). He claimed that it is individuals who have high scores on the personality dimension of anxiety who are most likely to become dysthymics. The reason for this is that such individuals are especially sensitive to punishment and are also likely to manifest certain innate fears, some of the most important of which arise during the course of social development in adolescence. According to this perspective, it may be true that conditioning is often involved in the development of neurotic fears, but this is merely one of the consequences of the great sensitivity to threat and punishment that neurotic introverts inherit.

These theoretical differences in explaining the genesis of dysthymia are reflected in the explanations offered for successful methods of treatment such as behavior therapy. According to H. J. Eysenck, the crucial element in the treatment of dysthymia is extinction—that is, the conditioned fear responses are extinguished. In contrast, Gray (1981) argued that if the behavior of dysthymics is affected by innate reactions to threatening stimuli the process of treatment involves habituation. In line with this hypothesis, there is some evidence that the efficacy of treatment of dysthymia by behavior theory depends on the total exposure time to the threatening stimulus (Teasdale, 1977; Watts, 1971).

A further difference between the theoretical positions of H. J. Eysenck and Gray concerns the interpretation of the finding that various physiological and other forms of treatment reduce dysthymic symptoms. If dysthymia is regarded as an amalgam of high neuroticism and low extraversion, then it is natural to assume that these treatments have had two effects, reducing the level of neuroticism and increasing that of extraversion. In contrast, Gray argues that these treatments simply reduce anxiety.

The differences between these theories may actually be less marked than they appear at first glance. First, it is the same group of vulnerable individuals that is being referred to whether we call them neurotic introverts or anxious. Second, genetically determined individual differences in personality are regarded as important in both theories. Third, both theorists agree that fear or anxiety is a crucial goal of treatment. However, H. J. Eysenck appears to allow more scope for situational factors to play a part in the development of dysthymia through his emphasis on conditioning. It is more difficult with Gray's theory to understand why it is that there are numerous people with high anxiety scores who nevertheless do not become dysthymics.

CONCLUSIONS

The most obvious conclusion based on the research reviewed in this chapter is that the personality dimensions of extraversion, neuroticism, and psychoticism all have predictive and explanatory power across a heterogeneous collection of real-life situations. Thus, we have seen that these three dimensions are relevant in accounting for social and sexual behavior, study and work preferences and achievements, and antisocial, criminal, and abnormal behavior. It is reassuring to discover that there are behavioral correlates of personality dimensions in everyday life as well as in the well-controlled and somewhat artificial confines of the laboratory.

It is also reassuring to have confirmatory evidence that behavior is interactively determined by personality and by situational factors. Although it seems obvious to many people that this should be so, others have attempted to account for all of the kinds of social behavior discussed in this chapter in environmental terms. We are now in a position to claim categorically that social phenomena such as criminality and mental illness depend in part on genetically determined individual differences in personality. This is not to deny the important role played by environmental factors; rather, it is a matter of correcting the undue emphasis placed by others on such factors.

Why does personality appear to have more impact on some kinds of social behavior than on others? This is an intriguing question but one that does not have any simple answer. However, it is probably relevant that the personality dimensions of extraversion, neuroticism, and psychoticism are all closely related to emotional and/or motivational factors. It is thus to be expected that those aspects of social behavior with a particularly strong emotional or motivational ingredient should be much affected by personality, whereas the more purely cognitive aspects of everyday functioning might be less affected.

A possible implication of the research findings discussed in this chapter is that some kinds of personality are better adapted than others to contemporary society. In particular, high scores on the neuroticism and psychoticism dimensions are associated with mental illness, psychopathy, and sexual dissatisfaction. However, it perhaps makes more sense to argue that people with all kinds of personality have a potentially useful contribution to make to society. This is supported by the fact that there is no genetic dominance associated with either end of the extraversion, neuroticism, and psychoticism dimensions. There are potential hazards stemming from high levels of psychoticism and neuroticism, but it is nevertheless the case that there are situations in which aggressive and hostile behavior (high psychoticism) or emotional lability (high neuroticism) are extremely useful. In similar fashion, society needs individuals who relish social interaction (extraverts), as well as others who prefer to work on their own (introverts).

In sum, the benefits of human diversity exceed the disadvantages. It is an urgent task within society to ensure that individuals find themselves in environments that they enjoy and that enable them to make optimal use of their talents. The personality theorist has an important role to play in this vital task of matching individuals and environments.

PART THREE

EPILOGUE

CHAPTER TWELVE

Is There a Paradigm in Personality Research?

It has often been suggested (e.g., by Kuhn, 1970, 1974; Barnes, 1982) that the difference between the hard sciences and the social sciences is the absence of *paradigms* in the latter. In his early work, Kuhn was by no means so clear or consistent as might be wished in his definition and use of the term *paradigm* (cf. Masterman, 1970), but later Kuhn (1974) recognized the problem and tried to eliminate it, without perhaps achieving complete success (Suppe, 1974). In his later formulation, Kuhn recognized two aspects of his original notion of a paradigm, called *disciplinary matrices* and *exemplars*. Exemplars Kuhn defined as concrete problem solutions, accepted by the scientific group involved as serving as useful paradigms for the pursuit of research; in other words, they are the accepted applications of symbolic generalizations to various concrete problems confined in the examples and solutions to the exercises in standard textbooks and laboratory manuals.

According to Kuhn, it is from the study of these exemplars that one learns to apply symbolic generalizations to nature. It is from the study of exemplars and attempts to solve problems that the student develops a similarity or resemblance relation which is used to model the application of symbolic generalizations to new experimental situations, and it is through the acquisition of this similarity relationship—and not through correspondence rules (as suggested by the Vienna School of logical positivists)—that the theoretical generalizations come to have empirical content.

By contrast, a disciplinary matrix is the common possession of a professional discipline, containing three kinds of elements: symbolic generalizations, models, and exemplars. In other words, the disciplinary matrix contains all those shared elements which make for relative full-ness of professional communication and unanimity of professional judg-ment. The main point about the disciplinary matrix is that is constitutes a *conceptual framework* (Suppe, 1974).

Paradigms, either by way of exemplars or disciplinary matrices, are notoriously lacking in social science in general and in psychology in par-ticular. Nowhere is this lack more obvious than in the personality field. Following the example of Hall and Lindzey (1970), most textbooks simply give a set of chapters organized around one particular author, explaining his theories, quoting a few examples of empirical work more or less rel-evant to it: but they eschew the scientifically important and indeed essential job of judging the *adequacy* of the theory in terms of the exper-imental work devoted to it and thus fail to compare the adequacy of one theory along these lines with that of all the others. Thus what we have is not the evolution of a paradigm, but a Dutch auction in ideas, alien to the spirit of science and conducive to arbitrary choice in terms of exist-ing prejudices on the part of the student. Not along these lines will we ever arrive at a paradigm (H. J. Eysenck, 1983b).

Yet it might be claimed that we do in fact have the beginnings at least of a paradigm in the personality field, in terms of a descriptive and causal system of concepts centered around the three major dimensions of personality we have discussed in this book P, E, and N. There is, as indicated in the relevant chapters, a growing realization that the results of literally hundreds of factor analytic studies, starting with very differ-ent premises and hypotheses, carried out by psychologists of quite dif-ferent theoretical orientation, located in many different countries and using different methods of analysis and rotation, have practically always found major dimensions corresponding to E and N and often to P as well. As already indicated, Royce and Powell (1983), after a thorough review of all the available evidence, have clearly come to the same con-clusion, although they use a slightly different nomenclature to identify these three major dimensions. There is thus a surprising degree of agree-ment on the descriptive side, amplified, as we have seen, by the fact that animal work as well as human work supports the same conclusion.

The fact that these major dimensions of personality are found in many different countries and cultures, from Hong Kong to Uganda and from Japan to India, exemplifies the universality of these descriptive parameters, and the fact that they have a firm foundation in genetics suggests their biological derivation and importance. When we note in addition that these descriptive dimensions also have a firm foundation in causal theories derived from well-established physiological and psy-chological theories and are supported by large-scale physiological and psychological experimental laboratory studies, it must, we think, be conceded that the system is beginning to resemble in outline what Kuhn conceives of as a proper paradigm in science.

One possible objection to the easy acceptance of this paradigm might be that there are anomalies and empirical failures of the theories in question to generate verified predictions. We shall deal with some of these objections presently, but note an answer given by Barnes (1982) to the question: "How does acceptance of a paradigm indicate problems for research; and how does the paradigm itself actually serve as a resource for scientists?" (p. 46). His reply was:

> The answer lies in the perceived inadequacy of a paradigm as it is initially formulated and accepted, in its crudity, its unsatisfactory predictive power, and its limited scope, which may in some cases amount to but a single replication. In agreeing upon a paradigm scientists do not accept a finished product, rather they agree to accept the basis for future work, and to treat as illusory or eliminable all its apparent inadequacies and defects. Paradigms are refined and elaborated in normal science. Their use in the development of further problem-solutions thus extends the scope of the scientific competencies and procedures. (p. 46)

In other words, a paradigm is not expected to be perfect in the verification of its predictions, but to be fallible; the work of what Kuhn calls "ordinary science" of the puzzle-solving kind is precisely that of looking at anomalies and trying to reconcile them with the theory, through parametric studies and in other ways.

Anomalies, if persistent and presenting insoluble problems, can of course lead to what Kuhn (1970) called "revolutions," such as the changes from Newtonian to Einsteinian theory concerning gravitation. It is always difficult to know when an apparent anomaly is a real one, of course, and there is no way of ensuring accurate prediction. Anomalies in the movements of planets could be explained in theory and were later shown in actual fact to be due to the influence of hitherto undiscovered planets; thus an apparent anomaly in fact served to strengthen the belief in Newton's theories. Failure to discover stellar parallax appeared to be an absolutely crucial anomaly as far as Copernicus's heliocentric theory was concerned, but it was explained away in terms of the immense distances of the stars and the inaccuracies of measurement, and after 250 years, when measurement was improved, stellar parallax was actually observed. However, the precession of the perihelion of Mercury, apparently a very minor anomaly, remained unexplained on Newtonian principles but was fully explained by Einstein's theory of relativity.

> As Kuhn himself has pointed out: Every problem that normal science sees as a puzzle can be seen from another viewpoint as a counter-instance, and

thus as a source of crisis. Copernicus saw as counter-instances what most of Ptolemy's other successors had seen as puzzles in the match between observation and theory. Lavoisier saw as a counter-instance what Priestley had seen as a successfully solved puzzle in the articulation of the phlogiston theory. And Einstein saw as counter-instances what Lorentz, Fitzgerald and others had seen as puzzles in the articulation of Newton's and Maxwell's theories. (1970, pp. 79-80)

Barnes (1982) comments, "What one scientist sees as an anomaly another sees as a puzzle for the same paradigm—even as a successfully solved puzzle" (p. 100). Thus the existence of an anomaly should not be a bar to the acceptance of the paradigm; the existence of such an anomaly should merely act as a spur for the puzzle-solving capacities of ordinary science. It is only when all such efforts have failed in a number of different instances that what Kuhn calls a "crisis" is reached, and a "revolution" may be in order. But while anomalies certainly exist in the field here discussed, they just as certainly do not amount to crisis proportions; ordinary science has hardly had time to come to grips with these anomalies!

In the text we have drawn attention consistently to some of the major anomalies that exist at the moment, for example, the exact position of impulsivity in the three dimensional framework, the contradictory results reported in some studies of the psychophysiological basis of extraversion, the curious findings concerning circadian rhythms, and many others. We see these anomalies not as occasions for crises and revolutionary developments in the theory, but rather as the typical kind of puzzle which in ordinary science calls for the puzzle-solving abilities of scientists to clarify the issues in question. Undoubtedly the results of such studies will profoundly affect the details of the theory and may require many alterations and improvements; scientific theories in their early stages seldom survive in their original form for any length of time, and the fact that they are adaptable and responsive to criticism and new data is generally regarded as speaking in their favor.

In considering theories and paradigms, it is interesting to consider their place in the development of a science. There is considerable argument among philosophers of science concerning the precise nature of science and what Popper has called "demarcation criteria," criteria that enable us to tell science from pseudoscience and nonscience. There is by no means uniform agreement (Suppe, 1974) among protagonists, but to us it seems that what the major theorists have to say is seldom universally correct but usually applies to a particular stage in the development of a science.

Eysenck (in press) has suggested a sequence that is illustrated in Figure 43. Near the beginning of the development of a science we have

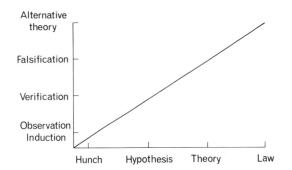

FIGURE 43. Different stages of theory making. (From *The Measurement of Personality* by H. J. Eysenck, Lancaster: Medical & Technical Publishers, 1976. Copyright 1976 by Medical & Technical Publishers. Reprinted by permission.)

what might be called the Baconian phase of observation and induction, linked with hunches as to the kind of results one might expect, and the kind of variable that might be important in producing a given effect. At a rather higher stage of development, we find the principle of *verification* as advocated by the Vienna School of logical positivists. This is linked with the postulation of hypotheses and their testing; if deductions from a hypothesis are verified, then this constitutes an important advance in the development of the science in question.

At a still higher stage, Popper, Lakatos, and others find verification insufficient as a principle because we can never in fact verify a hypothesis; however many deductions from it are tested, an infinite number remain untested, and these may provide evidence to overthrow the hypothesis. Popper has suggested that falsification is the critical step in the demarcation of science; a theory is scientific when it can be falsified. This is an acceptable and realistic proposition, but only when a science has already reached an advanced stage. Popper declares that astrology, for instance, is a pseudoscience, but on his criterion it would have to be accepted as a true science because astrology makes many testable predictions (H. J. Eysenck & Nias, 1982)! It is only when a science has passed through the stages of observation and induction and of verification that falsification as a principle becomes applicable.

Higher still in the development sequence we come to Kuhn's principle of revolution as an alternative to ordinary science. When a theory, such as Newton's, has withstood attempts of falsification for a long time, it becomes a law, and laws can be overthrown only by alternative theories, not by a simple process of falsification. Lakatos (1968, 1970) has indicated why this is so.

Paradigms begin to be feasible when we reach the second stage in this process, and we believe that verification at the moment in psychology is a much more important desideratum than falsification; this applies to a much higher stage of development, and it would be unfortunate if promising theories were cut short because of an erroneous philosophical assumption. Psychological theories in general are weak theories (Eysenck, 1960) and are to be distinguished from strong theories precisely by the differential stress on verification and falsification. Our conclusion would be that the theories discussed in this book have received sufficient verification to make it worthwhile to regard them as a paradigm and to try and iron out anomalies, make detailed studies of parameter values, and generally attempt to improve the theory before exposing it to the cold winds of falsification.

In this book we have stressed the historical roots of our personality theory, particularly E and N, stretching back over more than 2,000 years. From another point of view, however, it might be more instructive to view the theory as a revolution in the Kuhnian sense in that it produces a picture altogether different from the psychoanalytic-cum-projective alignment that has passed as orthodoxy in the past 50 years or so. A characteristic of such a revolution, according to Kuhn, is that it involves a change in the *character of research*. Speculation becomes more acceptable, and novel and radically different procedures in interpretations are tolerated more easily. When a new paradigm is accepted, a large-scale reordering of practice and perception occurs, reflecting the requirements exemplified in the new paradigm. Our theory of extraversion–introversion, as compared with the orthodox Jungian theory, is clearly a case in point. No Jungian would countenance a theory of extraversion–introversion which involved measures of EEG and evoked potentials, conditioning and extinction, figural aftereffects and CFF, alternation behavior and vigilance, salivary responses to lemon juice and circadian rhythm changes, tolerance of pain and sensory deprivation, sensory stimulation modulation and sensory thresholds, drug effects and reminiscence, memory retrieval and consolidation of learning, and many, many others. It is revealing that none of these tests, or their connections with personality, are mentioned in the typical undergraduate text on personality; they simply do not belong in the orthodox paradigm of personality research, and hence their inclusion constitutes a true Kuhnian revolution.

Another example of such a revolution is the recent work on evoked potential measures of intelligence, particularly the paradigms associated with Hendrickson, Schafer, and Robinson (H. J. Eysenck & Barrett, 1984). We have shown in our discussion that very high correlations can be obtained between IQ as measured by standard tests like the Wechsler or the Progressive Matrices and certain scores on the evoked potential. These correlations, in excess of .8, are impossible to reconcile with the traditional Binet-type paradigms, which postulate the IQ as a rather artificial mixture of separate and independent abilities, individual differences which are largely determined by environmental factors, education, socioeconomic status, the teaching of differential strategies, and so forth. None of these are very likely to play any part in the genesis of differential evoked potential patterns on the EEG, and hence a revolution in outlook is implicit in these findings, leading to altogether new and different expectations and experiments. What would now concern investigators would be parameter studies related to the positioning and types of electrodes used; the intensity and duration of the stimuli used, and the intertrial intervals of these stimuli; the type of analysis undertaken of the records, looking at variability as well as amplitude and latency; and many other variables that might help distinguish between the various paradigms suggested. At the same time, the factors usually discussed in connection with the topic of intelligence (Sternberg, 1983) are now seen to be relevant not to intelligence as such but to the social application of intelligence (intelligence B as opposed to intelligence A—Eysenck & Barrett, 1984). Thus a single fact, clearly unassimilable to current paradigms, is sufficient to produce a revolutionary change in perspective and to lead to entirely new avenues of research and of interpretation of existing data. This revolution in the field of intelligence is similar in many ways to the one already discussed in relation to temperament, that is, the noncognitive aspects of personality; both are reductionist in that they seek to link social behavior and its consistencies (in the form of traits or abilities) with genetically determined biological factors in the organism. That there are such relations and that they are very prominent is now hardly in doubt; what is in doubt, of course, and will perhaps remain so for a long time is the precise nature of the relationships involved.

It is here that the puzzle-solving aspects of normal science will have to take over and settle the issues remaining. The new paradigms do not do away with knowledge painfully acquired under the guidance of the old paradigms; they simply show the inefficiency of the old paradigms and add an entirely new dimension of causality to them, which marks the essential revolution that has taken place. Given the considerable amount of positive reinforcement that the general theory here considered has had from a large number of experimental and empirical investigations, it is interesting to ask why it has not in fact been generally accepted as a paradigm and as a basis for research along the lines of normal science. The main answer appears to be that psychologists (and other social scientists) are not on the whole aware of the demands for discipline that science exerts from all its followers: they prefer the free and easy atmosphere of arbitrary choice to the rigors of puzzle-solving within the well-defined context of a paradigm. Psychology grants every research worker the right to choose one from the numerous theories of personality and then among the many measuring instruments those which he prefers, without asking for justification along the lines of known reliability, validity, and experimental support. Theories and instruments alike are almost completely impervious to demonstrations of lack of reliability, lack of validity, and lack of such empirical support—hence the presentation of these theories in textbooks of personality in a personalized and eponymous form, as if there were no scientific grounds for choosing between the many different offerings.

This refusal to lay down and accept rules of decision between different theories has the disastrous consequence that ordinary science cannot function. Theories and paradigms demand rigorous testing, detailed investigation along parametric lines, and the experimental investigation of deductions; also required in many instances is the assessment of the ability of different theories to predict the experimental outcome actually determined. This means long continued investigations of detailed problems, a concentration on fundamental questions rather than easy applications to social problems more readily perceived as relevant, and a decision to try to complete the investigation of one particular deduction rather than jumping from one to another in the random fashion so much favored by modern researchers in this field. One particular point should be made in this connection because it concerns the optimal use of different measures of personality dimensions. If it be agreed that the three major dimensions of P, E, and N are relatively all-pervasive in the personality sphere and can be shown to be fundamental in terms of genetic research, animal work, cross-cultural studies, and so forth, certain consequences follow for the creation and use of other measures of personality. Almost every week sees the arrival of a new type of test, and the total number of tests, published and unpublished, runs into the hundreds and possible the thousands. It is a solemn thought that results achieved by the use of one of these tests can in no way be transferred to the personality space created by other tests, so that we do not have a general psychology of personality but instead individual psychologies created in terms of different testing procedures. That this is an absurdity will hardly need documentation or discussion; the question arises what can be done to obviate it.

The first step, or so we would suggest, would be to determine for each particular test (or score within a given test) the degree to which this is determined by and correlated with the major dimensions of P, E, and N. Once this has been done, the question arises as to whether any specific variance is still associated with the test or score or whether it merely measures to varying degree the fundamental personality traits of P, E, and N. It would then be possible to determine the position of the test of score within the three-dimensional space generated by P, E, and N, and in addition to say what proportion of the total variance of the test or score was specific to that test and outside the three-dimensional space in question. (Cf. our demonstration in connection with Cattell's 16PF in Chapter 4.) In this way different tests and scores would become comparable in a meaningful fashion, and many would indeed be shown to measure nothing but combinations of P, E, and N (H. J. Eysenck & S. B. G. Eysenck, in press). Such tests, clearly adding nothing to the fundamental dimensions, should be rejected outright for further use, and direct measures of P, E, and N substituted.

We can thus think of what Cattell would call primary traits of personality as clusters of item-points lying partly within and partly without the three-dimensional space generated by P, E, and N. Such clusters can be relatively tight and homogeneous or larger and less homogeneous; decisions on such points are of course purely arbitrary (recalling the battles among instinct psychologists who were either "splitters" or "lumpers," that is, preferred to subdivide instincts and end up with a large number, or else simply retain a very small number of relatively heterogeneous instincts, like those of self, sex, and society). The putative trait of impulsivity would thus be constituted of a cluster of item points lying in the $P+$, $E+$, and $N+$ octant; this fairly heterogeneous cluster can be shown to be divisible into four more homogeneous clusters (and these no doubt could be split again into more homogeneous clusters still). There is no *true* impulsivity, and the best way of looking at impulsive behavior in a causal manner would be by referring it to P, E, and N in combination.

It may be doubted whether anything survives of "impulsivity" after the contributions to its variance of P, E, and N have been summed. In other cases, the portion of the cluster lying outside the three-dimensional space defined by P, E, and N, may of course be much larger; this is an empirical problem that ought to be resolved in each case before any new trait is proposed and admitted to the science of individual differences. Basic to this approach is the belief that there is something more fundamental and special to P, E, and N than to other suggested personality dimensions and variables; the evidence strongly suggests that this is so, and if we do indeed have here the beginnings of a paradigm, then it seems logical that we must follow some such procedure as that outlined above. (One excellent example of how this procedure can be used is given by Lynn *et al.*, 1984).

The same procedure might with advantage be used in making predictions or in calculating correlations between personality and various experimental or social variables. The first step should be to see to what extent these variables can be predicted or are correlated with P, E, and N; it should then be established whether the particular trait or score to be added did in fact add a significant amount of variance to the interaction between personality test and criterion behavior. Again, only in this way can investigations using different instruments be brought down to a common denominator and hence compared in a meaningful manner. The importance of agreeing on such procedures cannot be exaggerated; in no other way can we achieve a unification of the field that hitherto has been so sadly lacking and change the belief that paradigms do not, and possible cannot, exist in the social sciences.

It is not suggested in what has been said so far that the whole science of personality, the measurement of traits and abilities, and the application to ordinary life of these scales can be encompassed by the small number of variables with which we have been dealing. Undoubtedly, for instance, social attitudes, values, vocational and avocational choices, and many other types of variables will be included in any picture of the total personality, and indeed many of these variables have already been found to be related to the major dimensions of personality (H. J. Eysenck, 1970c). But just as certainly they do contribute something over and above what is contributed by P, E, and N, and a full theory of personality will of course have to deal with that contribution as well as with the many primary traits that will contain variance not included in the major dimensions of personality. It is not our purpose in writing this book to restrict unduly the field of personality research; what we want to do is to map out the fundamental if as yet fragmentary model that can be used as a paradigm on which to build; it will make future work much easier if this aspect of the total picture, central as it is, is agreed upon among personality theorists.

It might of course be replied that the existing measures of P, E, and N are far from perfect and that their use in this manner would therefore be contraindicated. This does not appear to be a reasonable objection. Once it is agreed that the model here advocated is a fundamental paradigm in personality, research using the puzzle-solving propensities of scientists should not find it difficult to improve on existing scales and arrive finally at a set of scales both highly reliable and highly valid that could form the foundation for experimental work along the lines suggested.

One problem for research that has not been discussed and on which research has been rather spasmodic is the interaction of the various factors. A high-N, high-E individual who is highly intelligent is very different from a high-N, high-E individual who is very dull; the former may present an exciting glitter and a sparkling sociability overlying the less desirable aspects of such a character whereas the dull high-E, high-N scorer is likely to be distinguished by nothing but his immaturity, his antisocial behavior, and possibly his all-too-obvious hysterical symptoms.

The behavior of a high-P scorer who is introverted and neurotic will differ profoundly and in a predictable direction from that of a high-P scorer who is a stable extravert. In looking at any particular person or in making predictions for individuals, these interactions should always be borne in mind, and indeed from many points of view it would be most desirable if explicit predictions were made and tested in the laboratory. Unfortunately, such studies are complex and extensive; if we let each dimension be represented by something like ten extreme high and extreme low scorers, then, simply looking at P, E, N and g, we would need 160 highly selected subjects to fill all the different combinations created by these four dimensions. It would require a subject group of several thousand in order to find the various combinations in question, and even then there might be difficulties in filling all the positions.

What is more common is to test the effects of just one dimension, keeping scores on all the other dimensions near the mean, but that of course does not tell us anything about any interactions that may be occuring. That these reactions can be of vital importance is illustrated by a study (Frcka, Beyts, Levey, & Martin, 1983) of eyeblink conditioning in which a highly significant $P \times E$ interaction was observed. High-P, high-E scorers and low-P, low-E scorers both conditioned more strongly and rapidly than did other subjects. Findings of this kind suggest that interaction factors may be of vital importance in the study of personality.

We believe that the pursuit of some such methodology is vitally important for psychology and social science as a whole because we believe that a solution to the problem of personality research and measurement is fundamental to the development of a truly scientific psychology, whether in the experimental, social, industrial, educational, or clinical field (H. J. Eysenck, 1983b). Practically every main effect to be investigated in these various disciplines is moderated by personality factors, or correlated with them, and shows important interaction effects that may be much larger than the main effects normally studied. But adequate use of such procedures requires proper theories of personality and measuring instruments derived from these theories. The arbitrary use of multiphasic instruments of doubtful validity or unknown psychological import does not encourage the proper formulation of theories regarding interaction effects, and the likely failure of arbitrary selection of such multiphasic tests bids fair to destroy the belief in the necessity for amalgamating what Cronbach (1957) called the two disciplines of scientific psychology: the experimental and the correlational or personality side. Again we must reject the evil of arbitrariness and demand a much more disciplined approach, necessitating the justification of instruments used and theories employed. This is taken for granted in the hard sciences, and the absence of such discipline is one of the major reasons why psychology has not achieved scientific respect and repute in spite of its now quite lengthy history, extending over more than 100 years. It would not be meaningful or sensible to carry on along lines that have proved to be barren and unsuccessful. The creation and use of a proper paradigm is imperative if we want to make the study of personality a truly scientific discipline; nothing else will do.

References and Bibliography

Abraham, K. The influence of oral eroticism on character-formation. In *Selected papers*. London: Hogarth Press, 1952.

Achenback, T. M. The child behavior profile. I: Boys aged 6-11. *Journal of Consulting and Clinical Psychology*, 1978, *46*, 478-488.

Ackerson, L. *Children's behavior problems*. Chicago: University of Chicago Press, 1942.

Adcock, C. J. A factorial examination of Sheldon's types. *Journal of Personality*, 1948, *10*, 312-319.

Adcock, C. J. A comparison of the concepts of Eysenck and Cattell. *British Journal of Psychology*, 1965, *35*, 90-97.

Adcock, N. Testing the test: How adequate if the 16PF with NZ studied sample? *New Zealand Psychologist*, 1974, *3*, 11-15.

Adcock, N., & Adcock, C. The validity of the 16PF personality structure: A large New Zealand sample item analysis. *Journal of Behavioral Science*, 1977, *2*, 227-237.

Adcock, N., & Adcock, C. *Cultural, motivational and temporal problems with the 16PF test*. Paper presented at the I.A.A.P. Congress, Munich, July 1978.

Adkins, M.M., Cobb, E. A., Miller, R. B., Sanford, R. N., & Stewart, D. H. Physique, personality and scholarship. *Monographs of the Society for Research in Child Development*, 1943, *8*, 1-108.

Affleck, D. C., & Garfield, S. L. The prediction of psychosis with the MMPI. *Journal of Clinical Psychology*, 1960, *16*, 14-26.

Aiba, S. The suppression of the primary visual stimulus. In H. J. Eysenck (Ed.), *Experiments with drugs*. New York: Pergamon, 1963.

Alanen, Y. O. The family in the pathogenesis of schizophrenia and neurotic disorders. *Acta Psychiatrica Scandinavia, Supplementum*, 1966. *189*, 1.

Alker, H. & Gawin, F. On the intrapsychic specificity of happiness. *Journal of Personality*, 1978, *46*, 311-322.

Allen, B. P., & Potkay, C. R. On the arbitrary distinction between states and traits. *Journal of Personality and Social Psychology*, 1981, *41*, 916-928.

Allen, G. J. Effect of three conditions of administration on 'trait' and 'state' measures of anxiety. *Journal of Consulting and Clinical Psychology*, 1970, *34*, 355-359.

Allport, F. H., & Allport, G. W. Personality traits: their classification and measurement. *Journal of Abnormal and Social Psychology*, 1921, *16*, 6-40.

Allport, G. *Personality*. London: Constable, 1937.

Allport, G. W., & Odbert, H. S. Trait-names: A psycho-lexical study. *Psychological Monographs*, 1936, *47*, 171.

Allsopp, J. F., & Feldman, M. P. Personality and anti-social behavior in schoolboys. *British Journal of Criminology*, 1976, *16*, 337–351.

Amelang, M., & Borkenau, P. Über die faktorielle Struktur und externe Validität einiger Fragebogen-Skalen zur Erfassung von Dimensionen der Extraversion und emotionalen Labilität. *Zeitschrift für Differentielle und Diagnostische Psychologie*, 1982, *3*, 119–146.

Amelang, M., & Breit, C. Extraversion and rapid tapping: Reactive inhibition or general cortical activation as determinants of performance differences. *Personality and Individual Differences*, 1983, *4*, 103–106.

Anastasi, A. *Psychological testing*. London: Macmillan, 1982.

Andrews, F. M., & Withey, S. B. Social indicators of well-being: American's perceptions of life quality. New York: Plenum Press, 1976.

Angleitner, A. *Einführung in die Persönlichkeitspsychologie: Vol. 2. Faktorielle Ansätze*. Bern: Huber, 1982.

Angst, J., & Maurer-Groeli, Y. A. Blutgruppen und Persönlichkeit. *Archiv für Psychiatrie und Nervenkrankheiten*, 1974, *218*, 291–300.

Anthony, W. S. The development of extraversion, of ability, and of the relation between them. *British Journal of Educational Psychology*, 1973, *43*, 223–227.

Anthony, W. S. The development of extraversion and ability: An analysis of Rushton's longitudinal data. *British Journal of Educational Psychology*, 1977, *47*, 193–196.

Anthony, W. S. Extraversion and intelligence: Re-analysis of data of Crookes *et al*. *British Journal of Educational Psychology*, 1982, *52*, 119–120.

Argyle, M., Furnham, A., & Graham, J. A. *Social situations*. Cambridge: Cambridge University Press, 1981.

Arnhoff, F. N., & Leon, H. V. Personality factors related to success and failure in sensory deprivation subjects. *Perceptual and Motor Skills*, 1963, *16*, 46.

Atkinson, R. C., & Shiffrin, R. M. Human memory: A proposed system and its control processes. In K. W. Spence & J. T. Spence (Eds.), *The psychology of learning and motivation* (Vol. 2). London: Academic Press, 1968.

Backteman, G. Longitudinal stability of social relations. University of Stockholm: *Reports from the Department of Psychology*, 1978, No. 538.

Backteman, G. Longitudinal stability of personal characteristics: Questionnaire data. University of Stockholm: *Reports from the Department of Psychology*, 1979, No. 555.

Backteman, G., & Magnusson, D. Longitudinal stability of personality characteristics. *Journal of Personality*, 1981, *49*, 148–160.

Baddeley, A. D., & Hitch, G. Working memory. In G. H. Bower (Ed.), *The psychology of learning and motivation* (Vol. 8). London: Academic Press, 1974.

Baehr, M. A. *A factorial study of temperament*. Chicago: Psychometric Laboratory, 1951.

Baken, P. Extraversion–introversion and improvement in an auditory vigilance task. *British Journal of Psychology*, 1959, *50*, 325–332.

Bakan, P., Belton, J. A., & Toth, J. C. Extraversion–introversion and decrement in an auditory vigilance task. In D. N. Buckner & J. J. McGrath (Eds.), *Vigilance: A symposium*. New York: McGraw-Hill, 1963.

Baker, A. H., Mishara, B. L., Kostin, I. W., & Parker, L. Kinesthetic after-effect and personality: A case study of issues involved in construct validation. *Journal of Personality and Social Psychology*, 1976, *34*, 1–13.

Baker, H. D., Ryder, E. A., & Baker, N. H. *Temperature measurement in engineering* (2 vols.). Stamford: Omega, 1975.

Baltes, P. B., & Nesselroade, J. R. The developmental analysis of individual differences on multiple measures. In J. R. Nesselroade & H. W. Reese (Eds.), *Life-span developmental psychology: Methodological issues.* New York: Academic Press, 1973.

Banks, O., & Finlayson, D. *Success and failure in the secondary school.* London: Methuen, 1973.

Barker, R. G., & Wright, H. F. *Midwest and its children: The psychological ecology of an American town.* Evanston, Ill: Row, Peterson, 1955.

Barnes, B. *T. S. Kuhn and social science.* London: Macmillan, 1982.

Barnes, G. Extraversion and pain. *British Journal of Social and Clinical Psychology,* 1975, *14,* 303–308.

Barnes, G. E. Individual differences in perceptual reactance: A review of the stimulus intensity modulation individual difference dimension. *Canadian Psychological Review,* 1976, *17,* 29–52.

Barr, R. E., & McConaghy, N. A general factor of conditionability: A study of galvanic skin responses and penile responses. *Behaviour Research and Therapy,* 1972, *10,* 215–227.

Barratt, E. S. Psychophysiological correlates of classical differential eyelid conditioning among subjects selected on the basis of impulsiveness and anxiety. *Biological Psychiatry,* 1971, *3,* 339–346.

Barrat, E. S., Patton, J., Olsson, N. G., & Zuker, G. Impulsivity and paced tapping. *Journal of Motor Behavior,* 1981, *13,* 286–300.

Barrett, P., & Kline, P. The location of superfactors P, E, and N within a unexplored factor space. *Personality and Individual Differences,* 1980, *1,* 239–247. (a)

Barrett, P., & Kline, P. The observation to variable ratio in factor analyses. *Personality and Group Behavior,* 1980, *1,* 23–33. (b)

Barrett, P., Kline, P. Personality factors in the Eysenck Personality Questionnaire. *Personality and Individual Differences,* 1980, *1,* 317–333. (c)

Barrett, P., & Kline, P. An item and radial parcel analysis of the 16PF questionnaire. *Personality and Individual Differences,* 1982, *3,* 259–270. (a)

Barrett, P., & Kline, P. The itemetric properties of the Eysenck Personality Questionnaire: A reply to Helmes. *Personality and Individual Differences,* 1982, *3,* 73–80. (b)

Bartol, C. R., & Costello, N. Extraversion as a function of temporal duration of electrical shock: An exploratory study. *Perceptual and Motor Skills,* 1976, *42,* 1174.

Barton, K., & Cattell, R. B. An investigation of the common factor space of some well-known questionnaire scales: The Eysenck E.P.I., the Comrey Scales and the IPAT Central Trait-State Test (CST). *Journal of Multivariate Experimental Personality and Clinical Psychology,* 1975, *1,* 268–273.

Bartram, D., & Dale, H. C. A. The Eysenck Personality Inventory as a selection test for military pilots. *Journal of Occupational Psychology,* 1982, **55,** *287–296.*

Bass, B. M. Famous Saying Test: General Manual. *Psychological Reports,* 1958, *4,* 478–497.

Battig, W. F., & Montague, W. E. Category norms for verbal items in 56 categories: A replication and extension of the Connecticut category norms. *Journal of Experimental Psychology,* 1969, *80,* (No. 3, Pt. 2).

Beck, A. T., Laude, R., & Bohnert, M. Ideational components of anxiety neurosis. *Archives of General Psychiatry,* 1974, *31.* 319–325.

Bendig, A. W. The relation of temperament traits of social extraversion and emotionality to vocational interests. Journal of General Psychology, 1963, *69,* 311–318.

Berger, M. The "scientific approach" to intelligence: An overview of its history with special reference to mental speed. In H. J. Eysenck (Ed.), *A model for intelligence.* New York: Springer, 1982.

Bergius, R. Die Ablenkung von der Arbeit durch Lärm und Musik und ihre strukturpsychologischen Zusammenhänge. *Zeitschrift für Arbeitspsychologie und Praktische Psychologie im Allgemeinen*, 1939, *12*, 90–114.

Bergstrom, B., Gillberg, M., & Arnberg, P. Effects of sleep loss and stress upon radar watching. *Journal of Applied Psychology*, 1973, *58*, 158–162.

Berka, K. *Measurement: Its concepts, theories and problems*. London: Reidel, 1982.

Berlyne, D. E., Borsa, D. M., Craw, M. A., Gelman, R. S., & Mandell, E. E. Effects of stimulus complexity and induced arousal on paired-associate learning. *Journal of Verbal Learning and Verbal Behavior*, 1965, *4*, 291–299.

Binet, A. *L'étude expérimentale de l'intelligence*. Paris: Schleicher Frères, 1903.

Binet, A. *La psychologie du raisonnement*. Paris: F. Alcan, 1907.

Binet, A. Qu'est-ce qu'une émotion, qu'est-ce qu'un acte intellectuel? *L'Année Psychologique*, 1911, *17*, 1–47.

Bishop, D. V. M. The P scale and psychosis. *Journal of Abnormal Psychology*, 1977, *86*, 127–134.

Blackburn, R. The scores of Eysenck's criterion groups on some MMPI scales related to emotionality and extraversion. *British Journal of Social and Clinical Psychology*, 1968, *7*, 3–12.

Blake, M. J. F. Relationship between circadian rhythm of body temperature and introversion-extraversion. *Nature*, 1967, *215*, 896–897.

Blakemore, C., & Campbell, F. W. On the existence of neurones in the human visual system selectively sensitive to the orientation and size of retinal images. *Journal of Physiology*, 1969, *203*, 237–260.

Blanchard, R. J., Fukunaga, K., Blanchard, D. C., & Kelley, M. J. Conspecific aggression in the laboratory rat. *Journal of Comparative and Physiological Psychology*, 1975, *89*, 1204–1209.

Blanchard, R. J., Blanchard, D. C., Takahashi, T., & Kelley, M. J. Attack and defensive behaviour in the Albino rat. *Animal Behaviour*, 1977, *25*, 622–634.

Blinkhorn, S. F., & Henrickson, D. E. Averaged evoked responses and psychometric intelligence. *Nature*, 1982, *295*, 596–597.

Block, J. *The challenge of response sets*. New York: Appleton-Century-Crofts, 1965.

Block, J. The P scale and psychosis: Continued concern. *Journal of Abnormal Psychology*, 1977, *86*, 431–434. (a)

Block, J. The Eysencks and psychoticism. *Journal of Abnormal Psychology*, 1977, *86*, 653–654. (b)

Block, J., & Haan, N. *Lives through time*. Berkeley: Bancroft, 1971.

Block, J., Weiss, D. S., Thorne, A. How relevant is a semantic similarity interpretation of personality ratings? *Journal of Personality and Social Psychology*, 1979, *37*, 1055–1074.

Blunt, P. Personality characteristics of a group of white South African managers: Some implications for placement procedures. *International Journal of Psychology*, 1978, *13*, 139–146.

Bolton, B. Evidence for the 16 PF primary and secondary factors. *Multivariate Experimental Clinical Research*, 1977, *3*, 1–15.

Bone, R. N., & Eysenck, H. J. Extraversion, field dependence and the Stroop test. *Perceptual and Motor Skills*, 1972, *34*, 873–874.

Borgatta, E. F. The coincidence of subtests in four personality inventories. *Journal of Social Psychology*, 1962, *56*, 227–244.

Borgatta, E. F. The structure of personality characteristics. *Behavioral Science*, 1964, *9*, 8–17.

Bowers, K. Situationism in psychology: An analysis and a critique. *Psychological Review*, 1973, *80*, 307-336.

Bràdburn, N. M. *The structure of psychological well-being*. Chicago: Aldine, 1969.

Bradburn, N. M., & Caplowitz, D. *Reports on happiness: A pilot study of behavior related to mental health*. Chicago: Aldine, 1965.

Brady, A. T., & Walker, M. B. Interpersonal distance as a function of situationally induced anxiety. *British Journal of Social and Clinical Psychology*, 1978, *17*, 127-133.

Brand, C. R., & Deary, I. J. Intelligence and "inspection time" In H. J. Eysenck (Ed.), *A model for intelligence*. New York: Springer, 1982, pp. 133-148.

Brebner, J. A comment on Paisey and Mangan's neo-Pavlovian temperament theory and the biological bases of personality. *Personality and Individual Differences*, 1983, *4*, 229-230.

Brebner, J., & Cooper, C. The effect of a low rate of regular signals upon the reaction times of introverts and extraverts. *Journal of Research in Personality*, 1974, *8*, 263-276.

Brebner, J., & Cooper, C. Stimulus- or response-produced excitation: A comparison of the behavior of introverts and extraverts. *Journal of Research in Personality*, 1978. *12*, 306-311.

Brebner, J., & Flavel, R. The effect of catch-trials on speed and accuracy among introverts and extraverts in a simple RT task. *British Journal of Psychology*, 1978, *69*, 9-15.

Brengelmann, J. C. Kretschmer's zyklothymer und schizothymer Typus in Bereich der normalen Persönlichkeit. *Psychologische Rundschau*, 1952, *3*, 31-38.

Broadbent, D. E. *Decision and stress*. London: Academic Press, 1971.

Broadhurst, P. L. Determinants of emotionality in the rat: I. Situational factors. *British Journal of Psychology*, 1957, *48*, 1-12.

Broadhurst, P. L. Applications of biometrical genetics to the inheritance of behaviour. In H. J. Eysenck (Ed.), *Experiments in personality*. London: Routledge & Kegan Paul, 1960.

Broadhurst, P. L. Animal studies bearing on abnormal behaviour. In H. J. Eysenck (Ed.), *Handbook of abnormal psychology*. London: Pitman, 1973, pp. 721-754.

Broadhurst, P. L. The Maudsley Reactive and Non-reactive Strains of Rats: A Survey. *Behavior Genetics*, 1975, *5*, 299-319.

Broadhurst, P. L., & Eysenck, H. J. Emotionality in the rat: A problem of response specificity. In C. Banks & P. L. Broadhurst (Eds.), *Stephanos: Studies in psychology presented to Cyril Burt*. London: University of London Press, 1965, pp. 202-221.

Brody, n. *Personality: Research and theory*. New York: Academic Press, 1972.

Broen, V. E. *Schizophrenia: research and theory*. New York: Academic Press, 1968.

Bronson, W. C. Central orientations: A study of behavior organization from childhood to adolescence. *Child Development*, 1966, *37*, 125-155.

Bronson, W. C. Adult derivatives of emotional experiences and reactivity-control: Developmental continuities from childhood to adulthood. *Child Development*, 1967, *38*, 801-878.

Brook, F. R., & Johnson, R. W. Self-descriptive adjectives associated with a Jungian Personality Inventory. *Psychological Reports*, 1979, *44*, 747-750.

Browne, J. A. *Extraversion: In search of a personality dimension*. Unpublished Ph.D. thesis, University of Alberta, 1971.

Browne, J. A., & Howarth, E. A comprehensive factor analysis of personality questionnaire items: A test of twenty putative factor hypotheses. *Multivariate Behavioral Research*, 1977, *12*, 399-427.

Bruning, J. L., Capage, J. E., Kozuh, G. F., Young, P. F. & Young, W. E. Social induced drive and range of cue utilization. *Journal of Personaltiy and Social Psychology*, 1968, *9*, 242-244.

Bull, R. H., & Gale, M. A. The reliability of and interrelationships between various measures of electrodermal activity. *Journal of Experimental Research in Personality*, 1973, *6*, 300–306.

Burdsal, C., & Bolton, B. An item factoring of 16PF-E: Further evidence concerning Cattell's normal personality sphere. *Journal of General Psychology*, 1979, *100*, 103–109.

Burdsal, C. A., & Vaughn, D. S. A contrast of the personality structure of college students found in the questionnaire medium by items as compared to parcels. *Journal of Genetic Psychology*, 1974, *125*, 219–224.

Burgess, L. S. *Extraversion: Recovery from the effects of somatosensory stimulation.* Unpublished Ph.D. dissertation, University of Ottawa, 1973.

Burgess, P. K. Eysenck's theory of criminality: A new approach. *British Journal of Criminology*, 1972, *12*, 74–82.

Burt, C. Experimental tests of general intelligence. *British Journal of Psychology*, 1909, *3*, 94–177.

Burt, C. Factorial studies of personality and their bearing on the work of the teacher. *British Journal of Educational Psychology*, 1965, *35*, 368–378.

Burton, R. V. Generality of honesty reconsidered. *Psychological Review*, 1963, *70*, 481–499.

Buss, A. H. Two anxiety factors in psychiatric patients. *Journal of Abnormal and Social Psychology*, 1962, *65*, 426–427.

Buss, A. H. *Psychopathology.* New York: Wiley, 1966.

Buss, A. H., & Plomin, R. *A temperament theory of personality development.* New York: Wiley, 1975.

Byth, W. Extraversion and the visual figural after-effect with tachistoscopic presentation. *British Journal of Psychology*, 1972, *63*, 569–571.

Campbell, A., & Rushton, J. P. Bodily communication and personality. *British Journal of Social and Clinical Psychology*, 1978, *17*, 31–36.

Campbell, D. T. Recommendations for APA standards regarding construct, trait, and discriminant validity. *American Psychologist*, 1960, *15*, 546–553.

Campbell, D. T., & Fiske, D. W. Convergent and discriminant validation by the multitrait-multimethod matrix. *Psychological Bulletin*, 1959, *56*, 81–105.

Campbell, J. B., & Hawley, C. W. Study habits and Eysenck's theory of extraversion-introversion. *Journal of Research in Personality*, 1982, *16*, 139–146.

Campbell, J. B., & Reynolds, J. H. Interrelationships of the Eysenck Personality Inventory and the Eysenck Personality Questionnaire. *Educational and Psychological Measurement*, 1982, *42*, 1067–1073.

Campbell, J. B., & Reynolds, J. H. A comparison of the Guilford and Eysenck factors of personality. *Journal of Research in Personality*, in press.

Campbell, J. B., Baribeau-Bräun, J., & Braun, C. Neuroanatomical and physiological foundations of extraversion. *Psychophysiology*, 1981, *18*, 263–267.

Canter, H., & Loo, R. Relationships between field dependence and Eysenck's personality dimensions. *Journal of Psychology*, 1979, *103*, 45–49.

Carlier, M. Factor analysis of Strelau's questionnaire and attempt to validate some of the factors. In J. Strelau, F. Farley, & A. Gale (Eds.) *The biological foundations of personality and behavior.* New York: Hemisphere Press, 1982.

Carlson, R., & Levy, N. Studies of Jungian typology: I. Memory, social perception, and social action. *Journal of Personality*, 1973, *41*, 559–576.

Carlyn, M. An assessment of the Myers-Briggs Type Indicator. *Journal of Personality Assessment*, 1977, *41*, 461–473.

Carment, D. W., Miles, C. G., & Cervin, V. B. Persuasiveness and persuasibility as related to intelligence and extraversion. *British Journal of Social and Clinical Psychology*, 1965, *4*, 1-7.

Carr, G. Introversion-extraversion and vigilance performance. *Proceedings of the Annual Meeting of the American Psychological Association*, 1971, *79*. 379-380.

Carrigan, P. M. Extraversion-introversion as a dimension of personality: A reappraisal. *Psychological Bulletin*, 1960, *57*, 329-360.

Carson, R. C. Interpretive manual to the MMPI. In J. N. Butcher (Ed.), *MMPI: Research developments and clinical applications*. New York: McGraw-Hill, 1969.

Cattell, R. B. *Personality and motivation structure and measurement*. Yonkers: New World, 1957.

Cattell, R. B. The isopodic and equipotent principles for comparing factor scores across different populations. *British Journal of Mathematical and Statistical Psychology*, 1970, *23*, 23-41.

Cattell, R. B. The 16PF and basic personality structure: A reply to Eysenck. *Journal of Behavioral Science*, 1972, *1*, 169-188.

Cattell, R. B. *Personality and mood by questionnaire*. New York: Jossey-Bass, 1973.

Cattell, R. B. *The inheritance of personality and ability*. New York: Academic Press, 1982.

Cattell, R. B., & Bolton, L. G. What pathological dimensions lie beyond the normal dimensions of the 16PF? A comparison of MMPI and 16PF factor domains. *Journal of Consulting and Clinical Psychology*, 1969, *33*, 18-29.

Cattell, R. B., & Scheier, I. H. *The meaning and measurement of neuroticism and anxiety*. New York: Ronald, 1961.

Cattell, R. B., Eber, H. W. & Tatsouka, M. M. *Handbook for the Sixteen Personality Factor Questionnaire (16PF)*. Champaign, Ill.: Institute for Personality and Ability Testing, 1970.

Cattell, R. B., Schmidt, L., & Pavlik, K. Cross-cultural comparisons (USA, Japan, Austria) of the personality factor structures of 10-to 14-year olds in objective tests. *Social Behavior and Personality*, 1973, *1*, 182-211.

Chamove, A. S., Eysenck, H. J., & Harlow, H. F. Personality in monkeys: Factor analysis of rhesus social behaviour. *Quarterly Journal of Experimental Psychology*, 1972, *24*, 496-504.

Child, D. The relationships between introversion-extraversion, neuroticism and performance in school examinations. *British Journal of Educational Psychology*, 1964, *34*, 178-196.

Child, I. L. The relation of somatotype to self-ratings on Sheldon's temperamental traits. *Journal of Personality*, 1950, *18*, 440-453.

Cicero, M. T. *Tusculanarum disputationum*. London: Heinemann, 1971.

Claridge, G. S. *Personality and arousal*. Oxford: Pergamon, 1967.

Claridge, G. S. The schizophrenics as nervous types. *British Journal of Psychiatry*, 1972, *121*, 1-17.

Claridge, G. S. A nervous typological analysis of personality variation in normal twins. In G. S. Claridge (Ed.), *Personality differences and biological variations: A study of twins*. London: Pergamon Press, 1973.

Claridge, G. S. Psychoticism. In R. Lynn (Ed.), *Dimensions of personality* London: Pergamon Press, 1981. pp. 79-109.

Claridge, G. S. The Eysenck psychoticism scale. In J. D. Butcher & C. D. Spielberger (Eds.), *Advances in personality assessment* Vol. 2. Hillsdale, N. J.: Lawrence Erlbaum, 1983, pp. 71-114.

Claridge, G. S., & Birchall, P. Bishop, Eysenck, Berck and psychoticism. *Journal of Abnormal Psychology*, 1978, *87*, 664–668.

Claridge, G. S., & Chappa, H. J. Psychoticism: A study of its biological basis in normal subjects. *British Journal of Social and Clinical Psychology*, 1973, *12*, 175–187.

Claridge, G. S., & Clark, K. Covariation between two-flash thresholds and skin conductance level in first-breakdown schizophrenia: Relationships in drug-free patients and effects of treatment. *Psychiatry Research*, 1982, *6*, 371–380.

Claridge, G. S., & Herrington, R. N. An EEG correlate of the Archimedes spiral aftereffect and its relationship with personality. *Behaviour Research and Therapy*, 1963, *1*, 217–230. (a)

Claridge, G. S., & Herrington, R. N. Excitation-inhibition and the theory of neurosis: A study of the sedation threshold. In H. J. Eysenck (Ed.), *Experiments with drugs*. London: Pergamon Press, 1963, pp. 131–168. (b)

Claridge, G. S., & Ross, E. Sedative drug tolerance in twins. In G. S. Claridge, S. Carter, & W. I. Hume (Eds.), *Personality differences and biological variations*. Oxford: Pergamon, 1973.

Claridge, G. S., Donald, J. R., & Birchall, P. M. Drug tolerance and personality: Some implications for Eysenck's theory. *Personality and Individual Differences*, 1981, *2*, 153–166.

Claridge, G., Robinson, D., & Birchall, P. Characteristics of schizophrenics' and neurotics' relatives. *Personality and Individual Differences*, 1983, *4*, 651–664.

Clough, F. Staff ratings of children's behaviour in hospital: Comparability of factor structures. *British Journal of Psychology*, 1978, *69*, 59–68.

Coles, M. G. H., Gale, A., & Kline, P. Personality and habituation of the orienting reaction: Tonic and response measures of electrodermal activity. *Psychophysiology*, 1971, *8*, 54–63.

Collier, R., & Emch, M. Introversion-extraversion: The concepts and their clinical use. *American Journal of Psychiatry*, 1938, *94*, 1045–1075.

Colquhoun, W. P. Temperament, inspection efficiency, and time of day. *Ergonomics*, 1960, *3*, 377–378.

Comrey, A. L. A factor analysis of items on the MMPI hypochondriasis scale. *Educational and Psychological Measurement*, 1957, *17*, 568–577. (a)

Comrey, A. L. A factor analysis of items on the MMPI Depression Scale. *Educational and Psychological Measurement*, 1957, *17*, 578–585. (b)

Comrey, A. L. A factor analysis of items on the MMPI Hysteria Scale. *Educational and Psychological Measurement*, 1957, *17*, 586–592. (c)

Comrey, A. L. A factor analysis of items of the F scale of the MMPI. *Educational and Psychological Measurement*, 1958, *18*, 621–632. (a)

Comrey, A. L. A factor analysis of items on the MMPI Psychopathic Deviate Scale. *Educational and Psychological Measurement*, 1958, *18*, 91–98. (b)

Comrey, A. L. A factor analysis of items on the MMPI Paranoia Scale. *Educational and Psychological Measurement*, 1958, *18*, 99–107. (c)

Comrey, A. L. A factor analysis of items on the MMPI Psychasthenia Scale. *Educational and Psychological Measurement*, 1958, *18*, 293–300. (d)

Comrey, A. L. A factor analysis of items on the MMPI Hypomania Scale. *Educational and Psychological Measurement*, 1958, *18*, 313–323. (e)

Comrey, A. L. A factor analysis of items on the K Scale of the MMPI. *Educational and Psychological Measurement*, 1958, *18*, 633–639. (f)

Comrey, A. L. *Handbook of intepretations for the Comrey Personality Scales*. San Diego: Edits, 1980.

Comrey, A. L., & Duffy, K. E. Cattell and Eysenck factor scores related to Comrey personality factors. *Multivariate Behavioral Research*, 1968, *3*, 379-392.

Comrey, A. L., & Margraff, W. A factor analysis of items on the MMPI Schizophrenia Scale. *Educational and Psychological Measurement*, 1958, *18*, 301-311.

Comrey. A. L., & Soufi, A. Further investigation of some factors found in MMPI items. *Educational and Psychological Measurement*, 1960, *20*, 777-786.

Conklin, E. S. The definition of extraversion, introversion and allied concepts. *Journal of Abnormal and Social Psychology*, 1923, *17*, 367-382.

Conklin, E. S. The determination of normal extravert-introvert differences. *Journal of Genetic Psychology*, 1927, *34*, 28-32.

Conley, J. J. Longitudinal consistency of adult personality: Neuroticism and social introversion-extraversion over forty years. *Journal of Personality and Social Psychology*, 1984, in press.

Conley, J. J. The hierarchy of consistency: A review and model of longitudinal findings on adult individual differences in intelligence, personality and self-opinion. *Personality and Individual Differences*, 1984, *5*, 11-26.

Cook, D. L., Linden, J. D., & McKay, H. E. A factor analysis of teacher trainee responses to selected personality inventories. *Educational and Psychological Measurement*, 1961, *21*, 865-872.

Cooper, R., & Payne, R. Extraversion and some aspects of work behavior. *Personnel Psychology*, 1967, *20*, 45-57.

Corah, N. L. Neuroticism and extraversion in the MMPI: Empirical validation and exploration. *British Journal of Social and Clinical Psychology*, 1964, *3*, 168-174.

Corcoran, D. W. J. Noise and loss of sleep. *Quarterly Journal of Experimental Psychology*, 1962, *14*, 178-182.

Cortes, J. B., & Gatti, F. M. Physique and self-description of temperament. *Journal of Consulting Psychology*, 1965, *29*, 432-439.

Costa, P. T., & McCrae, R. R. Age differences in personality structure: A cluster analytic approach. *Journal of Gerontology*, 1976, *31*, 564-570.

Costa, P. T., & McCrae, R. R. *The relations between smoking motives, personality, and feelings. Progress Report III*. University of Massachusetts, 1977.

Costa, P. T., & McCrae, R. R. Objective personality assessment. In M. Storandt, I. C. Siegler, & M. F. Elias (Eds.), *The clinical psychology of aging*. New York: Plenum Press, 1978.

Costa, P. T., & McCrae, R. R. Influence of extraversion and neuroticism on subjective well-being: Happy and unhappy people. *Journal of Personality and Social Psychology*, 1980, *38*, 668-678. (a)

Costa, P. T., & McCrae, R. R. Still stable after all these years: Personality as a key to some issues in adulthood and old age. In P. B. Baltes & O. G. Brim (Eds.), *Life span development and behaviour*. (Vol. 3.) New York: Academic Press, 1980. (b)

Costa, P. T., McCrae, R. R., & Arenberg, D. Enduring dispositions in adult males. *Journal of Personality and Social Psychology*, 1980, *38*, 793-800.

Costa, P. T., McCrae, R. R., & Norris, A. H. Personal adjustment to aging: Longitudinal prediction from neuroticism and extraversion. *Journal of Gerontology*, 1981, *36*, 78-85.

Costa, P. T., McCrae, R. R., & Arenberg, D. Recent longitudinal research on personality and aging. In K. W. Shaie (Ed.), *Longitudinal studies of aging*. New York: Guilford Press, 1983.

Costello, C. G. The effects of meprobamate on the spiral after-effect. In H. J. Eysenck (Ed.), *Experiments with drugs*. New York: Pergamon, 1963.

Cowie, V. The incidence of neurosis in the children of psychotics. *Acta Psychiatrica Scandinavica*, 1961, *37*, 37–59.

Craig, M. J., Humphreys, M. S., Rocklin, T., & Revelle, W. Impulsivity, neuroticism, and caffeine: Do they have additive effects on arousal? *Journal of Research in Personality*, 1979, *13*, 404–419.

Craik, F. I. M., & Lockhart, R. S. Levels of processing: A framework for memory research. *Journal of Verbal Learning and Verbal Behavoir*, 1972, *11*, 671–684.

Crites, J. O., Bechtoldt, H. P., Goodstein, L. D., & Heilbrun, A. B. A factor analysis of the California Psychological Inventory. *Journal of Applied Psychology*, 1961, *45*, 408–414.

Cronbach, L. J. The two disciplines of scientific psychology. *American Psychologist*, 1957, *12*, 671–684.

Cronbach, L. J., & Meehl, P. E. Construct validity in psychological tests. *Psychological Bulletin*, 155, *52*, 381–402.

Crozier, W. R. Shyness as a dimension of personality. *British Journal of Social and Clinical Psychology*, 1979, *18*, 121–128.

Daly, S. Behavioral correlates of social anxiety. *British Journal of Social and Clinical Psychology*, 1978, *17*, 117–120.

D'Andrade, R. G. Trait psychology and componential analysis. *American Anthropologist*, 1965, *37*, 215–228.

Davidson, W. B., & House, W. J. On the relationship between reflection-impulsivity and field-dependence-independence. *Perceptual and Motor Skills*, 1978, *47*, 306.

Davies, D. R., & Hockey, G. R. J. The effects of noise and doubling the signal frequency on individual differences in visual vigilance performance. *British Journal of Psychology*, 1966, *57*, 381–389.

Davies, D. R., & Tune, G. S. *Human vigilance performance.* London: Staples, 1970.

Davies, D. R., Hockey, G. R., & Taylor, A. Varied auditory stimulation, temperament differences and vigilance performance. *British Journal of Psychology*, 1969, *60*, 453–457.

Davis, D. R. *Pilot error: Some laboratory experiments.* London: H.M.S.O., 1948.

Davis, H. What does the P scale measure? *British Journal of Psychiatry*, 1974, *125*, 161–167.

Deakin, J. F., & Exley, K. A. Personality and male-female influences on the EEG alpha rhythm. *Biological Psychology*, 1979, *8*, 285–290.

De Bonis, M. Étude factorielle de la systomatologie subjective de l'anxiété pathologique. *Revue de Psychologie Appliquée*, 1968, *18*, 177–187.

Deffenbacher, J. L. Worry, emotionality, and task-generated interference in test anxiety: An empirical test of attentional theory. *Journal of Educational Psychology*, 1978, *70*, 248–254.

Delay, J., Daniker, P., & Green, A. Le milieu familial des schizophrènes. *Encephale*, 1957, *46*, 189–196.

Demangeon, M. Évolution de la personalité à l'adolescence. *Orientation Scolaire et Professionelle*, 1977, *6*, 137–159.

Denenberg, V. H. Open-field in the rat: What does it mean? *Annals of the New York Academy of Science*, 1969, *159*, 831–851.

Di Loreto, A. O. *Comparative psychotherapy: An experimental analysis.* Chicago: Aldine-Atherton, 1971.

Dombrose, L. A., & Slobin, M. S. The IES Test. *Perceptual and Motor Skills*, 1958, *8*, 347–389.

Dornic, S. Mental load, effort, and individual differences. *Reports from the Department of Psychology*. The University of Stockholm, No. 509, 1977.

Dornic, S. Efficiency vs. effectiveness in mental work: The differential effect of stress. *Reports from the Department of Psychology*, University of Stockholm, No. 568, 1980.

Doyle, A. C. *The Memoirs of Sherlock Holmes*. London: J. Murray & J. Cape, 1974.

Doyle, J. A. Self-actualization, neuroticism and extraversion revisited. *Psychological Reports*, 1976, *39*, 1081-1082.

Drake, L. E. A social IE scale for the MMPI. *Journal of Applied Psychology*, 1946, *30*, 51-54.

Dreger, R. M. First-, second- and third-order factors from the children's behavioral classification project instrument and an attempt at rapprochement. *Journal of Abnormal Psychology*, 1981, *90*, 242-260.

Duckitt, J., & Broll, T. Personality factors as moderators of the psychological impact of life stress. *South African Journal of Psychology*, 1982, *12*, 76-80.

Duffy, E. *Activation and behaviour*. London: Wiley, 1962.

Dunnette, M. D., Kirchner, W. K., & De Gidio, S. A. Relations among scores on Edwards Personal Preference Schedule, California Psychological Inventory, and Strong Vocational Interest Blank for an industrial sample. *Journal of Applied Psychology*, 1958. *42*, 178-181.

Dunstone, J. J., Dzendolet, G., & Henckeruth, O. Effect of some personality variables on electrical vestibular stimulation. *Perceptual and Motor Skills*, 1964, *18*, 689-695.

Duran, N. *Flindar de dolor i diferencies individuals en rates*. Barcelona: Unpublished M.A. thesis. Universidad Autonoma de Barcelona, 1978.

Dworkin, R. H., Burke, B. W., Maher, B. A., & Gottesmann, I. I. A longitudinal study of the genetics of personality. *Journal of Personality and Social Psychology*, 1976, *34*, 510-518.

Dworkin, R. H., Burke, B. W., Maher, B. A., & Gottesmann, I. I. Genetic influences on the organization and development of personality. *Developmental Psychology*, 1977, *13*, 164-165.

Easterbrook, J. A. The effect of emotion on cue utilization and the organization of behavior. *Psychological Review*, 1959, *66*, 183-201.

Eaves, L. The structure of genotype and environmental covariation of personality measurements: an analysis of the PEN. *British Journal of Social and Clinical Psychology*, 1973, *12*, 275-282.

Eaves, L. Twins as a basis for the causal analysis of personality. In W. E. Nance (Ed.), *Twin research: Psychology and methodology*. New York: Allan, 1978.

Eaves, L., & Eysenck, H. J. The nature of extraversion: A genetical analysis. *Journal of Personality and Social Psychology*, 1975, *32*, 102-112.

Eaves, L. J., & Eysenck, H. J. Genetic and environmental components of inconsistency and unrepeatability in twins' responses to a neuroticism questionnaire. *Behavoir Genetics*, 1976, *6*, 145-160. (a)

Eaves. L., & Eysenck, H. J. Genotype × age interaction for neuroticism. *Behavior Genetics*, 1976, *6*, 359-362. (b)

Eaves, L. J., & Eysenck, H. J. A genotype-environmental model for psychoticism. *Advances in Behaviour Research and Therapy*, 1977. *1*, 5-26.

Eaves, L. J., & Eysenck, H. J. The genetics of smoking. In H. J. Eysenck (Ed.), *The causes and effects of smoking*. London: Maurice Temple Smith, 1980. Part II, pp. 140-314.

Eaves, L. J., & Young, P. A. Genetical theory and personality differences. In R. Lynn (Ed.), *Dimensions of personality*. London: Pergamon Press, 1981, pp. 129-179.

Eaves. L. J., Last, K. A., Martin, N. G., & Jinks, J. L. A progressive approach to non-additivity and genotype-environmental covariance in the analysis of human differences. *British Journal of Mathematical and Statical Psychology*, 1977, *30*, 1–42.

Eaves, L. J., Martin, N. G., & Eysenck, S. B. G. An application of the analysis of covariance structures to the psychogenetical study of impulsiveness. *British Journal of Mathematical and Statistical Psychology*, 1977, *30*, 185–197.

Eaves, L. J., Young, P. A., Last, K. A., & Martin, N. G. Model fitting approaches to the analysis of human behaviour. *Heredity*, 1978, *41*, 249–320.

Edman, G., Schalling, D., & Rissler, A. Interaction effects of extraversion and neuroticism on detection thresholds. *Biological Psychology*, 1979, *9*, 41–47.

Edwards, A. L. *Manual for the Edwards Personal Preference Schedule*. New York: Psychological Corporation, 1957.

Edwards, A. L., & Klockars, A. J. Significant others and self-evaluation: Relationships between perceived and actual evaluations. *Personality and Social Psychology Bulletin*, 1981, *7*, 244–251.

Edwards, A. L., Abbott, R. D., & Klockars, A. J. A factor analysis of the EPPs and PRF personality inventories. *Educational and Psychological Measurement*, 1972, *32*, 23–29.

Eich, J. E. The cue-dependent nature of state-dependent retrieval. *Memory and Cognition*, 1980, *8*, 157–173.

Eichorn, D. H. The Berkeley longitudinal studies: Continuities and correlates of behaviour. *Canadian Journal of Behavioural Science*, 1973, *5*, 297–320.

Eichorn, D. H., Clausen, J. A., Haan, N., Honzik, M. P., & Mussen, P. H. *Present and past in middle life*. New York: Academic Press, 1981.

Ekman, G. On the number and definition of dimensions in Kretschmer's and Sheldon's constitutional systems. In G. Ekman (Ed.), *Essays in psychology dedicated to David Katz*. Uppsala, Sweden: Rijkman, 1951. (a)

Ekman, G. On typological and dimensional systems of reference in describing personality. *Acta Psychologica*, 1951, *8*, 1–24. (b)

Eliatamby, A. (1984). *Anxiety and anagram solving*. Unpublished manuscript.

Elliott, A. G. P. Some implications of lie scale scores in real-life selection. *Journal of Occupational Psychology*, 1981, *54*, 9–16.

Elliott, C. D. Noise tolerance and extraversion in children. *British Journal of Psychology*, 1971, *62*, 375–380.

Ellis, A. The validity of personality questionnaires. *Psychological Bulletin*, 1946, *43*, 385–440.

Ellis, A. *Reason and emotion in psychotherapy*. New York: Lyle Stuart, 1962.

Ellis, A., & Conrad, H. S. The validity of personality inventories in military practice. *Psychological Bulletin*, 1948, *45*, 385–426.

Emmerick, W. Continuity and stability in early social development. *Child Development*, 1964, *35*, 311–332.

Emmerick, W. Continuity and stability in early social development: Teacher ratings. *Child Development*, 1966, *36*, 17–27.

Emmerick, W. Stability and change in early personality development. In W. W. Hartup & N. L. Smothergill, *The young child: Review of research*. Washington: National Association for the Education of Young Children, 1967.

Endler, N. S., Magnusson, D., Ekehammar, B., & Okada, M. The multi-dimensionality of state and trait anxiety. *Scandinavian Journal of Psychology*, 1976, *17*, 81–96.

Entwistle, N. J., & Cunningham, S. Neuroticism and school attainment—A linear relationship *British Journal of Educational Psychology*, 1968, *38*, 123–132.

Epstein, S. Trains are alive and well. In D. Magnusson & N. S. Endler (Eds.), *Personality at the crossroads: Current issues in international psychology*. Hillsdale, N. J.: Lawrence Erlbaum, 1977.

Ertl, J. P. Fourier analysis of evoked potentials and human intelligence. *Nature*, 1971, *230*, 525-526.

Ertl, J. P., & Schafer, E. W. P. Brain response correlates of psychometric intelligence. *Nature*, 1969, *223*, 421-422.

Essen-Möller, E. The concept of schizoidia. *Psychiatry, Neurology and Medical Psychology*, 1946, *112*, 258-271.

Evans, F. J. Field dependence and the Maudsley Personality Inventory. *Perceptual and Motor Skills*, 1967, *24*, 526.

Eye, A., & Krampen, G. Zu den teststatistischen Eigenschaften der deutschsprachigen Version des Eysenck-Persönlichkeits-Inventars EPI, *Diagnostica*, 1979, *25*, 327-328.

Eysenck, H. J. Primary mental abilities. *British Journal of Educational Psychology*, 1939, *9*, 260-265.

Eysenck, H. J. Types of personality—A factorial study of 700 neurotics. *Journal of Mental Science*, 1944, *90*, 851-861.

Eysenck, H. J. *Dimensions of personality*. London: Routledge & Kegan Paul, 1947.

Eysenck, H. J. Criterion analysis: An application of the hypothetico-deductive method to factor analysis. *Psychological Review*, 1950, *57*, 38-53.

Eysenck, H. J. Schizothymia-cyclothymia as a dimension of personality: II. Experimental. *Journal of Personality*, 1952, *20*, 345-384. (a)

Eysenck, H. J. *The scientific study of personality*. London: Routledge & Kegan Paul, 1952. (b)

Eysenck, H. J. The logical basis of factor analysis. *American Psychologist*, 1953, *8*, 105-114.

Eysenck, H. J. *The psychology of politics*. London: Routledge & Kegan Paul, 1954.

Eysenck, H. J. Cortical inhibition, figural after-effect, and the theory of personality. *Journal of Abnormal and Social Psychology*, 1955, *51*, 94-106. (a)

Eysenck, H. J. Psychiatric diagnosis as a psychological and statistical problem. *Psychological Reports*, 1955, *1*, 3-17. (b)

Eysenck, H. J. The inheritance of extraversion-introversion. *Acta Psychologica*, 1956, *12*, 95-110. (a)

Eysenck, H. J. The questionnaire measurement of neuroticism and extraversion. *Rivista di Psichologia*, 1956, *50*, 113-140. (b)

Eysenck, H. J. *The dynamics of anxiety and hysteria*. London: Routledge & Kegan Paul, 1957.

Eysenck, H. J. *Manual of the Maudsley Personality Inventory*. London: University of London Press, 1959 (now Hodder & Stoughton). (a)

Eysenck, H. J. Personality and problem solving. *Psychological Reports*, 1959, *5*, 592. (b)

Eysenck, H. J. (Ed.) *Experiments in personality*, (Vol. 2). London: Routledge & Kegan Paul, 1960. (a)

Eysenck, H. J. The place of theory in psychology. In H. J. Eysenck (Ed.), Experiments in personality (Vol. 2). London: Routledge & Kegan Paul, 160, pp. 303-315. (b)

Eysenck, H. J. Symposium: The development of moral values in children: VII. The contribution of learning theory. *British Journal of Educational Psychology*, 1960, *30*, 11-21. (c)

Eysenck, H. J. Reminiscence, drive and personality—Revision and extension of a theory. *British Journal of Social and Clinical Psychology*, 1962, *1*, 127-140.

Eysenck, H. J. (Ed.). *Experiments with drugs.* London: Pergamon, 1963.

Eysenck, H. J. The biological basis of criminal behaviour. *Nature,* 1964, *203,* 952–953. (a)

Eysenck, H. J. Biological factors in neurosis and crime. *Scientia,* 1964, Dec., 1–11. (b)

Eysenck, H. J. Involuntary rest pauses in tapping as a function of drive and personality. *Perceptual and Motor Skills,* 1964, *18,* 173–174. (c)

Eysenck, H. J. Extraversion and the acquisition of eyeblink and GSR conditioned responses. *Psychological Bulletin,* 1965, *63,* 258–270.

Eysenck, H. J. *The biological basis of personality.* Springfield, Ill.: Charles C Thomas, 1967. (a)

Eysenck, H. J. Intelligence assessment: A theoretical and experimental approach. *British Journal of Educational Psychology,* 1967, *37,* 81–98. (b)

Eysenck, H. J. Personality patterns in various groups of businessmen. *Occupational Psychology,* 1967, *41,* 249–250. (c)

Eysenck, H. J. A dimensional system of psychodiagnostics. In A. R. Mahrer (Ed.), *New approaches to personality classification.* London: Columbia University Press, 1970, pp. 169–208. (a)

Eysenck, H. J. Explanation and the concept of personality. In R. Barger & F. Cioffi (Eds.). *Explanation in the Behavioural Sciences.* Cambridge: Cambridge University Press, 1970, pp. 389–410. (b)

Eysenck, H. J. The structure of human personality (3rd. ed.). London: Methuen, 1970. (c)

Eysenck, H. J. On the choice of personality tests for research and prediction. *Journal of Behaviour Science,* 1971, *1,* 85–89. (a)

Eysenck, H. J. *Readings in extraversion–introversion: 2. Fields of application.* London: Staples, 1971. (b)

Eysenck, H. J. *Readings in extraversion–introversion: 3. Bearings on basic psychological processes.* London: Staples, 1971. (c)

Eysenck, H. J. Relation between intelligence and personality. *Perceptual and Motor Skills,* 1971, *32,* 637–638. (d)

Eysenck, H. J. An experimental and genetic model of schizophrenia. In A. R. Kaplan (Ed.), *Genetic factors in schizophrenia.* Springfield, Ill.: Charles C Thomas, 1972, pp. 504–515. (a)

Eysenck, H. J. Primaries or second-order factors: A critical consideration of Cattell's 16PF battery. *British Journal of Social and Clinical Psychology,* 1972, *11,* 265–269. (b)

Eysenck, H. J. (Ed.). *Handbook of abnormal psychology.* London: Pitman, 1973. (a)

Eysenck, H. J. (Ed.). *The measurement of intelligence.* Lancaster: Medical & Technical Publishers, 1973. (b)

Eysenck, H. J. *Personality, learning, and anxiety.* In H. J. Eysenck (Ed.), *Handbook of abnormal psychology* (2nd ed.) London: Pitman, 1973. (c)

Eysenck, H. J. Who needs a random sample? *Bulletin of the British Psychological Society,* 1975, *28,* 195–198.

Eysenck, H. J. The learning theory model of neurosis—A new approach. *Behaviour Research and Therapy,* 1976, *14,* 251–267. (a)

Eysenck, H. J. *The measurement of personality.* Lancaster: Medical & Technical Publishers, 1976. (b)

Eysenck, H. J. *Sex and personality.* London: Open Books, 1976. (c)

Eysenck, H. J. *Crime and personality* (3rd ed.). London: Routledge & Kegan Paul, 1977. (a)

Eysenck, H. J. Personality and factor analysis: A reply to Guilford. *Psychological Bulletin,* 1977, *84,* 405–411. (b)

Eysenck, H. J. National differences in personality as related to ABO blood groups polymorphism. *Psychological Reports*, 1977, *41*, 1257-1258. (c)

Eysenck, H. J. The development of personality and its relation to learning. In S. Murray-Smith (Ed.), *Melbourne studies in education*. Melbourne: University Press, 1978, pp. 143-181. (a)

Eysenck H. J. Superfactors P, E and N in a comprehensive factor space. *Multivariate Behavioral Research*, 1978, *13*, 475-482. (b)

Eysenck, H. J. Personality factors in a random sample of the population. *Psychological Reports*, 1979, *44*, 1023-1027.

Eysenck, H. J. *A model for personality*. New York: Springer, 1981.

Eysenck, H. J. The biological basis of cross-cultural differences in personality: Blood group antigens. *Psychological Reports*, 1982, *51*, 531-540. (a)

Eysenck, H. J. *A model for intelligence*. New York: Springer, 1982. (b)

Eysenck, H. J. The psychophysiology of intelligence. In C. Spielberger & J. Butcher (Eds.). *Advances in personality assessment*. Hillsdale, N.J.: Lawrence Erlbaum, 1982. (c)

Eysenck, H. J. A biometrical-genetical analysis of impulsive and sensation-seeking behavior. In M. Zuckerman (Ed.), *Biological bases of sensation-seeking, impulsivity and anxiety*. Hillside, N.J.: Lawrence Erlbaum, 1983. (a)

Eysenck, H. J. Is there a paradigm in personality research? *Journal of Research in Personality*, 1983, *17*, 369-397. (b)

Eysenck, H. J. Psychopharmacology and personality. In W. Janke (Ed.), *Response variability to psychotropic drugs*. London: Pergamon Press, 1983. (c)

Eysenck, H. J. The theory of intelligence and the psychophysiology of cognition. In R. J. Sternberg (Ed.), *Advances in the psychology of human intelligence* (Vol. 3). Hillsdale, N.J.: Lawrence Erlbaum, 1984.

Eysenck, H. J. The place of theory in a world of facts. *International Journal of Theoretical Psychology*, in press.

Eysenck, H. J, & Barrett, P. Psychophysiology and the measurement of intelligence. In C. R. Reynolds & V. Willson (Ed.), *Methodological and statistical advances in the study of individual differences*. New York: Plenum Press, 1984.

Eysenck, H. J., & Broadhurst, P. L. Experiments with animals. In H. J. Eysenck (Ed.), *Experiments in motivation*. Oxford: Pergamon Press, 1965, pp. 285-291.

Eysenck, H. J., & Cookson, D. Personality in primary school children: I. Ability and achievement. *British Journal of Educational Psychology*, 1969, *39*, 109-122.

Eysenck, H. J., & Easterbrook, J. A. Drugs and personality: VIII. The effects of stimulant and depressant drugs on visual after-effects of a rotating spiral. *Journal of Mental Science*, 1960, *106*, 842-844.

Eysenck, H. J., & Eysenck, S. B. G. *Manual of the Eysenck Personality Inventory*. London: Hodder & Stoughton, 1965, (San Diego: Edits, 1965.)

Eysenck, H. J., & Eysenck, S. B. G. *Personality structure and measurement*. London: Routledge & Kegan Paul, 1969.

Eysenck, H. J., & Eysenck, S. B. G. *Manual of the Eysenck Personality Questionnaire*. London: Hodder & Stoughton, 1975. (San Diego: Edits, 1975.)

Eysenck, H. J., & Eysenck, S. B. G. *Psychoticism as a dimension of personality*. London: Hodder & Stoughton, 1976.

Eysenck, H. J., & Eysenck, S. B. G. Recent advances: The cross-cultural study of personality. In C. D. Spielberger & J. N. Butcher (Eds.), *Advances in personality assessment* (Vol. 2.) Hillsdale, N. J.: Lawrence Erlbaum, 1983, pp. 41-69.

Eysenck, H. J., & Frith, C. *Reminiscence, motivation and personality*. New York: Plenum Press, 1977.

Eysenck, H. J., & Fulker, D. The components of Type A behaviour and its genetic determinants. *Personality and Individual Differences*, in 1983, *4*, 499–505.

Eysenck, H. J., & Furneaux, D. Primary and secondary suggestibility: An experimental
and statistical study. *Journal of Experimental Psychology*, 1945, *35*, 485–503.

Eysenck, H. J., & Holland, H. C. Length of spiral after-effect as a function of drive. *Perceptual and Motor Skills, 1960*, 11, *129–130*.

Eysenck, H. J., & Levey, A. B. Konditionierung, Introversion-Extraversion und die Stärke des Nervensystems. *Zeitschrift für Psychologie*, 1967, *174*, 96–106.

Eysenck, H. J., & Levey, A. Conditioning, introversion-extraversion and the strength of
the nervous system. In V. D. Neblitsyn & J. A. Gray (Eds.), *Biological basis of individual behaviour.* London: Academic Press, 1972.

Eysenck, H. J., & Nias, D. *Astrology: Science or superstition?* London: Maurice Temple
Smith, 1982.

Eysenck, H. J., & Prell, D. B. The inheritance of neuroticism: An experimental study.
Journal of Mental Science, 1951, *97*, 441–465.

Eysenck, H. J., & Wilson, G. D. *The psychological basis of ideology.* Lancaster: Medical
& Technical Publishers, 1978. (Baltimore: University Park Press, 1978.)

Eysenck, H. J., & Wilson, G. D. *The psychology of sex.* London: Dent, 1979.

Eysenck, H. J., Holland, H. C., & Trouton, D. S. Drugs and personality: III. The effects of
stimulant and depressant drugs on visual after-effect. *Journal of Mental Science,*
1957, *103*, 650–655.

Eysenck, H. J., Willett, R. A., & Slater, P. Drive, direction–rotation, and massing of practice as determinants of the duration of the after-effect from the rotating spiral.
American Journal of Psychology, 1962, *75*, 127–133.

Eysenck, H. J., Nias, D., & Cox, D. N. Sport and personality. *Advances in Behaviour
Research and Therapy*, 1982, *1*, 1–56.

Eysenck, M. W. Extraversion, arousal, and retrieval from semantic memory. *Journal of
Personality*, 1974, *42*, 319–331. (a)

Eyseneck, M. W. ndividual differences in speed of retrieval from semantic memory. *Journal of Research in Personality*, 1974, 8, 307–323. (b)

Eysenck, M. W. Arousal and speed of recall. *British Journal of Social and Clinical Psychology*, 1975, *14*, 269–277.

Eysenck, M. W. Arousal, learning, and memory. *Psychological Bulletin*, 1976, *83*, 389–404.
(a)

Eysenck, M. W. Individual differences in speed of retrieval from semantic memory. *Jour-
83*, 75–90. (b)

Eysenck, M. W. *Human memory: Theory, research and individual differences.* Oxford:
Pergamon, 1977.

Eysenck, M. W. Anxiety, learning, and memory: A reconceptualization. *Journal of
Research in Personality*, 1979, *13*, 363–385.

Eysenck, M. W. Learning, memory and personality. In H. J. Eysenck (Ed.), *A model for
personality.* London: Springer, 1981.

Eysenck, M. W. *Attention and arousal: Cognition and performance.* Berlin: Springer,
1982.

Eysenck, M. W. Anxiety and individual differences. In G. R. J. Hockey (Ed.), *Stress and
fatigue in human performance.* Chichester: Wiley, 1983.

Eysenck, M. W. *Trait anxiety and cognitive task performance.* Manuscript in
preparation.

Eysenck, M. W., & Eysenck, H. J. Mischel and the concept of personality. *British Journal
of Psychology*, 1980, *71*, 191–209.

Eysenck, M. W., & Eysenck, M. C. Memory scanning, introversion—extraversion, and levels of processing. *Journal of Research in Personality*, 1979, *13*, 305–315.

Eysenck, M. W., & Folkard, S. Personality, time of day, and caffeine: Some theoretical and conceptual problems in Revelle *et al. Journal of Experimental Psychology: General*, 1980, *109*, 32–41.

Eysenck, S. B. G. Neurosis and psychosis: An experimental analysis. *Journal of Mental Science*, 1956, *102*, 517–529.

Eysenck, S. B. G. *Manual of the Junior Eysenck Personality Inventory*. London: Hodder & Stoughton, 1965.

Eysenck, S. B. G. Impulsiveness and antisocial behavior in children. *Current Psychological Research*, 1981, *1*, 31–37.

Eysenck, S. B. G., & Eysenck, H. J. On the dual nature of extraversion. *British Journal of Social and Clinical Psychology*, 1963, *2*, 46–55.

Eysenck, S. B. G., & Eysenck, H. J. Salivary response to lemon-juice as a measure of introversion. *Perceptual and Motor Skill*, 1967, *24*, 1047–1051.

Eysenck, S. B. G., & Eysenck, H. J. Crime and personality: Item analysis of questionnaire responses. *British Journal of Criminology*, 1971, *11*, 49–62.

Eysenck, S. B. G., & Eysenck, H. J. The place of impulsiveness in a dimensional system of personality description. *British Journal of Social and Clinical Psychology*, 1977, *16*, 57–68.

Eysenck, S. B. G., & Eysenck, H. J. Impulsiveness and venturesomeness: Their position in a dimensional system of personality description. *Psychological Reports*, 1978, *43*, 1247–1255.

Eysenck, S. B. G., & Eysenck, H. J. Impulsiveness and venturesomeness in children. *Personality and Individual Differences*, 1980, *1*, 73–78.

Eysenck, S. B. G., & McGurk, B. J. Impulsiveness and venturesomeness in a detention center population. *Psychological Reports*, 1980, *47*, 1299–1306.

Faergemann, D. M. *Psychogenic psychoses*. London: Butterworth, 1963.

Fagerstrom, K. O., & Lisper, H. O. Effects of listening to car radio, experience, and personality of the driver on subsidiary reaction time and heart rate in a long-term driving task. In R. R. Mackie (Ed.), *Vigilance*. New York: Plenum Press, 1977.

Fahrenberg, J., & Selg, H. *Das Freiburger Persönlichkeits Inventar FPI*. Göttingen: University Press, 1973.

Farley, F. H. Individual differences in solution time in error-free problem-solving. *British Journal of Social and Clinical Psychology*, 1966, *5*, 306–309.

Farley, F. H. On the independence of extraversion and neuroticism. *Journal of Clinical Psychology*, 1967, *23*, 154–156.

Farley, F. H. *Personality and reminiscence*. Paper read at the International Congress of Psychology, London, 1969.

Farley, F. H. Further investigation of the two personae of extraversion. *British Journal of Social and Clinical Psychology*, 1970, *9*, 377–379.

Feather, N. T. Some personality correlates of external control. *Australian Journal of Psychology*, 1967, *19*, 253–260.

Feldman, M. P. Response reversal performance as a function of drive level. In H. J. Eysenck (Ed.), *Experiments in motivation*. Oxford: Pergamon, 1964.

Feldman, M. P. *Criminal behaviour: A psychological analysis*. New York: Wiley, 1977.

Fiedler, F. E., Lodge, J. A., Jones, R. E., & Hutchings, E. B. Interrelations among measures of personality adjustment in nonclinical populations. *Journal of Abnormal and Social Psychology*, 1958, *56*, 345–351.

Fine, B. Field-dependent introverts and neuroticism: Eysenck and Witkin united. *Psychological Reports*, 1972, *31*, 939–956.

Fine, B. J. Field-dependence, extraversion, Eysenck and autarky. *Personality and Individual Differences*, 1983, *4*, 359–360.

Fine, B. J., & Danforth, A. V. Field-dependence, extraversion and perception of the vertical: Empirical and theoretical perspectives of the rod-and-frame test. *Perceptual and Motor Skills*, 1975, *40*, 683–693.

Fine, B. J., & Kobrick, J. L. Note on the relationship between introversion-extraversion, field-dependence–independence and accuracy of visual target detection. *Perceptual and Motor Skills*, 1976, *42*, 763–766.

Fine, B. J., & Kobrick, J. L. Field-dependence, practice, and low illumination as related to the Farnsworth-Mussell 100-Hue test. *Perceptual and Motor Skill*, 1980, *51*, 1167–1177.

Fischer, R., Griffin, F., & Rockey, M. L. Gustatory chemoreception in man: Multi-disciplinary aspects and perspectives. *Perspectives in Biological Medicine*, 1966, *9*, 549–577.

Fiske, D. W. A study of relationships to somatotype. *Journal of Applied Psychology*, 1944, *28*, 504–519.

Fiske, D. W. Consistency of the factorial structures of personality ratings from different sources. *Journal of Abnormal and Social Psychology*, 1949, *44*, 329–344.

Fitch, W. M., & Margoliash, E. Construction of phylogenetic trees. *Science*, 1967, *155*, 279–284.

Flemming, E. G. The 'halo' around personality, *Teachers College Record*. 1942, *43*, 564–569.

Floderus-Myrhed, B., Pedersen, N., & Rasmuson, I. Assessment of heritability for personality based on a short form of the Eysenck Personality Inventory. *Behavior Genetics*, 1980, *10*, 153–162.

Folkard, S. The nature of diurnal variations in performance and their implications for shift work studies. In P. Colquhoun, S. Folkard, P. Knauth, J. Rutenkrary (Eds.), *Experimental studies of shift work*. Opladen: Westdeutsche Verlag, 1975.

Foulds, G. A., & Caine, T. M. Personality factors and performance on timed tests of ability. *Occupational Psychology*, 1958, *32*, 102–105.

Fowles, D. C., Roberts, R., & Nagel, K. The influence of introversion–extraversion on the skin conductance response to stress and stimulus intensity. *Journal of Research in Personality*, 1977, *11*, 129–146.

Francis, L., Person, R. R., Carter, M., & Kay, W. K. The relationship between neuroticism and religiosity among English 15-and 16-year olds. *Journal of Social Psychology*, 1981, *114*, 99–102. (a)

Francis, L., Pearson, P. R., Carter, M., & Kay, W. K. Are introverts more religious? *British Journal of Social Psychology*, 1981, *20*, 101–104. (b)

Francis, R. D. Introversion and isolation tolerance. *Perceptual and Motor Skills*, 1969, *28*, 534.

Frankenhaeuser, A., Nordeheden, B., Myrsten, A. L., & Post, B. Psychophysiological reactions to under-stimulation and over-stimulation. *Acta Psychologica*, 1971, *35*, 298–308.

Franks, C. Différences déterminées par la personalité dans la perception visuelle de la verticalité. *Revue de la Psychologie Appliquée*, 1956, *6*, 235–246.

Franks, C. M., & Laverty, S. G. Sodium amytal and eyelid conditioning. *Journal of Mental Science*, 1955, *101*, 654–663.

Franks, C. M., & Trouton, D. Effects of amobarbital sodium and dexamphetamine sulfate on the conditioning of the eyeblink response. *Journal of Comparative and Physiological Psychology*, 1958, *51*, 220–222.

Frcka, G., Beyts, J., Levey, A. B., & Martin, I. The role of awareness in human conditioning. *Pavlovian Journal of Biological Science*, 1983, *18*, 69–76.

Freud, S. *General introduction to psychoanalysis*. New York: Liveright 1920.

Freud, S. Libidinal types. In J. Strachey (Ed.), *Complete psychological works of Sigmund Freud* (Vol. 21). New York: Norton, 1976. (Originally published 1931)

Freud, S. Three contributions to a theory of sex. In *The basic writings of Sigmund Freud*. New York: Modern Library, 1938.

Freyd, M. Introverts and extraverts. *Psychological Review*, 1924, *31*, 74–87.

Friedman, M., & Rosenman, R. H. *Type A behavior and your heart*. London: Wildwood House, 1974.

Frith, C. D. The interaction of noise and personality with critical fusion performance. *British Journal of Psychology*, 1967, *58*, 127–131.

Frith, C. D. Strategies in rotary pursuit tracking. *British Journal of Psychology*, 1971, *62*, 187–197.

Frith, C. D. Nature of rest pauses in a single tapping task. *Perceptual and Motor Skills*, 1973, *36*, 437–438.

Frith, C. D. *Habituation of the pupil size and light responses to sound*. Paper presented at the meeting of the American Psychological Association, San Francisco, 1977.

Frost, B. P. A note on extraversion and aggression. *Western Psychologist*, 1970, *1*, 101–112.

Frost, B. P. On the relation between extraversion and agression. *Psychological Reports*, 1981, *49*, 1009–1010.

Fulker, D. W. The genetic and environmental architecture of psychoticism, extraversion and neuroticism. In H. J. Eysenck (Ed.), *A model for personality*. Berlin and New York: Springer-Verlag, 1981, p. 88–122.

Fulker, D. W., & Eysenck, H. J. Nature and nurture: Heredity. In H. J. Eysenck (Ed.), *The structure and measurement of intelligence*. New York: Springer, 1979.

Furneaux, W. D. *Report to Imperial College of Science and Technology*, London, 1957.

Furneaux, W. D. Intellectual abilities and problem solving behaviour. In H. J. Eysenck (Ed.), *Handbook of abnormal psychology* (1st ed.). New York: Basic Books, 1961.

Furnham, A. Personality and activity preference. *British Journal of Social Psychology*, 1981, *20*, 57–68.

Furnham, A. Extraversion, sensation seeking, stimulus screening and 'A' type behaviour patterns. *Personality and Individual Differences*, 1984, *5*, 133–140.

Gabrys, J. B. Stability of scores on the Junior Eysenck Personality Inventory in an outpatient population. *Perceptual and Motor Skills*, 1980, *51*, 743–746.

Gale, A. "Stimulus hunger": Individual differences in operant strategy in a button-pressing task. *Behaviour Research and Therapy*, 1969, *7*, 265–274.

Gale, A. The psychophysiology of individual differences: Studies of extraversion and the EEG. In P. Kline (Ed.), *New approaches in psychological measurement*. New York: Wiley, 1973.

Gale, A. Electroencephalographic studies of extraversion-introversion: A case study in the psychophysiology of individual differences. *Personality and Individual Differences*, 1983, *4*, 371–380.

Gale, A., Coles, M., & Blaydon, J. Extraversion-introversion and the EEG. *British Journal of Psychology*, 1969, *60*, 209–223.

Gale, A., Harpham, B., & Lucas, B. Time of day and the EEG: Some negative results. *Psychonomic Science*, 1972, *28*, 269–271.

Gale, A., Morris, P. E., Lucas, B., & Richardson, A. Types of imagery and imagery types: An EEG study. *British Journal of Psychology*, 1972, *63*, 523–531.

Galton, F. *Inquiries into human faculty and its development*. London: Macmillan, 1883.

Galton, F. *Memories of my life*. London: Methuen, 1908.

Gange, J. J., Geen, R. G., & Harkins, S. G. Autonomic differences between extraverts and introverts during vigilance. *Psychophysiology*, 1979, *16*, 392–397.

Garau, A. *Deambulazio al campo abierto: Postulats dels farmaci d'Eysenck*. Barcelona: Unpublished M.A. thesis, Universidad Autonoma de Barcelona, 1976.

Garuau, A., Tobena, A., & Garcia, L. Studies on an analogue of extraversion in the rat. *Proceedings of the Second World Congress of Biological Psychiatry*, Elsevier, North Holland, 1980.

Garcia, L., & Garau, A. Extraversion y deambulacion de la rata en el campo abierto. *Revue Latino-Americana de Psicologia*, 1978, *10*, 211–222.

Gardner, R. W. Individual differences in figural after-effects and response to reversible figures. *British Journal of Psychology*, 1961, *52*, 269–272.

Garnett, J. C. M. General ability, cleverness and purpose. *British Journal of Psychology*, 1918, *9*, 345–366.

Gary, A. L., & Glover, J. *Eye color, sex, and children's behavior*. Chicago: Nelson-Hall, 1976.

Gasser, T., Mocks, J., Lenard, N. G., Bocher, D., & Verleger, R. The EEG of mildly retarded children: Developmental, classificatory, and topographic aspects. *Electroencephalography and Clinical Neurophysiology*, 1983, *55*, 131–144. (a)

Gasser, T., Lucadon-Muller, I., Verleger, R., & Backer, P. Correlating EEG and IQ: A new look at an old problem using computerized EEG parameters. *Electroencephalography and Clinical Neurophysiology*, 1983, *55*, 493–504. (b)

Gattaz, W. F. HLA-B27 as a possible genetic marker of psychoticism. *Personality and Individual Differences*, 1981, *2*, 57–60.

Gattaz, W. F., & Seitz, M. A. possible association between HLA 13–27 and the vulnerability to schizophrenia. *Personality and Individual Differences*, in press, 1984.

Gattaz, W. F., Ewald, R. W., & Beckmann, H. The HLA System and schizophrenia: A study in a German population. *Archiv für Psychiatrie und Nervenkrankheiten*, 1980, *228*, 205–211.

Geen, R. G. Test anxiety, observation, and range of cue utilization. *British Journal of Social and Clinical Psychology*, 1976, *15*, 253–259.

Ghodsian, M., Fogelman, K., Lambert, L., & Tibbenham, A. Changes in behaviour ratings of a national sample of children. *British Journal of Social and Clinical Psychology*, 1980, *19*, 247–256.

Gibson, H. B. Relations between performance on the Advanced Matrices and the EPI in high-intelligence subjects. *British Journal of Social and Clinical Psychology*, 1975, *14*, 363–369.

Gidwani, D. G. S. *The effect of previously learned habits and personality variables in verbal conditioning*. Unpublished Ph.D. thesis, University of London, 1971.

Giedt, F. H., & Downing, L. An extraversion scale for the MMPI. *Journal of Clinical Psychology*, 1961, *17*, 156–159.

Giel, R., Ten Horn, M., Ormel, J., Schudel, W., & Wiersma, D. Mental illness, neuroticism and life events in a Dutch village sample: A follow-up. *Psychological Medicine*, 1978, *8*, 235–243.

Giese, H., & Schmidt, A. *Studenten Sexualität.* Hamburg: Rowohlt, 1968.

Gifford, R. Sociability: Traits, settings, and interactions. *Journal of Personality and Social Psychology,* 1981, *41,* 340–347.

Gill, S. *The effect of personality and cortical excitants and depressants on perceptual, psychomotor and vigilance tasks.* Unpublished Ph.D. thesis, Panjan University, Chandigarh, 1979.

Gillespie, C. R., & Eysenck, M. W. Effects of introversion–extraversion on continuous recognition memory. *Bulletin of the Psychonomic Society,* 1980, *15,* 233–235.

Gilliland, A. R., & Morgan, J. J. B. An objective measure of introversion–extraversion. *Journal of Abnormal Psychology,* 1931, *26,* 296–303.

Gillis, J. S., & Cattell, R. B. Comparison of second-order personality structures at 6–8 years with later patterns. *Multivariate Experimental Clinical Research,* 1979, *4,* 93–99.

Giugannio, B. M., & Hindley, C. B. Stability of individual differences in personality characteristics from 3 to 5 years. *Personality and Individual Differences,* 1982, *3,* 287–301.

Glover, E. The significance of the month in psychoanalysis. *British Journal of Medical Psychology,* 1924, *4,* 135–155.

Goggin, J., Flemenbaum, A., & Anderson, D. Field dependence and extraversion neuroticism in an inpatient psychiatric service. *Journal of Clinical Psychology,* 1979, *35,* 538–541.

Goh, D. S., & Farley, F. H. Personality effects on cognitive test performance. *Journal of Psychology,* 1977, *96,* 111–122.

Goh, D. S., & Moore, C. Personality and academic achievement in three educational levels. *Psychological Reports,* 1978, *43,* 71–79.

Goldberg, L. R., & Hase, H. D. Strategies and tactics of personality inventory construction: An empirical investigation. *Oregon Research Institute Research Monograph,* 1967, *7*(1).

Goldman-Eisler, F. Breastfeeding and character-formation. *Journal of Personality,* 1948, *17,* 83–103.

Goldman-Eisler, F. Breastfeeding and character-formation: The etiology of the oral character in psychoanalytic theory. *Journal of Personality,* 1950, *19,* 189–196.

Goldman-Eisler, F. The problem of orality and its origin in early childhood. *Journal of Mental Science,* 1951, *97,* 765–781.

Goma, M. *Fiabilitat i validesa de les mesures del camp obert.* Barcelona: unpublished thesis, Universidad Autonoma, 1977.

Goorney, A. B. MPI and MMPI scores, correlations and analysis for a military aircrew population. *British Journal of Social and Clinical Psychology,* 1970, *9,* 164–170.

Gothschaldt, K. Das Problem der Phanogenetik der Persönlichkeit. In P. Lersch & M. Thomae (Eds.), *Handbuch der Psychologie* (Bd. 4). Göttingen: Hogrefe, 1960.

Götz, K. O., & Götz, K. Introversion–extraversion and neuroticism in gifted and ungifted art students. *Perceptual and Motor Skills,* 1973, *36,* 675–678.

Gough, H. G. Diagnostic patterns on the Minnesota Multiphasic Personality Inventory. *Journal of Clinical Psychology,* 1946, *2,* 23–27.

Gough, H. G. *Manual for the California Psychological Inventory.* Palo Alto: Consulting Psychologist Press, 1957.

Gough, H. G., & Heilbrun, A. B. *The Adjective Check List Manual.* Palo Alto: Consulting Psychologists Press, 1965.

Granger, G. W. Abnormalities of sensory perception. In H. J. Eysenck (Ed.), *Handbook of abnormal psychology.* London: Pitman, 1960.

Gray, J. A. Strength of the nervous system and levels of arousal: A re-interpretation. In J. A. Gray (Ed.), *Pavlov's typology*. Oxford: Pergamon, 1964.

Gray, J. A. Strength of the nervous system, introversion-extraversion, conditionability and arousal. *Behaviour Research and Therapy*, 1967, *5*, 151-169.

Gray, J. A. *Level of arousal and length of rest as determinants of pursuit rotor performance*. Unpublished Ph.D. thesis, University of London, 1968.

Gray, J. A. The psychophysiological basis of introversion-extraversion. *Behaviour Research and Therapy*, 1970, *8*, 249-266.

Gray, J. A. The psychophysiological nature of introversion-extraversion: A modification of Eysenck's theory. In V. D. Neblitsyn & J. A. Gray (Eds.), *Biological bases of individual behaviour*. London: Academic Press, 1972.

Gray, J. A. Causal theories of personality and how to test them. In J. R. Royce (Ed.), *Multivariate analysis of psychological theory*. New York: Academic Press, 1973.

Gray, J. A. A critique of Eysenck's theory of personality. In H. J. Eysenck (Ed.), *A model for personality*. Berlin: Springer, 1981.

Green, D. E., & Walkey, F. H. A nonmetric analysis of Eysenck's Personality Inventory. *Multivariate Behavioral Research*, 1980, *15*, 157-163.

Grief, S. Untersuchungen zur deutschen Übersetzung der 16PF Fragebogen. *Psychologische Beiträge*, 1970, *12*, 180-213.

Gross, O. *Die cerebrale Sekundärfunktion*. Leipzig: 1902.

Gross, O. *Über psychologische Minderwertigkeiten*. Leipzig: 1909.

Grygier, T. J. *The Dynamic Personality Inventory*. Windsor: NFER, 1961.

Guilford, J. P. Unitary traits of personality and factor theory. *American Journal of Psychology*, 1936, *48*, 673-680.

Guilford, J. P. Factors and factors of personality. *Psychological Bulletin*, 1975, *82*, 802-814.

Guilford, J. P. Will the real factor of extraversion-introversion please stand up! A reply to Eysenck. *Psychological Bulletin*, 1977, *84*, 412-416.

Guilford, J. P., & Hoepfner, R. *The analysis of intelligence*. New York: McGraw-Hill, 1971.

Guilford, J. S., Zimmerman, W. S., & Guilford, J. P. *The Guilford-Zimmerman Temperament Survey Handbook*. San Diego: Edits, 1976.

Gupta, B. S. The effects of stimulant and depressant drugs on verbal conditioning. *British Journal of Psychology*, 1973, *64*, 553-557.

Gupta, B. S. Extraversion and reinforcement in verbal operant conditioning. *British Journal of Psychology*, 1976, *67*, 47-52.

Gupta, B. S., & Kaur, S. The effects of dextroamphetamine on kinesthetic figural after-effects. *Psychopharmacology*, 1978, *56*, 199-204.

Gupta, B. S., & Nagpal, M. Impulsivity/sociability and reinforcement in verbal operant conditioning. *British Journal of Psychology*, 1978, *69*, 203-206.

Haier, R. J., Robinson, D. L., Braden, W., & Williams, D. Electrical potentials of the cerebral cortex and psychometric intelligence. *Personality and Individual Differences*, 1984, *4*, 591-599.

Hall, C. S. The inheritance of emotionality. In W. F. Martin & C. B. Stendler (Eds.), *Readings in child development*. New York: Harcourt Brace, 1938, pp. 58-68.

Hall, C. S., & LIndzey, G. *Theories of personality* (2nd ed.). New York: Wiley, 1970.

Hallam, R. S. The Eysenck personality scales: Stability and change after therapy. *Behaviour Research and Therapy*, 1976, *14*, 369-372.

Hamilton, M. Diagnosis and rating of anxiety. In M. H. Lader (Ed.), *Studies of anxiety*. *British Journal of Psychiatry Special Publication No. 3*, 1969.

Hamilton, P., Hockey, G. R. J., & Quinn, J. G. Information selection, arousal and memory. *British Journal of Psychology*, 1972, *63*, 181–190.

Hamilton, V. *The cognitive analysis of personality related to information-processing deficits with stress and anxiety.* Paper presented at the British Psychological Society meeting, London, 1978.

Hammond, D. Effects of visual and thermal stimulation upon reminiscence in rotary pursuit tracking. *Irish Journal of Psychology*, 1972, *3*, 177–184.

Hampel, R., & Wittmann, W. *Eine Methodenstudie zum Rotationsproblem.* Quoted by Fahrenberg & Selg, 1973.

Hare, R. D. Psychopathy and the personality dimensions of psychoticism, extraversion and neuroticism. *Personality and Individual Differences*, 1982, *3*, 35–42.

Hare, R. D., & Schalling, D. (Eds.). *Psychopathic behavior: Approaches to research.* Chichester: Wiley, 1978.

Harkins, S., & Geen, R. G. Discriminability and criterion differences between extraverts and introverts during vigilance. *Journal of Research in Personality*, 1975, *9*, 335–340.

Hartley, L. R., & Carpenter, A. Comparison of performance with headphone and free-field noise. *Journal of Experimental Psychology*, 1974, *103*, 377–380.

Hartog, P., & Rhodes, E. C. *An examination of examinations.* London: Macmillan, 1936.

Hartshorne, H., & May, M. A. *Studies in deceit.* New York: Macmillan, 1928.

Hartshorne, H., & May, M. A. *Studies in service and self control.* New York: Macmillan, 1929.

Hartshorne, H., & Shuttleworth, F. K. *Studies in the organization of character.* New York: Macmillan, 1930.

Hasenfus, N., & Magaro, P. Creativity and schizophrenia: An equality of empirical constructs. *British Journal of Psychiatry*, 1976, *129*, 346–349.

Häseth, K., Shagass, C., & Straumanis, J. J. Perceptual and personality correlates of EEG and evoked response measures. *Biological Psychiatry*, 1969, *1*, 49–60.

Haslam, D. R. Individual differences in pain threshold and level of arousal. *British Journal of Psychology*, 1967, *58*, 139–142.

Hayes, K. J. Exploration and fear. *Psychological Reports*, 1960, *6*, 91–93.

Heidbreder, E. Measuring introversion and extraversion. *Journal of Abnormal and Social Psychology*, 1926, *21*, 120–134.

Helmes, E. A psychometric investigation of the Eysenck Personality Questionnaire. *Applied Psychology Measurement*, 1980, *4*, 43–55.

Hendrickson, A. E. An integrated molar/molecular model of the brain. *Psychological Reports*, 1972, *30*, 343–368.

Hendrickson, A. The biological basis of intelligence, Part I: Theory. In H. J. Eysenck (Ed.), *A model for intelligence.* New York: Springer, 1982.

Hendrickson, A. E. & White, P. O. A method for the relation of higher-order factors. *British Journal of Mathematical and Statistical Psychology*, 1966, *19*, 97–103.

Hendrickson, D. E. *An examination of individual differences in cortical evoked response.* London: Unpublished Ph.D. thesis, University of London, 1972.

Hendrickson, D. E. The biological basis of intelligence, Part 2: Measurement. In H. J. Eysenck (Ed.), *A model for intelligence.* New York: Springer, 1982.

Hendrickson, D. E., & Hendrickson, A. E. The biological basis of individual differences in intelligence. *Personality and Individual Differences*, 1980, *1*, 3–34.

Herbert, G. W. Teachers' ratings of classroom behaviour: Factorial structure. *British Journal of Educational Psychology*, 1974, *44*, 233–240.

Hernandez, S. K., & Manager, P. A. Assertiveness, aggressiveness and Eysenck's personality variables. *Personality and Individual Differences*, 1980, *1*, 143-149.

Heskin, K. J., Smith, F. V., Bannister, P. A., & Belton, N. Psychological correlates of long term imprisonment: II. Personality variables. *British Journal of Criminology*, 1973, *13*, 323-330.

Heston, L. L. Psychiatric disorders in foster home reared children of schizophrenic mothers. *British Journal of Psychiatry*, 1960, *112*, 819-828.

Hewitt, L. E., & Jenkins, R. L. *Fundamental patterns of adjustment.* Illinois: D. H. Green, 1946.

Heymans, G. Über einige psychische Korrelationen. *Zeitschrift für Angewandte Psychologie*, 1908, *1*, 313-381.

Heymans, G., & Brugman, H. Intelligenz Prüfungen mit Studierenden. *Zeitschrift für Angewandte Psychologie*, 1913, *7*, 317-331.

Heymans, G., & Wiersma, E. Beiträge zur speziellen Psychologie auf Grund einer Massenuntersuchung. *Zeitschrift für Psychologie*, 1906, *42*, 81-127 (a); 1906, *43*, 321-373 (b); 1907, *45*, 1-42; 1908, *46*, 321-333 (a); 1908, *49*, 414-439 (b); 1909, *51*, 1-72.

Hick, W. On the rate of gain of information. *Quarterly Journal of Experimental Psychology*, 1952, *4*, 11-26.

Hill, A. B. Extraversion and variety-seeking in a monotonous task. *British Journal of Psychology*, 1975, *66*, 9-13.

Hindley, C. B., & Giuganino, B. M. Continuity of personality patterning from 3 to 15 years in a longitudinal sample. *Personality and Individual Differences*, 1982, *3*, 127-144.

Hitch, G. J., & Baddeley, A. D. Verbal reasoning and working memory. *Quarterly Journal of Experimental Psychology*, 1976, *28*, 603-621.

Hobi, V. Ein Vergleich zwischen MMPI 16PF und FPI an Kleinen Stichproben. *Zeitschrift für Klinische Psychologie und Psychotherapie*, 1973, *21*, 129-139.

Hobi, V., & Klar, A. Ein Beitrag zur Faktorenstruktur des FPI. *Diagnostica*, 1973, *19*, 88-96.

Hodges, W. F. Effects of ego threat and threat of pain on state anxiety. *Journal of Personality and Social Psychology*, 1968, *8*, 364-372.

Hogan, M. J. Influence of motivation on reactive inhibition in extraversion-introversion. *Perceptual and Motor Skills*, 1966, *22*, 187-192.

Holland, H. C. *An experimental investigation of the influence of personality variables upon perception.* Unpublished Ph.D. thesis, University of London, 1959.

Holland, H. C. Judgments and the effects of instructions. *Acta Psychologica*, 1961, *18*, 229-238.

Holland, H. C. "Visual masking" and the effects of stimulant and depressant drugs. In H. J. Eysenck (Ed.), *Experiments with drugs.* New York: Pergamon, 1963.

Holland, H. C. *The spiral after-effect.* London: Pergamon, 1965.

Holland, H. C., & Gomez, B. H. The effects of stimulant and depressant drugs upon visual figural after-effects. In H. J. Eysenck (Ed.), *Experiments with drugs.* New York: Pergamon, 1963.

Holmes, D. S. Pupillary response, conditioning and personality. *Journal of Personality and Social Psychology*, 1967, *5*, 95-103.

Howarth, E. Personality differences in serial learning under distraction. *Perceptual and Motor Skills*, 1969, *28*, 379-382. (a)

Howarth, E. Extraversion and increased interference in paired-associate learning. *Perceptual and Motor Skills*, 1969, *29*, 403-406. (b)

Howarth, E. A factor analysis of selected markers for objective personality factors. *Multivariate Behavioral Research*, 1972, *7*, 431-476. (a)

Howarth, E. A source of independent verification: Consequences and divergencies in the work of Cattell and Eysenck. In R. M. Dreger (Ed.), *Multivariate Personality Research*. Baton Rouge: Calitor, 1972, pp. 122-160. (b)

Howarth, E. An hierarchical oblique factor analysis of Eysenck's rating study of 700 neurotics. *Social Behavior and Personality*, 1973, *1*, 81-87.

Howarth, E. Were Cattell's "Personality Sphere" factors correctly identified in the first instance? *British Journal of Psychology*, 1976, *67*, 213-230.

Howarth, E. The Comrey Personality Inventory. In O. K. Buros (Ed.), *Mental measurements year book*. Lincoln: University of Nebraska Press, 1978.

Howarth, E. Interrelations between state and trait: Some new evidence. *Perceptual and Motor Skills*, 1980, *51*, 613-614. (a)

Howarth, E. A test of some old concepts by means of some new scales: Anality or psychoticism, oral optimism or extraversion, oral pessimism or neuroticism. *Psychological Reports*, 1980, *47*, 1039-1042. (b)

Howarth, E. Comment on Kline's note on "A test of some old concepts by means of some new scales". *Psychological Reports*, 1981, *49*, 178.

Howarth, E., & Browne, J. A. An item-factor-analysis of the 16PF. *Personality*, 1971, *2*, 117-139.

Howarth, E., & Eysenck, H. J. Extraversion, arousal, and paired-associate recall. *Journal of Experimental Research in Personality*, 1968, *3*, 114-116.

Huckabee, M. W. Introversion-extraversion and imagery. *Psychological Reports*, 1974, *34*, 453-454.

Hull, J., & Zubek, J. P. Personality characteristics of successful and unsuccessful sensory isolation subjects. *Perceptual and Motor Skills*, 1962, *14*, 231-240.

Hundleby, J. D., & Connor, W. H. Interrelationships between personality inventories: The 16PF, the MMPI and the MPI. *Journal of Consulting and Clinical Psychology*, 1968, *32*, 152-157.

Hunter, S., Overall, J. E., & Butcher, J. P. Factor structure of the MMPI in a psychiatric population. *Multivariate Behavior Research*, 1974, *9*, 283-301.

Iacomo, W. G., & Lykken, D. T. Eye tracking and psychopathology. *Archives of General Psychiatry*, 1979, *36*, 1361-1369.

Irvine, J., Lyle, R., & Allen, R. Type A personality as psychopathology. *Journal of Psychosomatic Research*, 1982, *26*, 183-189.

Jackson, D. N. *Manual for the Personality Research Forms*. London, Ontario: University of Western Ontario, 1967.

Jackson, D. N., & Messick, S. Content and style in personality assessment. *Educational and Psychological Measurement*, 1961, *21*, 771-790.

Janet, P. *L'état mental des hystériques*. Paris: Rueff, 1894.

Jarvik, L. F., & Chadwick, S. B. Schizophrenia and survival. In M. Hammer, K. Salzinger, & S. Sutton (Eds.), *Psychopathology: Contributions from the social, behavioral, & biological sciences*. New York: Wiley, 1973.

Jenkins, C. D., Zyzanski, S. J., & Rosenman, R. H. Progress towards validation of a computer-scored test of type A coronary-prone behavior patterns. *Psychosomatic Medicine*, 1971, *33*, 192-202.

Jensen, A. *Individual differences in learning: Interference factor*. United States Department of Health, Education and Welfare, Project Report, No. 1867, Washington, D.C., 1964.

Jensen, A. R. The measurement of reactive inhibition in humans. *Journal of General Psychology*, 1966, *75*, 85–94.

Jensen, A. R. Reaction time and psychometric *g*. In H. J. Eysenck (Ed.), *A model for intelligence*. New York: Springer, 1982. (a)

Jensen, A. R. The chronometry of intelligence. In R. J. Sternberg (Ed.), *Advances in research in intelligence*. Hillsdale, N. J.: Lawrence Erlbaum, 1982, pp 242–267. (b)

Jensen, A. R., & Rohwer, W. D. The Stroop colour-word test: A review. *Acta Psychologica*, 1966, *25*, 36–93.

Jessup, G., & Jessup, H. Validity of the Eysenck Personality Inventory in pilot selection. *Occupational Psychology*, 1971, *45*, 111–123.

Jogawar, V. V. Individuality correlations in human blood groups. *Personality and Individual Differences*, 183, *4*, 215–216.

Johnson, J. H., & Overall, J. E. Factor analysis of the Psychological Screening Inventory. *Journal of Consulting and Clinical Psychology*, 1973, *41*, 57–60.

Johnson, R. C., Ackerman, J. M., Frank, H., & Fionda, A. J. Resistance to temptation, and guilt following yielding, and psychopathology. *Journal of Consulting and Clinical Psychology*, 1968, *32*, 169–175.

Jones, J., Eysenck, H. J., Martin, I., & Levey, A. B. Personality and the topography of the conditioned eyelid response. *Personality and Individual Differences*, 1981, *2*, 61–84.

Joubert, C. E. Some correlations between famous sayings test and Eysenck Personality Inventory variables. *Psychological Reports*, 1977, *40*, 697–698.

Joubert, C. E. Multidimensionality of locus of control and the Eysenck Personality Inventory. *Psychological Reports*, 1978, *43*, 338.

Jung, C. *Psychologische typus*. Zurich: Rascher, 1921.

Jung, C. G. *Psychological Types*. London: Routledge & Kegan Paul, 1923.

Jurkovic, G. J., & Prentice, N. M. Relation of moral and cognitive development to dimensions of juvenile delinquency. *Journal of Abnormal Psychology*, 1977, *86*, 414–420.

Kahneman, D. *Attention and effort*. Englewood Cliffs, N.J.: Prentice-Hall, 1973.

Kaiser, H., Hunka, S., & Bianchini, J. Relating factors between studies based upon different individuals. In H. J. Eysenck & S. B. G. Eysenck (Eds.), *Personality structure and measurement*. London: Routledge & Kegan Paul, 1969, pp. 333–343.

Kallmann, F. J., & Sander, G. Twin studies in senescence. *American Journal of Psychology*, 1949, *106*, 29.

Kant, I. *Anthropologie in pragmatischer Hinsicht*. Berlin: Bresser Cassiner, 1912. (Originally published, 1798.)

Kantorowitz, D. A. Personality and conditioning of tumescence and detumescence. *Behaviour Research and Therapy*, 1978, *16*, 117–123.

Karlsson, J. L. Genealogical studies of schizophrenia. In D. Rosenthal & S. S. Ketty (Eds.), *The transmission of schizophrenia*. London: Pergamon Press, 1968.

Kassebaum, G. G., Couch, A. S., & Slater, P. E. The factorial dimension of the MMPI. *Journal of Consulting Psychology*, 1959, *23*, 226–234.

Katsikitis, M., & Brebner, J. Individual differences in the effects of personal space invasion: A test of the Brebner-Cooper model of extraversion. *Personality and Individual Differences*, 1981, *2*, 5–10.

Keister, M. E., & McLaughlin, R. J. Vigilance performance related to extraversion-introversion and caffeine. *Journal of Experimental Research in Personality*, 1972, *6*, 5–11.

Kelly, D., & Martin, I. Autonomic reactivity, eyelid-conditioning and their relationship to neuroticism and extraversion. *Behaviour Research and Therapy*, 1969, *7*, 233–244.

Kelly, D. H. Visual responses to time-dependent stimuli. *Journal of the Optical Society of America*, 1971, *61*, 537–546.

Kelly, D. H. Flicker. In D. Jameson L. M. Huerich (Eds.), *Handbook of sensory physiology, Vol. 7: Sensory Psychophysics.* Berlin: Springer Verlag, 1972.

Kempf, E. J. *The autonomic function and the personality.* Washington: Nervous and Mental Disease Pub. Co., 1921.

Kendon, A. & Cook, M. The consistency of gage patterns in social interaction. *British Journal of Psychology.* 1969, *60*, 481-494.

Kimble, G. A. An experimental test of a two-factor theory of inhibition. *Journal of Experimental Psychology*, 1949, *39*, 15-23.

Kishimoto, Y. Visual vigilance performance of extraverts and introverts under two conditions of signal frequency. *Japanese Journal of Psychology*, 1977, *48*, 53-57.

Kjellberg, A. Sleep deprivation and some aspects of performance: Lapses and other attentional aspects. *Waking and Sleeping*, 1977, *1*, 145-148.

Kleiber, S., Veldman, D., & Menaker, S. L. The multidemensionality of areas of control. *Journal of Clinical Psychology*, 1973, *29*, 411-416.

Kleinsmith, L. J., & Kaplan, S. Paired associate learning as a function of arousal and interpolated interval. *Journal of Experimental Psychology*, 1963, *65*, 190-193.

Kline, P. Extraversion, neuroticism and academic performance among Ghanaian university students. *British Journal of Educational Psychology*, 1966, *36*, 92-94.

Kline, P. The status of the anal character: A methodological and empirical reply to Hill. *British Journal of Medical Psychology*, 1978, *51*, 87-90.

Kline, P. *Psychometrics and psychology.* London: Academic Press, 1979.

Kline, P. *Fact and fantasy in Freudian theory.* London: Methuen, 1981.

Kline, P., & Barrett, P. The factors in personality questionnaires among normal subjects. *Advances in Behaviour, Research and Therapy*, 1983, *5*, 141-202.

Kline, P., & Storey, R. A factor analytic study of the anal character. *British Journal of Social and Clinical Psychology*, 1977, *16*, 317-328.

Kline, P., & Storey, R. The Dynamic Personality Inventory: What does it measure? *British Journal of Psychology*, 1978, *69*, 375-383.

Knapp, R. R. Relationship of a measure of self-actualisation to neuroticism and extraversion. *Journal of Consulting Psychology*, 1965, *29*, 108-172.

Knowles, J. B., & Krasner, L. Extraversion and duration of the Archimedes spiral aftereffect. *Perceptual and Motor Skills* 1965, *20*, 997-1000.

Knutson, A. Personal security as related to stations in life. *Psychological Monographs*, 1952, *66*, 1-31.

Köhler, W., & Wallach, H. Figural after-effects. *Proceedings of the American Philosophical Society*, 1944, *88*, 265-357.

Kohn, M., & Rosman, B. Z. Cross-situational and longitudinal stability of social-emotional functioning in young children. *Child Development.* 1973, *44*, 721-727.

Korchin, S. Anxiety and cognition. In C. Scheeser (Ed.), *Cognition: Theory, research, promise.* New York: Harper & Row, 1964.

Koskenvuo, M., Langinvainio, H., Kapriv, J., Rantasalr, I., Sarna, S. *The Finnish Twin Registry: Baseline Characteristics, Section 700: Occupational and Psychosocial Factors.* Helsinki: Helsinkin Yliopston Kausanterreysteen Laitos, 1979.

Koss, M. P. MMPI item content: Recurring issues. In J. N. Butcher (Ed.), *New developments in the use of the MMPI.* Minneapolis: University of Minnesota Press, 1979.

Kraulis, W. Zur Klinik der Erbpsychosen. *Allgemeine Zeitschrift der Psychiatrie*, 1939, *113*, 32-44.

Kretschmer, E. *Körperbau und Charakter.* Berlin: Springer, 1925, 1948.

Kringlen, E. Hereditary and social factors in schizophrenic twins. In J. Romano (Ed.), *Origins of schizophrenia.* Amsterdam: Excerpta Medica Foundation, 1967.

Krishnamoorti, S. R., & Shagass, C. Some psychological test correlates of sedation threshold. In J. Wortis (Ed.), *Recent advances in biological psychiatry*. New York: Plenum Press, 1963.

Kristjansson, M., & Brown, R. I. F. Instruction effects and relation between extraversion and the spiral after-effects. *Perceptual and Motor Skills*, 1973, *36*, 1323-1326.

Krug, R. E., & Mayer, K. E. An analysis of the F scale: II. Relationship to standardized personality inventories. *Journal of Social Psychology*, 1961, *53*, 293-301.

Krupski, A., Raskin, D., & Bakan, P. Physiological and personality correlates of commission errors in an auditory vigilance task. *Psychophysiology*, 1971, *8*, 304-311.

Kuhn, T. S. *The structure of scientific revolutions*. Chicago: University of Chicago Press, 1962, 1970.

Kuhn, T. S. Logic of discovery or psychological research, In I. Lakatos & A. Musgrave (Eds.), *Criticism and the growth of knowledge*. Cambridge: Cambridge University Press, 1970.

Kuhn, T. S. Second thoughts in paradigms. In F. Suppe (Ed.), *The structure of scientific theories*. London: University of Illinois Press, 1974, pp. 459-982.

Lacey, J. I. Individual differences in somatic response patterns. *Journal of Comparative and Physiological Psychology*, 1950, *43*, 338-350.

Lacey, J. I. Somatic response patterning and stress: Some revisions of activation theory. In M. H. Appley & R. Turnbull (Eds.), *Psychological stress*. New York: Appleton-Century-Crofts, 1967.

Lacey, J. I., & Lacey, B. C. Verification and extension of the principle of autonomic response-stereotype. *American Journal of Psychology*, 1958, *71*, 50-73.

Lader, M., & Wing, L. *Physiological measures, sedative drugs and morbid anxiety*. Oxford: Maudsley Monograph No. 14, 1966.

Lagerspetz, K. M., & Lagerspetz, K. Y. H. Changes in the aggressiveness of mice resulting from selection breeding, learning and social isolatives. *Scandinavian Journal of Psychology*, 1971, *12*, 241-248.

Laird, D. A. Detecting abnormal behaviour. *Journal of Abnormal and Social Psychology*, 1925, *20*, 128-141.

Lakatos, I. Criticism and the methodology of scientific research programmes. *Proceedings of the Aristotelian Society*, 1968, *69*, 149-186.

Lakatos, I. Fabrification and the methodology of scientific research programmes. In I. Lakatos & D. Musgrave (Eds.), *Criticism and the growth of knowledge*. Cambridge: Cambridge University Press, 1970, pp. 91-196.

Lally, M., Nettlebeck, T. Intelligence, reaction time and inspection time. *American Journal of Mental Deficiency*, 1977, *82*, 273-281.

Langan, R. I. *Manual of the Psychological Screening Inventory*. New York: Research Psychology, 1970. (a)

Langan, R. I. Development and validation of a psychological screening inventory. *Journal of Consulting and Clinical Psychology Monograph*, 1970, *35*, 1-24. (b)

Langan, R. I. Factor structure of Psychological Screening Inventory scales. *Psychological Reports*, 1978, *42*, 383-386.

Langan, R. I., Johnson, J. H., & Overall, J. E. Factor structure of the Psychological Screening Inventory items in a normal population. *Journal of Consulting and Clinical psychology*, 1974, *42*, 219-223.

Lannay, G., & Slade, P. The measurement of hallucinatory predisposition in male and female prisoners. *Personality and Individual Differences*, 1981, *2*, 221-234.

Laverty, S. G. Sodium amytal and extraversion. *Journal of Neurology, Neurosurgery and Psychiatry*, 1958, *21*, 50-54.

Lay, C. H., & Jackson, D. N. Analysis of the generality of trait-inferential relationships. *Journal of Personality and Social Psychology*, 1969, *12*, 12-21.

Lazzerini, A. J., Cox, T, & Mackay, C. J. Perceptions of and reactions to stressful situations: The ability of a general anxiety trait. *British Journal of Social and Clinical Psychology*, 1979, *18*, 363-369.

Lefton, L. A. Metacontrast: A review. *Perception and Psychophysics*, 1973, *13*, 161-171.

Lehrl, S. Subjectives Zeitquant und Intelligenz. *Grundlagenstudien aus Kybernetik und Geisteswissenschaften*, 1979, *15*, 91-96.

Lehrl, S. Subjectives Zeitquant als missing link zwischen Intelligenzpsychologie und Neuropsychologie. *Grundlagenstudien aus Kybernetik und Geisteswissenschaften*, 1980, *21*, 107-116.

Lehrl, S. Intelligenz, informations psychologische Grundlagen. *Enzyklopädie Naturwissenschaft und Technik*, Jahresband 1983. Landsberg, West Germany: Moderne Industrie, 1983.

Lehrl, S., & Erzigkeit, H. Determiniert der Kurzspeichen das allgemeine Intelligenzniveau? *Grundlagenstudien aus Kybernetik und Geisteswissenschaften*, 1976, *17*, 19-119.

Lehrl, S. Straub, P., & Straub, R. Informations-psychologische Elementarbausteine der Intelligenz. *Grundlagenstudien aus Kybernetik und Geisteswissenschaften*, 1975, *10*, 41-50.

Lehrl, S. Gallwitz, S., & Blaha, L. *Kurztest für Allegemeine Intelligenz NSI.* Vaterstetten: Vless, 1980.

Leigh, G. O. M., & Wisdom, B. An investigation of the effects of error-making and personality on learning. *Programmed Learning*, 1970, *7*, 120-126.

Leipold, W. D. *Psychological distance in a dyadic interview as a function of introversion-extraversion, anxiety, social desirability and stress.* Ph. D. thesis, University of North Dakota, 1963.

Leon, G. R. Gillenn, B., Gillenn, R., & Ganze, M. Personality, stability and change over a 30-year period—Middle age to old age. *Journal of Consulting and Clinical Psychology*, 1979, *47*, 517-524.

Lester, D. *A physiological basis for personality traits.* Springfield, Ill.: Charles C Thomas, 1974.

Lester, D. Eye color, extraversion and neuroticism. *Perceptual and Motor Skills*, 1977, *44*, 1162.

Levenson, H. Multidimensional locus of control in psychiatric patients. Journal of Consulting and Clinical Psychology, 1973, *41*, 397-404.

Levey, A. B., & Martin, I. Personality and conditioning. In H. J. Eysenck (Ed.), *A model for personality.* Berlin: Springer, 1981.

Levonian, E. A statistical analysis of the 16 Personality Factor Questionnaire. *Educational and Psychological Measurement*, 1961, *21*, 589-596.

Levy, P., & Lang, P. J. Activation, control, and the spiral aftermovement. *Journal of Personality and Social Psychology*, 1966, *3*, 105-112.

Lichtenstein, E., & Kentzer, C. Further normative and correlational data in the Internal-External Control of Reinforcement Scale. *Psychological Reports*, 1967, *21*, 1014-1016.

Liebert, R. M., & Morris, L. W. Cognitive and emotional components of test anxiety: A distinction and some initial data. *Psychological Reports*, 1967, *20*, 975-978.

Lindsley, D. B. The reticular system and perceptual dissemination. In H. H. Jasper (Ed.), *Reticular formation of the brain.* London: Churchill, 1957.

Line, W., & Griffin, J. D. M. The objective determination of factors underlying mental health. *American Journal of Psychiatry*, 1935, *19*, 833-842.

Lobel, T. G. Personality correlates of assertive behaviour. *Personality and Individual Differences*, 1981, *2*, 252–254.

Locke, K. D., Locke, E. A., Morgan, G. A., & Zimmerman, R. R. Dimensions of social interaction among infant rhesus monkeys. *Psychological Reports*, 1964, *15*, 339–349.

Loehlin, J. C., & Nichols, R. C. *Heredity, environment and personality*. Austin: University of Texas Press, 1976.

Loevinger, J. *Ego development: Conceptions and theories*. San Francisco: Jossey-Bass, 1976.

Loo, R. Field dependence and the Eysenck Personality Inventory. *Perceptual and Motor Skills*, 1976, *43*, 614.

Loo, R. Field dependence and Eysenck's Neuroticism Scale. *Perceptual and Motor Skills*, 1978, *47*, 522.

Loo, R. A psychometric investigation of the Eysenck Personality Questionnaire. *Journal of Personality Assessment*. 1979, *43*, 54–58. (a)

Loo, R. Role of primary personality factors in the perception of traffic signs and driver violations and accidents. *Accident Analysis and Prevention*, 1979, *11*, 125–127. (b)

Loo, R., & Townsend, O. J. Components underlying the relationship between field dependence and extraversion. *Perceptual and Motor Skills*, 1977, *45*, 528–530.

Lorr, M., & Manning, T. T. Higher-order personality factors of the ISI. *Multivariate Behavioral Research*, 1978, *13*, 39.

Lorr, M., Youniss, R. P. An inventory of interpersonal style. *Journal of Personality Assessment*, 1973, *37*, 165–173.

Lorr, M., O'Connor, J. P., & Seifert, R. F. A comparison of four personality inventories. *Journal of Personality Assessment*, 1977, *41*, 520–526.

Lovalls, W., Pishkin, V. Types of behaviour, self-involvement, autonomic activity and the traits of neuroticism and extraversion. *Psychosomatic Medicine*, 1980, *42*, 329–334.

Lowenthal, M. F., Thurner, M., & Chiriboga, D., *Four stages of life*. San Francisco: Jossey-Bass, 1975.

Ludvigh, E. J., & Happ, D. Extraversion and preferred level of sensory stimulation. *British Journal of Psychology*, 1974, *65*, 359–365.

Lykken, D. T. Research with twins: The concept of emergenesis. *Psychophysiology*, 1982, *19*, 361–373.

Lynn, R. Two personaltiy characteristics related to academic achievement. *British Journal of Educational Psychology*, 1959, *29*, 213–216.

Lynn, R. Extraversion, reminiscence, and satiation effects. *British Journal of Psychology*, 1960, *51*, 319–324.

Lynn, R. Personality characteristics of a group of entrepreneurs. *Occupational Psychology*, 1969, *43*, 151–152.

Lynn, R. *Personality and national character*. London: Pergamon, 1971.

Lynn, R. Cross-cultural differences in neuroticism, extraversion and psychoticism. In R. Lynn (Ed.), *Dimensions of personality*. London: Pergamon, 1981.

Lynn, R., & Gordon, I. E. The relation of neuroticism and extraversion to intelligence and educational attainment. *British Journal of Educational Psychology*, 1961, *31*, 194–203.

Lynn, R., & Hampson, S. National differences in extraversion and neuroticism. *British Journal of Social and Clinical Psychology*, 1975, *14*, 223–240.

Lynn, R., & Hampson, S. Fluctuations in national levels of neuroticism and extraversion, 1935–1970. *British Journal of Social and Clinical Psychology*, 1977, *16*, 131–137.

Lynn, R., Devane, S., & O'Neill, B. Extending the boundaries of psychoticism: Health care and self-sentiment. *Personality and Individual Differences*, 1984, *5*, 397–402.

Mabille, O. Revue de morpho. *Physiologie Humaine*, 1951, *4*, 29–41.

MacCorquodale, K., & Meehl, P. E. On a distinction between hypothetical constructs and intervening variables. *Psychological Review*, 1948, *55*, 95–107.

Mach, E. *History and root of the principle of the conservation of energy*. Chicago: Open Court, 1911.

Macht, M. L., Spear, N. E., & Levis, D. J. State-dependent retention in humans induced by alterations in affective state. *Bulletin of the Psychonomic Society*, 1977, *10*, 415–418.

MacKenzie, D. Why is a hand not a foot? *New Scientist*, 1982, Dec. 2, 558–561.

MacKinnon, D. W. The structure of personality. In J. McV. Hunt (Ed.), *Personality and the behavior disorders*. New York: Ronald, 1944.

Mackworth, N. H. *Researches in the measurement of human performance*. Medical Research Council Special Reports Series, 268, 1950.

MacLean, M. E. Chronic welfare dependency: A multivariate analysis of personality factors. *Multivariate Experimental and Clinical Research*, 1977, *3*, 83–93.

Madlung, K. Über den Einfluss der typologischen Veranlagung auf die Flimmergrenzen. *Untersuchungen für Psychologie, Philosophie, und Pädagogie*, 1936, *10*, 70–78.

Magnusson, D. Trait–state anxiety: Quote on conceptual and empirical relationships. University of Stockholm: *Reports from the Department of Psychology*, 1979, No. 557.

Magnusson, D. Wanted: A psychology of situations. Stockholm: *Reports from the Department of Psychology* 1980, No. 49.

Magnusson, D. (Ed). *Toward a Psychology of situations: An interactional perspective*. Hillsdale, N. N.: Lawrence Erlbaum, 1981.

Magoun, H. *The waking brain*. Springfield, Ill.: Charles C Thomas, 1963.

Malapert, G. Les Éléments du charactère et leurs lois de combinaison. Paris: Alcan, 1897.

Maller, D. General and specific factors in character. *Journal of Social Psychology*, 1934, *5*, 97–102.

Maller, J. B., & Zubin, J. The effect of motivation upon intelligence test scores. *Journal of Genetic Psychology*, 1932, *41*, 136–151.

Mangan, G. L. Personality and conditioning: Some personality, cognitive and psychophysiological parameters of appetitive (sexual) GSR conditioning. *Pavlovian Journal of Biological Science*, 1974, *9*, 125–135.

Mangan, G. L. *The biology of human conduct*. Oxford: Pergamon, 1982.

Mangan, G. L., & O'Gorman, J. G. Initial amplitude and rate of habituation of orienting reaction in relation to extraversion and neuroticism. *Journal of Experimental Research in Personality*, 1969, *3*, 275–282.

Marchman, J. N., & Fowles, D. C. *Orienting responses as a function of neuroticism, extraversion and stimulus intensity*. Paper presented at the meeting of the Society of Psychophysiological Research, Galveston, Texas, 1973.

Marston, L. R. The emotions of young children. *Iowa Studies in Child Welfare*, 1925, *3*.

Martin, N. G., & Eysenck, H. J. Genetic factors in sexual behaviour. In H. J. Eysenck (Ed.), *Sexual personality*. London: Open Books, 1976, pp. 192–219.

Martin, N. G., Eaves, L. J., & Fulker, D. V. The genetical relationship of impulsiveness and sensation seeking to Eysenck's personality dimensions. *Acta Geneticae Medicae et Gemellologiae*, 1979, *28*, 197–210.

Martiny, M. *Essai de biotypologie humaine*. Paris: Peyronnet, 1948.

Martuza, V. R., & Kallstrom, D. W. Validity of the State–Trait Anxiety Inventory in an academic setting. *Psychological Reports*, 1974, *35*, 363–366.

Maslow, A. H. *Towards a psychology of being*. New York: Van Nostrand, 1962.

Masterman, M. The nature of the paradigm. In I. Lakatos & A. Musgrave (Eds.), *Criticism and the growth of knowledge.* Cambridge: Cambridge University Press, 1970.

Materanz, A., & Hampel, R. Eine interkulturelle Vergleichsstudie mit den FPI am deutschen und spanischen probanden. *Zeitschrift für Experimentelle und Angewandte Psychologie*, 1978, *25*, 218-229.

Mather, K., Jinks, J. L. *Biometrical genetics: The study of continuous variation.* London: Chapman and Hall, 1971.

Mather, K., & Jinks, J. L. *Introduction to biometrical genetics.* London: Chapman and Hall, 1977.

Matthews, K. S., & Krantz, D. S. Resemblances of twins and their parents in patterns of behaviour. *Psychosomatic medicine*, 1976, *38*, 140-144.

Mayer, R. E. Problem-solving performance with task overload: Effects of self-pacing and trait anxiety. *Bulletin of the Psychonomic Society*, 1977, *9*, 283-286.

Mayo, P. R. Personality traits and the retrieval of positive and negative memories. *Personality and Individual Differences*, 1983, *4*, 465-472.

McCloy, C. H. A factor analysis of personality-traits to underlie character education. *Journal of Eductional Psychology*, 1936, *27*, 375-387.

McCord, R. R., & Wakefield, J. A. Jr. Arithmetic achievement as a function of introversion-extraversion and teacher-presented reward and punishment. *Personality and Individual Differences*, 1981, *2*, 145-152.

McCrae, R. R. Consensual validation of personality traits: Evidence from self reports and ratings. *Journal of Personality of Social Psychology*, 1982, *43*, 293-303.

McCrae, R. R., & Costa, P. T. Openness to experience and ego level in Loevinger's sentence completion test: Disparational contribution to developemtnal modes of personality. *Journal of Personality and Social Psychology*, 1980, *39*, 1179-1190.

McCrae, R. R., & Costa, P. T. Joint factors in self reports and ratings: Neuroticism, extraversion and openness to experience. *Personality and Individual Differences*, 1983, *4*, 245-256. (a)

McCrae, R. R., & Costa, P. T. Psychological maturity and subjective well-being: Toward a new synthesis. *Personality and Individual Differences*, 1983, *4*, 245-264. (b)

McCrae, R. R., & Costa, P. T. Joint factors in self reports and ratings: Neuroticism, extraversion, and openness to experience. *Personality and Individual Differences*, in press.

McCrae, R. R., Costa, P. T., & Grenlevy, D. Constancy of adult personality structure in males: Longitudinal, cross-sectional and times of measurement analyses. *Journal of Gerontology*, 1980, *35*, 877-883.

McDougall, W. *Is America safe for democracy?* New York: Scribners, 1921.

McDowell, I., & Praught, E. On the measurement of happiness. *American Journal of Epidemiology*, 1982, *110*, 949-958.

McEwan, A. W. Eysenck's theory of criminality and the personality types and offences of young delinquents. *Personality and Individual Differences*, 1983, *4*, 201-204.

McGuffin, P., Farmer, A. E., & Yonace, A. H. HLA antigens and subtypes of schizophrenia. *Psychiatric Research*, 1981, *5*, 115-122.

McGurk, B. J., & Bolton, V. A comparison of the Eysenck Personality Questionnaire and the psychological screening inventory in a delinquent sample and a comparison group. *Journal of Clinical Psychology*, 1981, *37*, 876-879.

McGurk, B. J., & McDougall, C. A new approach to Eysenck's theory of criminality. *Personality and Individual Differences*, 1981, *2*, 338-340.

McKinnon, D. W. *The violation of prohibitives in the sociology of problems.* Ph.D. thesis, Harvard University, 1933.

McLaughlin, R. J., & Eysenck, H. J. Visual masking as a function of personality. *British Journal of Psychology*, 1966, *57*, 393–396.

McLaughlin, R. J., & Kary, S. K. Amnesic effects in free recall with introverts and extraverts. *Psychonomic Science*, 1972, *29*, 250–252.

McLean, P. D. *Paired-associate learning as a function of recall interval, personality and arousal.* Unpublished Ph.D. thesis, University of London, 1968.

McLean, P. D. Induced arousal and time of recall as determinants of paired-associate recall. *British Journal of Psychology*, 1969, *60*, 57–62.

McPherson, F. M., Presley, A. S., Armstrong, J., & Curtis, R. H. 'Psychoticism' and psychotic illness. *British Journal of Psychiatry*, 1974, *125*, 152–160.

Medow, W. Zur Erblichkeit in der Psychiatrie. *Zeitschrift für die Gesamte Neurologie und Psychiatrie*, 1914, *26*, 493–534.

Medwick, S. A. A learning theory approach to research in schizophrenia. *Psychological Bulletin*, 1958, *55*, 316–327.

Medwick, S. A. The associational basis of the creative process. *Psychological Review*, 1962, *69*, 220–232.

Medwick, M. T., Medwick, S. A., & Jung, C. C. Continual association as a function of level of creativity and type of verbal stimulus. *Journal of Abnormal and Social Psychology*, 1964, *69*, 511–515.

Meehl, P. E. Probile analysis of the MMPI in differential diagnosis. In G. S. Welsh & W. G. Dahlstrom (Eds.), *Basic reactions on the MMPI in psychology and medicine.* Minneapolis: University of Minnesota Press, 1956.

Meggendorfer, F. Klinische und genealogische Untersuchungen über 'moral insecurity': *Zeitschrift für die Gesamte Neurologie und Psychiatrie*, 1921, *66*, 208–234.

Mehrabian, A., & Russell, J. A. *An approach to environmental psychology.* Cambridge, Mass: MIT Press, 1974.

Mehryar, A. H., Khayari, F., & Hebmat, J. H. Comparison of Eysenck's PEN and Langan's Psychological Screening Inventory in a group of American students. *Journal of Consulting and Clinical Psychology*, 1975, *93*, 9–12.

Mendelsohn, G. A., & Griswold, B. B., Differential use of incidental stimuli in problem solving as a funciton of creativity. *Journal of Abnormal and Social Psychology*, 1964, *68*, 431–436.

Messer, S. C. Reflection-impulsivity: A review. *Psychological Bulletin*, 1976, *83*, 1026–1052.

Messick, S. The standard problem: Meaning and values in measurement and evaluation. *American Psychologist*, 1975, *3*, 955–966.

Messick, S. Test validity and the ethics of assessment. *American Psychologist*, 1980, *35*, 1012–1027.

Metzner, R. Correlations between Eysenck's, Jung's and Sheldon's typologies. *Psychological Reports*, 1980, *47*, 343–348.

Michael, C. M. Follow-up studies of introverted children: IV Relative incidence of criminal behaviour. *Journal of Criminal Law and Criminality*, 1956, *43*, 412–422.

Michaelis, W., & Eysenck, H. J. The determination of personality inventory factor patterns and intercorrelations by change in real life motivation. *Journal of General Psychology*, 1971, *118*, 223–234.

Mikulka, P., Kandall, P., Constantine, S., & Posterfield, L. The effect of Pavlovian CS+ and CS− on explorative behaviour. *Psychonomic Science*, 1973, *27*, 308–310.

Miller, N. E. Experimental studies of conflict. In J. McV. Hunt (Ed.), *Personality and the behavior disorders.* New York: Ronald Press, 1944, pp. 431–466.

Mischel, W. *Personality and Assessment.* London: Wiley, 1968.

Mischel, W. Continuity and change in pesonality. *American Psychologist,* 1969, *24,* 112–1018.

Mischel, W. Toward a cognitive social learning reconceptualization of personality. *Psychological Review,* 1973, *80,* 252–283.

Mischel, W. The interaction of person and situation. In D. Magnusson & N. S. Endler (Eds.), *Personality at the crossroads: Current issues in international psychology.* Hillsdale, N.J.: Lawrence Erlbaum, 1977.

Mischel, W., & Peake, P. K. Beyond *Déjà vu* in the search for cross-situational consistency. Psychological Review, 1982, *89,* 730–755.

Mitchell, J. V., & Pierce-Jones, J. A factor analysis of Gough's California Psychological Inventory. *Journal of Consulting Psychology,* 1960, *24,* 453–456.

Mobbs, N. A. Eye contact in relation to social interaction/extraversion. *British Journal of Social and Clinical Psychology,* 1968, *7,* 305–306.

Mohan, V., & Dharmani, I. The effect of intelligence and personality on verbal conditioning. *Psychologica Belgica,* 1976, *16,* 223–232.

Mohan, V., & Gill, S. Effect of extraversion and drugs on letters cancellation. *Proceedings of the Indian Science Congress,* 1979.

Mohan, V., & Kumar, D. Qualitative analysis of the performance of introverts and extraverts on Standard Progressive Matrices. *British Journal of Psychology,* 1976, *67,* 391–397.

Monson, T. C., Hesley, J. W., & Chernick, L. Specifying when personality traits can and cannot predict behavior: An alternative to abondoning the attempt to predict single-act criteria. *Journal of Personality and Social Psychology,* 1982, *43,* 385–389.

Montag, I. *The Tel-Aviv-MMPI: Some validation studies.* Paper presented at the Fourth International Symposium on Personality Assessment at Haifa, June, 1977.

Montag, I., & Comrey, A. L. Comparison of certain MMPI, Eysenck and Comrey Personality constructs. *Multivariate Behavioral Research,* 1982, *17,* 93–97.

Morelli, G., & Andrews, L. Rationality and its relation to extraversion and neuroticism. *Psychological Reports,* 1980, *47,* 1111–1114.

Morelli, G., Krotinger, H., & Moore, S. Neuroticism and Levenson's Locus of Control Scale. *Psychological Reports,* 1979, *44,* 153–154.

Morgenstern, F. S., Hodgson, R. J., & Law, L. Work efficiency and personality: A comparison of introverted and extraverted subjects exposed to conditions of distraction nd distortion of stimulus in a learning task. *Ergonomics,* 1974, *17,* 211–220.

Morris, L. W. *Extraversion and introversion.* New York: Wiley. 1979.

Morris, L. W., & Liebert, R. M. Relationships of cognitive and emotional components of test anxiety to physiological arousal and academic performance. *Journal of Consulting and Clinical Psychology,* 1970, *35,* 332–337.

Morris, L. W., & Liebert, R. M. Effects of negative feedback, threat of shock and level of trait anxiety on the arousal of two components of anxiety. *Journal of Counseling Psychology,* 1973, *20,* 321–326.

Morris, L. W., Brown, N. R., & Halbert, B. L. Effects of symbolic modelling on the arousal of cognitive and affective components of anxiety in pre-school children. In C. D. Spielberger & I. G. Sarason (Eds.), *Stress and anxiety,* (Vol. 4.) London: Halsted, 1977.

Morris, L. W., Davis, M. A., & Hutchings, C. H. Cognitive and emotional components of anxiety: Literature review and a revised worry–emotionality scale. *Journal of Educational Psychology,* 1981, *73,* 541–555.

Morris, P. E., & Gale, A. A correlational study of variables related to imagery. *Perceptual and Motor Skills,* 1974, *38,* 659–665.

Moruzzi, G., & Magoun, H. W. Brain stem reticular formation and activation of the EEG. *Electroencephalography and Clinical Neurophysiology*, 1949, *1*, 455–473.

Moss, H. J., & Susman, E. J. Constancy and change in personality development. In O. G. Brim & J. Mayan (Eds.), *Constancy and change in human development.* Cambridge: Harvard University Press, 1980.

Mourant, A. E., Kopec, S. C., & Domaniewska-Soblezak, K. *The distribution of the human blood groups.* London: Oxford University Press, 1976.

Mueller, J. H. Anxiety and cue utilization in human learning and memory. In M. Zuckerman & C. D. Spielberger (Eds.), *Emotions and anxiety: New concepts, methods and applications.* Hillsdale, N.J.: Lawrence Erlbaum, 1976.

Mueller, J. H. Test anxiety, input modality, and levels of organization in free recall. *Bulletin of the Psychonomic Society*, 1977, *9*, 67–69.

Mueller, J. H. The effects of individual differences in test anxiety and type of orienting task on levels of organization in free recall. *Journal of Research in Personality*, 1978, *12*, 100–116.

Mueller, J. H. Test anxiety and the encoding and retrieval of information. In I. G. Sarason (Ed.), *Test anxiety: Theory, research, and applications.* Hillsdale, N.J.: Lawrence Erlbaum, 1979.

Mueller, J. H., Carlomusto, M., & Marler, M. Recall as a function of method of presentation and individual differences in test anxiety. *Bulletin of the Psychonomic Society*, 1977, *10*, 447–450.

Mueller, J. H., Bailis, K. L., & Goldstein, A. G. *Depth of processing and anxiety in facial recognition.* Paper read at the annual convention of the Midwestern Psychological Association, Chicago, 1978.

Mullin, J., Corcoran, D. W. J. Interaction of task amplitude with circadian variation in auditory vigilance performance. *Ergonomics*, 1977, *20*, 193–200.

Murray, H. A. (Ed.). *Explorations in personality.* New York: Oxford University Press, 1938.

Mussen, P., Eichorn, D. H., Hanzik, M. P., Bieher, S. L., & Meredith, W. Continuity and change in women's characteristics over four decades. *International Journal of Behavioral Development*, 1980, *3*, 333–347.

Myers, I. B. *The Myers-Briggs type indicator: Normal.* Princeton: Educational Testing Service, 1962.

Nagpal, M., & Gupta, B. S. Personality, reinforcement and verbal operant conditioning. *British Journal of Psychology*, 1979, *70*, 471–476.

Nebylitsyn, V. D. Individual differences in the strength and sensitivity of both Visual and Auditory analyses. In N. O'Connor (Ed.), *Recent Soviet Psychology.* Oxford: Pergamon, 1961.

Nelkon, M., & Parker, P. *Advanced level physics.* London: Heinemann. 1968.

Newman, H. H., Freeman, F. N., & Holzinger, K. J. *Twins.* Chicago: University of Chicago Press, 1937.

Newton, I. *Principia Mathematica.* Berkeley: University of California Press, 1934.

Neymann, C. A., & Kohlstedt, K. D. A new diagnostic test for introversion-extraversion. *Journal of Abnormal Psychology*, 1939, *23*, 482–487.

Nichols, R. C., & Schnell, R. R. Factor scales for the California Psychological Inventory. *Journal of Consulting Psychology*, 1963, *27*, 228–235.

Nicoll, M. *Dream psychology.* London: Macmillan, 1921.

Nideffer, R. M. Test of attentional and interpersonal style. *Journal of Personality and Social Psychology*, 1976, *34*, 394–404.

Nielsen, T. C., & Petersen, K. E. Electrodermal correlates of extraversion, trait anxiety and schizophrenia. *Scandinavian Journal of Psychology*, 1976, *17*, 73-80.

Norman, W. T. Toward an adequate taxonomy of personality attributes: Replicated factor structure in peer nomination personality ratings. *Journal of Abnormal and Social Psychology*, 1963, *60*, 574-583.

Norman, W. T. 'To see ourselves as others see us!' Relations among self-perceptions, peer perceptions, and expected peer perceptions and personality attributes. *Multivariate Behavioral Research*, 1969, *4*, 417-443.

Norman, W. T., & Goldberg, L. R. Raters, ratees and random-men in personality structure. *Journal of Personality and Social Psychology*, 1960, *4*, 681-691.

Oates, D. V. An experimental study of temperament. *British Journal of Psychology*, 1929, *19*, 1-30.

Ödegard, O. The psychiatric disease entities in the light of a genetic investigation. *Acta Psychiatrica Scandinavica*, Suppl., 1963, *169*, Whole no. 94.

O'Gorman, J. G. Individual differences in habituation of human physiological responses: A review of theory, method, and findings in the study of personality correlates in non-clinical populations. *Biological Psychology*, 1977, *5*, 257-318.

O'Hanlon, J. Adrenaline and noradrenaline: Relation to performance in a visual vigilance task. *Science*, 1965, *150*, 507-509.

Okaue, M., Nakamura, M., & Niwa, K. Personality characteristics of pilots on EPPS, MPI and DOSEFU Test: 1. Comparisons of pilots with Japanese male adults, and of fighter with carrier pilots. *Reports of Aeromedical Laboratory*, 1977, *18*, 83-93.

Olweus, D. Stability of aggressive reaction patterns in males: A review. *Psychological Bulletin*, 1979, *86*, 852-875.

Olweus, D. The consistency in a personality revisited, with special reference to aggression. *British Journal of Social and Clinical Psychology*, 1980, *19*, 377-390.

Opollot, J. A. *Paired-associate performance as a function of extraversion and variations in intra-task interference*. Unpublished Ph.D. thesis, University of Birmingham, 1970.

Osborne, J. W., & Steeves, L. Relation between self-actualization, neuroticism and extraversion. *Perceptual and Motor Skills*, 1981, *53*, 996-998.

Overall, J. E., Hunter, S., & Butcher, J. N. Factor structure of the MMPI-168 in a psychiatric population. *Journal of Consulting and Clinical Psychology*, 1973, *41*, 284-286.

Pagano, D. F., & Katahn, M. Effects of test anxiety on acquisition and retention of psychological terms and single letter stimulus-response pairs. *Journal of Experimental Research in Personality*, 1967, *2*, 260-267.

Paisey, T. J. H., & Mangan, G. L. Neo-Pavlovian temperament theory and the biological bases of personality. *Personality and Individual Differences*, 1982, *3*, 189-203.

Paisey, T. J. H., & Mangan, G. L. A developmental model of conditioning and personality: Acitivity and reactivity constructs In J. Strelan (Ed.), *Contemporary approaches to temperament: An East-West dialogue*. Oxford: Pergamon, in press.

Pallares, A. T. *Intensitat de l'estimul incondicionat i differencies individuals, en condicionament d'evitacio 'shuttle'*. Barcelona: Publicaciones de la Universidad Autonoma de Barcelona, 1978.

Palmore, E. *Social patterns in normal ageing: Findings from the Duke Longitudinal Study*. Durham, N.C.: Duke University Press, 1981.

Paramesh, C. R. Introversion-extraversion and figural aftereffects. *Proceedings of the Indian Science Congress*, 1963.

Parker, G. V., & Veldman, D. J. Item factor structure of the Adjective Checklist. *Educational Psychological Measurement*, 1969, *29*, 605-613.

Parnell, R. W. *Behaviour and physique.* London: Arnold, 1958.

Passingham, R. E. Crime and personality: A review of Eysenck's theory. In V. D. Neby-litzyn & J. A. Gray (Eds.), *Biological bases of individual behavior.* New York: Academic Press, 1972, pp. 342-371.

Passini, F. T., & Norman, W. T. A universal conception of personality structure. *Journal of Personality and Social Psychology,* 1966, *4,* 44-49.

Patterson, D. G. *Physique and intellect.* New York: Century, 1930.

Patrick, J. R. Studies in rational behavior and emotional excitement: I. Rational behavior in human subjects. *Journal of Comparative Psychology,* 1934, *18,* 1-22. (a)

Patrick, J. R. Studies in rational behavior and emotional excitement: II. The effect of emotional excitement on rational behavior in human subjects. *Journal of Comparative Psychology,* 1934, *18,* 153-195 (b).

Patterson, A. R. *Coercive family process.* Eugene, Ore.: Castalia, 1982.

Patterson, M. L. An arousal model of interpersonal intimacy. *Psychological Review,* 1976, *83,* 235-245.

Patterson, M., & Holmes, D. S. Social interaction, correlates of the MPI extraversion-introversion scale. *American Psychologist,* 1966, *21,* 724-725.

Pawlik, K. & Cattell, R. B., Third order factors in objective personality tests. *British Journal of Psychology,* 1964, *55,* 1-18.

Pavlov, I. P. *Conditioned reflexes.* London: Oxford University Press, 1927.

Payne, R., & Hewlett, J. H. Thought disorder in psychotic patients. In H. J. Eysenck (Ed.), *Experiments in personality* (Vol. 2). London: Routledge & Kegan Paul, 1960.

Petrie, A *Individuality in pain and suffering.* Chicago: University of Chicago Press, 1978.

Piers, E. V., & Kirchner, E. P. Eyelid conditioning and personality: Positive results from nonpartisans. *Journal of Abnormal Psychology,* 1969, *74,* 336-339.

Planansky, K. *Phenotypic boundaries of schizophrenia in twins: Mutation in populations.* Proceedings of a symposium held in Prague, 1965. Prague: Academia, 1966. (a)

Planansky, K. Schizoidness in twins. *Acta Geneticae Medicae et Gemellolgiae,* 1966, *15,* 151-175. (b)

Planansky, K. Conceptual boundaries of schizoidness: Suggestion of epidemiological and genetic research. *Journal of Nervous and Mental Disorders,* 1966, *142,* 318-329. (c)

Planansky, K. Phenotypic boundaries in genetic specificity in schizophrenia. In S. R. Kaplan (Ed.), *Genetic factors in schizophrenia.* Springfield, Ill.: Charles C. Thomas, 1972, pp. 441-572.

Platt, J. J., Pomeranz, D., & Eiseman, R. Validation of the EPI by the MMPI and internal-external control scale. *Journal of Clinical Psychology,* 1971. *27,* 104-105.

Plouffe, L, & Stelmack, R. M. Neuroticism and the effect of stress on the pupillary light reflex. *Perceptual and Motor Skills,* 1979, *49,* 635-642.

Posner, M. I. Abstraction and the process of recognition. In C. H. Bower & T. Spence (Eds.), *The Psychology of learning and motivation* (Vol. 3.) New York: Academic Press, 1969, pp. 43-100.

Powell, G. E. *Personality and conformity in normal children.* Unpublished Ph.D. thesis, University of London, 1976.

Powell, G. E. Psychoticism and social deviancy in children. *Advances in Behavior Research and Therapy,* 1977, *1,* 27-56.

Powell, G. E., & Stewart, R. A. The relationship of personality to antisocial and neurotic behaviors as observed by teachers. *Personality and Individual Differences,* 1983, *4,* 97-100.

Prentky, R. A. *Creativity and psychopathology.* New York: Praeger, 1980.

Putnins, A. L. The Eysenck personality questionnaire and delinquency prediction. *Personality and Individual Differences*, 1982, *3*, 339–340.

Queyrat, P. *Les caractères et l' éducation morale*. Paris: Alcan, 1896.

Rachman, S. *Psychomotor behaviour and personality with special reference to conflict*. Unpublished Ph.D. thesis, University of London, 1961.

Rachman, S. Extraversion and neuroticism in childhood. In H. J. Eysenck & S. B. G. Eysenck (Eds.), *Personality structure and measurement*. London: Routledge & Kegan Paul, 1967.

Rachman, M. A., & Eysenck, S. B. G. Psychoticism and response to treatment in neurotic patients. *Behavior Research and Therapy*, 1978, *16*, 183–189.

Rachman, S., & Grassi, J. Reminiscence, inhibition and consolidation. *British Journal of Psychology*, 1965, *56*, 157–162.

Raine, A. & Venables, P. H. Classical conditioning and socialization—A biosocial interaction. *Personality and Individual Differences*, 1981, *2*, 273–284.

Rakos, R. F. Content consideration in the distinction between assertive and aggressive behavior. *Psychological Reports*, 1979, *44*, 767–773.

Rathus, S. A. Factor structure of the MMPI-168 with and without repression weights. *Psychological Reports*, 1978, *42*, 643–646.

Ray, W. J., Katahn, M., & Snyder, C. R. Effects of test anxiety on acquisition, retention, and generalisation of a complex verbal task in a classroom situation. *Journal of Personality and Social Psychology*, 1971, *20*, 147–154.

Rechtschaffen, A. Neural satiation, reactive inhibition and introversion—extraversion. *Journal of Abnormal and Social Psychology*, 1958, *57*, 283–291.

Rees, L. Body build, personality and neurosis in women. *Journal of Mental Science*, 1950, *96*, 426–434.

Rees, L. Constitutional factors and abnormal behaviour. In H. J. Eysenck (Ed.), *Handbook of abnormal psychology*. London: Pitman, 1973, pp. 487–539.

Rees, L, & Eysenck, H. J. A factorial study of some morphological and psychological aspects of human constitution. *Journal of Mental Science*, 1945, *91*, 1–21.

Reid, D. W., & Ware, E. E. Multidimensionality of internal-external control: Implications for past and future research. *Canadian Journal of Behavioral Science*, 1973, *5*, 264–271.

Reinhardt, R. F. The outstanding jet pilot. *American Journal of Psychiatry*, 1970, *127*, 732–736.

Revelle, W., Humphreys, M. S., Simon, L., & Gilliland, K. The interactive effect of personality, time of day and caffeine: A test of the arousal model. *Journal of Experimental Psychology: General*, 1980, *109*, 1–31.

Revendsdorff, D. Vom unsinnigen Aufwand. *Archiv fürPsychologie*, 1978, *130*, 1–36.

Review Panel on Coronary-prone Behavior and Coronary Heart Disease Genetical review. *Circulation*, 1981, *63*, 1199–1215.

Reyburn, M. S., & Taylor, J. G. Some aspects of personality. *British Journal of Psychology*, 1939, *30*, 151–165.

Reynolds, C. N., & Nichols, R. C. Factor rates for the CPI: Do they capture the valid variance? *Educational and Psychological Measurement*, 1977, *37*, 907–915.

Ribot, T. *La psychologie des sentiments*. Paris: Alcan, 1892.

Richardson, J. T. E. *Mental imagery and human memory*. London: Macmillan, 1980.

Riding, R. J. The effect of extraversion and detail content on the recall of prose by eleven year-old children. *British Journal of Educational Psychology*, 1979, *49*, 297–303.

Riding, R. J., & Dyer, V. A. The relationship between extraversion and verbal imagery learning style in twelve year-old children. *Personality and Individual Differences,* 1980, *1,* 273–280.

Riedel, H. Zur empirischen Erbprognose der Psychopathie. *Zeitschrift für die Gesamte Neurologie und Psychiatrie,* 1937, *159,* 597–625.

Rim, Y. Significance of work and personality. *Journal of Occupational Psychology,* 1977, *50,* 135–138.

Roback, A. A. *The psychology of character.* London: Kegan Paul, French, & Frabner, 1931.

Robinson, D. L. Properties of the diffuse thalamocortical system and human personality: A direct test of Pavlovian/Eysenckian theory. *Personality and Individual Difference,* 1982, *3,* 1–16.

Robinson, D. L., Haier, R. J., Braden, W., & Krengel, M. Evoked potential augmenting and reducing: II. Methodological and theoretical significance of new electrophysiological observations. *Personality and Individual Differences,* in press.

Robinson, R. N. Eye-color, sex and personality: A case of negative findings for Woty's sociability hypothesis. *Perceptual and Motor Skills,* 1981, *52,* 855–863.

Robinson, T. N., & Zahn, T. P. Covariations of two-flash threshold and autonomic arousal for high and low scorers in a measure of psychoticism. *British Journal of Social and Clinical Psychology,* 1979, *18,* 431–441.

Rocklin, T., & Revelle, W. The measurement of extraversion: A comparison of the Eysenck Personality Inventory and the Eysenck Personality Questionnaire. *British Journal of Social Psychology,* 1981, *20,* 279–289.

Rodnight, E., & Gooch, R. N. A new method for the determination of individual differences in susceptibility to a depressant drug. In H. J. Eysenck (Ed.), *Experiments with drugs.* Oxford: Pergamon, 1963.

Rohracher, H. *Kleine Charakterkunde.* Wien: Urban & Schwarzenberg, 1965.

Rosenbaum, R. Stimulus generalization as a function of level of experimentally-induced anxiety. *Journal of Experimental Psychology,* 1953, *45,* 35–43.

Rosenbaum, R. Stimulus generalization as a function of clinical anxiety. *Journal of Abnormal and Social Psychology,* 1956, *53,* 281–285.

Rosenthal, D. (Ed.). *The Genain Quadruplets.* New York: Basic Books, 1963.

Roseveare, N. T. *Mercury's Perihelion from de Verrier to Einstein.* Oxford: Clarendon Press, 1982.

Roth, E. Die Geschwindigheit der Verarbeitung von Information und ihr Zusammenhang mit Intelligenz. *Zeitschrift für Experimentelle und Angewante Psychologie,* 1964, *11,* 616–622.

Rotter, J. B. Generalized expectancies for internal V–S external control of reinforcement. *Psychological Monographs,* 1960, *80,* No. 1.

Rousel, C. H., & Edwards, C. N. Some development antecedents of psychopathology. *Journal of Personality,* 1971, *39,* 362–377.

Rowe, D. C. Monozygotic twin cross-correlation as a validation of personality: A test of the semantic bias hypothesis. *Journal of Personality and Social Psychology,* 1982, *43,* 1072–1079.

Royce, J. R. The conceptual framework for a multi-factor theory of individuality. In J. R. Royce (Ed.), *Multivariate analysis and psychological theory.* London: Academic Press, 1973, pp. 305–407.

Royce, J. R., & Powell, S. *Theory of personality and individual differences: Factors, systems, and processes.* Englewood Cliffs, N.J.: Prentice-Hall, 1983.

Royce, J. R., Poley, W., & Yendall, L. T. Behavior genetic analysis of mouse emotionality:

I. Factor analysis. *Journal of Comparative and Physiological Psychology*, 1973, *83*, 36–47. (a)

Royce, J. R., Poley, W., & Yendall, L. T. Behavior genetic analysis of mouse emotionality: III. The clinical analysis. *Journal of Comparative and Physiological Psychology*, 1973, *83*, 347–354. (b)

Rüdin, E. *Zur Vererbung und Neuentstehung der Dementia Praecox*. Berlin: Springer, 1916.

Ruesch, J., & Bowman, K. Prolonged post-traumatic syndrome following head injury. *American Journal of Psychiatry*, 1945, *102*, 145–163.

Rushton, J. *A longitudinal study of the relationship between some personality variables and some measures of academic attainment*. Unpublished Ph.D. thesis, University of Manchester, 1969.

Rushton, J. P., & Chrisjohn, R. D. Extraversion, neuroticism, psychoticism and self-reported delinquency: Evidence from eight separate samples. *Personality and Individual Differences*, 1981, *2*, 11–20.

Rushton, J. P., Brainerd, C. J., & Preisley, M. Behavioral development and construct validity: The principle of aggregation. *Psychological Bulletin*, 1983, *94*, 18–38.

Russell, J. A. Affective space is bipolar. *Journal of Personality and Social Psychology*, 1979, *37*, 345–356.

Russell, P. A. Open-field defecation in rats: Relationship with body weight and basal defecation level. *British Journal of Psychology*, 1973, *64*, 109–114. (a)

Russell, P. A. Relationship between exploratory behaviour and fear: A review. *British Journal of Psychology*, 1973, *64*, 413–433. (b)

Rust, J. Cortical evoked potential, personality, and intelligence. *Journal of Comparative and Physiological Psychology*, 1975, *89*, 1220–1226.

Rutter, D. R., & Stephenson, G. M. Visual interaction in a group of schizophrenic and depressive patients. *British Journal of Social and Clinical Psychology*, 1972, *11*, 57–65.

Saklofske, D. H. Personality and behavior problems of schoolboys. *Psychological Reports*, 1977, *41*, 445–446.

Saltz, E. Manifest anxiety: Have we misread the data? *Psychological Review*, 1970, *77*, 568–573.

Saltz, E., & Hoehn, A. J. A test of the Taylor-Spence theory of anxiety. *Journal of Abnormal and Social Psychology*, 1957, *54*, 114–117.

Sarason, I. G., Smith, R. E., & Diener, E. Personality research: Components of variance attributable to the person and the situation. *Journal of Personality and Social Psychology*, 1975, *32*, 199–204.

Savage, R. D., & Eysenck, H. J. The definition and measurement of emotionality. In H. J. Eysenck (Ed.), *Experiments in motivation*. London: Pergamon Press, 1964., pp. 292–314.

Saville, P., & Blinkhorn, S. *Undergraduate personality by factored scores*. London: NFER Publishing Company, 1976.

Saville, P., & Blinkhorn, S. Reliability, homogeneity and the construct validity of Cattell's 16PF. *Personality and Individual Differences*, 1981, *2*, 325–333.

Schachter, S. *The psychology of affiliation*. Stanford: Stanford University Press, 1959.

Schafer, E. W. P. Neural adaptability: A biological determinant of behavioral intelligence. *International Journal of Neuroscience*, 1982, *13*, 183–191.

Schalling, D., Cronholm, B., Asberg, M., & Espmark, S. Ratings of psychic and somatic anxiety indicants. *Acta Psychiatrica Scandinavica*, 1973, *49*, 353–368.

Schalling, D., Cronholm, B., & Asberg, M. Components of state and trait anxiety as related to personality and arousal. In L. Levi (Ed.), *Emotions: Their parameters and measurement*. New York: Raven, 1975.

Schenk, J. Die neurotische Persönlichkeit des Haschischkonsumenten. *Zeitschrift für Klinische Psychologie und Psychotherapie*, 1974, *22*, 340–351.

Schenk, J., Rausche, A., & Steege, F. W. Zur Struktur des Freiburger Persönlichkeits-Inventars (FPI). *Zeitschrift für experimentelle und Angewandte Psychologie*, 1977, *24*, 492–509.

Schlegel, W. S. *Körper und Seele*. Stuttgart: Enke, 1957.

Schlegel, W. S. Genetic foundation of social behaviour. *Personality and Individual Differences*, 1983, *4*, 483–490.

Schneewind, K. A. Enswickelung einer dentschsprachigen Versian des 16PF Tests von Cattell. *Diagnostica*, 1977, *23*, 188–191.

Schuerger, J. M., Taid, E., & Taveruelli, M. Temporal stability of personality by questionnaire. *Journal of Personality and Social Psychology*, 1982, *43*, 176–182.

Schulz, B. Kinder manisch-depressiwer und anderer affektivpsychotischen Elterpaare. *Zeitschrift für die Gesamte Neurologie und Psychiatrie*, 1940, *169*, 311–324.

Schwartz, S. Individual differences in cognition: Some relationships between personality and memory. *Journal of Research in Personality*, 1975, *9*, 217–225.

Schwartz, S. Differential effects of personality on access to various long-term memory codes. *Journal of Research in Personaity*, 1979, *13*, 396–403.

Scott, J. P., & Fuller, J. L. *Genetics and the social behavior of the dog*. Chicago: University of Chicago Press, 1965.

Sekuler, R. W., & Ganz, L. Aftereffects of seen motion with a stabilized retinal image. *Science*, 1963, *139*, 419–420.

Sells, S. B., Demaree, R. G., & Will, D. P. *A taxonomic investigation of personality: Conjoint factor structure of Guilford and Cattell trait markers*. Texas: Christian Institute of Behavioral Research, 1968.

Sells, S. B., Demaree, R. G. & Will, D. P. Dimensions of Personality: I. Conjoint factor structure of Guilford and Cattell trait markers. *Multivariate Behavioral Research*, 1970, *5*, 391–422.

Seunath, O. H. M. *Strategy and consolidation in pursuit motor performance*. Unpublished Ph.D. thesis, University of London, 1973.

Sevilla, L. G. *Extincio de RF50, inhibicio i personalitat en rates mascles Wistar*. Unpublished Ph.D. thesis, Universidad Autonoma de Barcelona, 1974.

Sevilla, L. G. Extincio de RF 50, inhibicio i personalitat en rates mascles Wistar. *Temas Monograficos de Psicologia*, Departamento de Psicologia, Universidad Autonoma de Barcelona, 1975, *1*, 1–56.

Sevilla, L. G. Extraversion and neuroticism in rats. *Personality and Individual Differences*, in press.

Sevilla, L. G., & Garau, A. Extraversion y deambulacion de la rata en el campo abierto. *Revista Latino-americana de psicologia*, 1978, *10*, 211–226.

Shadbolt, D. R. Interactive relationships between measured personality and teaching strategy variables. *British Journal of Educational Psychology*, 1978, *48*, 227–231.

Shagass, C. A. The sedation threshold. A method for estimating tension in psychiatric patients. *EEG and Clinical Neurophysiology*, 1954, *6*, 221–233.

Shagass, C. A. A measurable neurophysiological factor of psychiatric significance. *EEG and Clinical Neurophysiology*, 1957, *9*, 101–108.

Shagass, C., & Jones, A. L. A neurophysiological test for psychiatric diagnosis: Results in 750 patients. *American Journal of Psychiatry*, 1958, *114*, 1002–1009.

Shagass, C., & Kerenyi, A. B. Neurophysiologic studies of personality. *Journal of Nervous and Mental Diseases*, 1958, *126*, 141–147.

Shagass, C., & Schwartz, M. Age, personality and somatosensory evoked responses. *Science*, 1965, *148*, 1359–1361.

Shanmugan, T. E., & Santhanam, M. C. Personality differences in serial learning when interference is presented at the marginal visual level. *Journal of the Indian Academy of Applied Psychology*, 1964, *1*, 25–28.

Shaplan, J., Rushton, J. P., & Campbell, A. Crime and personality: Further evidence. *Bulletin of the British Psychological Society*, 1975, *28*, 66–68.

Shaw, L., & Sichel, H. *Accident proneness*. Oxford: Pergamon, 1970.

Shedletsky, R., & Endler, N. S. Anxiety: The state-trait model and the interaction model. *Journal of Personality*, 1974, *42*, 511–527.

Sheldon, W. H. *The varieties of human physique*. New York: Harper, 1940.

Sheldon, W. H. *The varieties of human temperament*. New York: Harper, 1942.

Shields, J. *Monozygotic twins*. Oxford: Oxford University Press, 1962.

Shigenhisa, T., & Symons, J. R. Effects of intensity of visual stimulation on auditory sensitivity in relation to personality. *British Journal of Psychology*, 1973, *64*, 205–213.

Shigehisa, T., Shigehisa, P. M. J., & Symons, J. R. Effect of interval between auditory and preceding visual stimuli on auditory sensitivity. *British Journal of Psychology*, 1973, *64*, 367–373.

Shiomi, K. Relations of pain threshold and pain tolerance in cold water with scores on Maudsley Personality Inventory and Manifest Anxiety Scale. *Perceptual and Motor Skills*, 1978, *47*, 1155–1158.

Shiomi, K. Performance differences between extraverts and introverts on exercises using an ergonometer. *Perceptual and Motor Skills*, 1980, *50*, 356–358.

Shorkey, C. T., & Whiteman, V. L. Development of the Rational Behavior Inventory: Initial validity and reliability. *Educational and Psychological Measurement*, 1977, *37*, 527–534.

Shostrom, E. L. A test for the measurement of self-actualization. *Educational and Psychological Measurement*, 1964, *24*, 207–218.

Shostrom, E. L., & Knapp, R. R. The relationship of a measure of self-actualization (POI) to a measure of pathology (MMPI) and to therapeutic growth. *American Journal of Psychotherapy*, 1966, *20*, 193–202.

Shrauger, J. S., & Schoenenman, T. J. Symbolic interactionist view of self-concept: Through the looking glass darkly. *Psychological Bulletin*, 1979, *86*, 549–573.

Shucard, D. W., & Horn, J. L. Evoked cortical potential and measurement of human abilities. *Journal of Comparative and Physiological Psychology*, 1972, *78*, 59–68.

Shweder, R. A. How relevant is an individual difference theory of personality? *Journal of Personality*, 1975, *43*, 455–489.

Shweder, R. A., & D'Andrade, R. G. Aggregate reflection or systematic distortion? A reply to Berck, Vein & Thorne. *Journal of Personality and Social Psychology*, 1979, *37*, 1075–1089.

Siddle, D. A. T. The orienting response and distraction. *Australian Journal of Psychology*, 1971, *23*, 261–265.

Siddle, D. A. T., Morrish, R. B., White, K. D., & Mangan, G. L. Relation of visual sensitivity to extraversion. *Journal of Experimental Research in Personality*, 1969, *3*, 264–267.

Sigal, J. J., Star, K. H., & Franks, C. M. Hysterics and dysthymics as criterion groups in the study of introversion—extraversion. *Journal of Abnormal and Social Psychology*, 1958, *57*, 143–148.

Simonson, E., & Brozek, J. Flicker fusion frequency. *Physiological Review*, 1952, *32*, 349–378.

Singer, K., Lieh-Mak, F., & Ng, M. L. Physique, personality and mental illness in Southern Chinese women. *British Journal of Psychiatry*, 1976, *129*, 243–247.

Singh, V. K., & Gupta, B. S. Personality and drugs in visual figural after-effects. *International Journal of Psychology*, 1981, *16*, 35–44.

Skanthakumari, S. R. Personality differences in the rate of forgetting. *Journal of the Indian Academy of Applied Psychology*, 1965, *2*, 39–47.

Skinner, B. F. *The behavior of organisms: An experimental analysis.* New York: Appleton-Century-Crofts, 1938.

Skinner, H. J., & Jackson, D. G. A model of psychopathology based on an interpretation of MMPI actuarial systems. *Journal of Consulting and Clinical Psychology*, 1978, *46*, 231–238.

Slade, P. An investigation of psychological factors involved in the predisposition in auditory hallucination. *Psychological Medicine*, 1976, *6*, 123–132.

Slater, E. The neurotic constitutional statistical study of 2,000 neurotic soldiers. *Journal of Neurology, Neurosurgery and Psychiatry*, 1943, *6*, 1–16.

Slater, E. Genetical causes of schizophrenic symptoms. *Monatsschrift für Psychiatrie und Neurologie*, 1947, *113*, 50–62.

Slater, E. *Psychotic and neurotic illnesses in twins.* London: H. M. Stationery Office, 1953.

Slater, E., & Slater, P. A heuristic theory of neurosis. *Journal of Neurology, Neurosurgery and Psychiatry*, 1944, *7*, 49–55.

Smith, B. D. Usefulness of peer ratings of personality in education research. *Educational and Psychological Measurement*, 1967, *27*, 963–984.

Smith, B. D. Extraversion and electrodermal activity: Arousability and the inverted-U. *Personality and Individual Differences*, 1983, *4*, 411–419.

Smith, B. D., & Wigglesworth, M. J. Extraversion and neuroticism in orienting reflex dishabituation. *Journal of Research in Personality*, 1978, *12*, 284–296.

Smith, B. D., Rypma, C. B., & Wilson, R. J. Dishabituation and spontaneous recovery of the electrodermal orienting response: Effects of extraversion, impulsivity, sociability and caffeine. *Journal of Research in Personality*, 1981, *15*, 233–240.

Smith, B. D., Wilson, R. J., & Jones, B. E. Extraversion and multiple levels of caffeine-induced arousal: Effects of overhabituation and dishabituation. *Psychophysiology*, 1983, *20*, 29–34.

Smith, J. M., & Misiak, H. The effect of iris color on critical flicker frequency (CFF). *Journal of General Psychology*, 1973, *89*, 91–95.

Smith, P. J. *Personality in infants.* Unpublished doctoral thesis, University of Newcastle on Tyne, 1974.

Smith, S. L. Extraversion and sensory threshold. *Psychophysiology*, 1968, *5*, 293–299.

Sokal, R. R., & Sneath, P. H. *Principles of numerical taxonomy.* London: Freeman, 1963.

Sokolov, E. N. *Perception and the conditioned reflex.* London: Pergamon, 1963.

Sparrow, N. H., & Ross, J. The dual nature of extraversion: A replication. *Australian Journal of Psychology*, 1969, *16*, 214–218.

Spearman, C. E. General intelligence objectively determined and measured. *American Journal of Psychology*, 1904, *15*, 201–293.

Spearman, C. *The abilities of man.* London: Macmillan, 1927.

Spence, J. T., & Spence, K. W. The motivational components of manifest anxiety: Drive and drive stimuli. In C. D. Spielberger (Ed.), *Anxiety and behaviour.* London: Academic Press, 1966.

Spence, K. W. Anxiety (drive) level and performance in eyelid conditioning. *Psychological Bulletin*, 1964, *61*, 129–139.

Spence, K. W., Taylor, J., & Ketchel, R. Anxiety (drive) level and degree of competition in paired-associates learning. *Journal of Experimental Psychology*, 1956, *52*, 306–310.

Spiegler, M. D., Morris, L. W., & Liebert, R. M. Cognitive and emotional components of test anxiety: Temporal factors. *Psychological Reports*, 1968, *22*, 451–456.

Spielberger, C. D. The effects of anxiety on complex learning and academic achievement. In C. D. Spielberger (Ed.) *Anxiety and behavior*. New York: Academic Press, 1966.

Spielberger, C. D. Conceptual and methodological issues in anxiety research. In C. D. Spielberger (Ed.), *Anxiety: Current trends in theory and research* (Vol. 1). New York: Academic Press, 1971.

Spielberger, C. D. *Preliminary manual for the Stall-Treish Personality Inventory (STPI)*. Palo Alto, Calif.: Consulting Psychologists Press, 1980.

Spielberger, C. D., Gorsuch, R., & Lushene, R. *The State-Trait Anxiety Inventory (STAI) Test Manual Form X*. Palo Alto, Calif.: Consulting Psychologists Press, 1970.

Spielberger, C. D., O'Neil, H. F., & Hansen, D. N. Anxiety, drive theory, and computer-assisted learning. In B. A. Maher (Ed.), *Progress in experimental personality research* (Vol. 6). London: Academic Press, 1972.

Spielberger, C. D., Gonzalez, H. P., Taylor, C. J., Algaze, B., & Anton, W. D. Examination stress and test anxiety. In C. D. Spielberger & I. G. Sarason (Eds.), *Stress and anxiety* (Vol. 5). London: Halsted, 1978.

Spielman, J. *The relation between personality and the frequency and duration of involuntary rest pauses during massed practice*. Unpublished Ph.D. thesis, University of London, 1963.

Spiller, D., & Guski, R. Langfristiger Drogenkonsum und Persölichkeitsmerkmale. *Zeitschrift für Sozialforschung* 1975, *6*, 31–42.

Spitz, H. H., & Lipman, R. S. Reliability and intercorrelation of individual differences on visual and kinesthetic figural aftereffects. *Perceptual and Motor Skills*, 1960, *10*, 159–166.

Stagner, R. *Psychology of personality*. New York: McGraw-Hill, 1948.

Stallo, D. B. *The concepts and theories of modern physics*. New York: Appleton, 1881.

Steele, R. S., & Kelly, T. J. Eysenck Personality Questionnaire and Jungian Myers-Briggs Type Indicator correlation of extraversion–introversion. *Journal of Consulting and Clinical Psychology*, 1976, *44*, 690–691.

Stein, K. B. The TSC scales: The outcome of a cluster analysis of the 550 MMPI items. In P. McReynolds (Ed.), *Advances in psychological measurement* (Vol. 1). Palo Alto: Science and Behavior Books, 1968.

Stelmack, R. M. The psychophysiology of extraversion and neuroticism. In H. J. Eysenck (Ed.) *A model for personality*. Berlin: Springer, 1981.

Stelmack, R. M., & Campbell, K. B. Extraversion and auditory sensitivity to high and low frequency. *Perceptual and Motor Skills*, 1974, *38*, 875–879.

Stelmack, R. M., & Mandelzys, N. Extraversion and pupillary response to affective and taboo words. *Psychophysiology*, 1975, *12*, 536–540.

Stelmack, R. M., & Wilson, K. G. Extraversion and the effects of frequency and intensity on the auditory brain-stem evoked response. *Personality and Individual Differences*, 1982, *3*, 373–380.

Stelmack, R. M., Achorn, E., & Michaud, A. Extraversion and individual differences in auditory evoked response. *Psychophysiology*, 1977, *14*, 368–374.

Stelmack, R. M., Bourgeois, R. P., Chian, J. Y. C., & Pickard, C. W. Extraversion and the orienting reaction habituation rate to visual stimuli. *Journal of Research in Personality*, 1979, *13*, 49–58.

Steptoe, A. *Psychological factors in cardiovascular disorders*. London: Academic Press, 1981.

Stern, W. *Differentielle Psychologie*. Berlin: Rorer, 1921.

Sternberg, R. J. *Intelligence, information processing and analogical reasoning: The componential analysis of human abilities*. Hillsdale, N.J.: Lawrence Erlbaum, 1977.

Sternberg, R. *A handbook of human intelligence*. Cambridge: Cambridge University Press, 1983.

Sternberg, R. J., & Gardner, M. K. A componential interpretation of the general factor in human intelligence. In H. J. Eysenck (Ed.), *A model for intelligence*. New York: Springer, 1982, pp. 231–254.

Sternberg, S. High speed reasoning in human memory. *Science*, 1966, *153*, 652–654.

Stevenson-Hinde, J., Stillwell-Barnes, R., & Zunz, M. Subjective assessment of Rhesus monkeys over four successive years. *Primates*, 1980, *21*, 66–82.

Strelau, J. Nervous system type and extraversion-introversion: A comparison of Eysenck's theory with Pavlov's typology. *Polish Psychological Bulletin*, 1970, *1*, 17–24.

Strelau, J. A diagnosis of temperament by nonexperimental techniques. *Polish Psychological Bulletin*, 1972, *3*, 97–105.

Stricker, L. J. Personality research form: Factor structure and response style involvement. *Journal of Consulting and Clinical Psychology*, 1974, *42*, 529–533.

Stricker, L. J., & Ross, J. An assessment of some structural properties of the Jungian personality typology. *Journal of Abnormal and Social Psychology*, 1964, *68*, 67–71. (a)

Stricker, L. J., & Ross, J. Some correlates of a Jungian personality inventory. *Psychological Reports*, 1964, *14*, 623–543. (b)

Stricker, L. J., Jacob, P. I., & Kegan, N. Trait interrelations in implicit personality theories and questionnaire data. *Journal of Personality and Social Psychology*, 1974, *30*, 198–209.

Stumpfl, F. *Erbanlage und Verbrechen*. Berlin: Springer, 1935.

Suppe, F. Exemplars, theories and disciplinary matrices. In F. Suppe (Ed.), *The structure of scientific theories*. London: University of Illinois Press, 1974, pp. 483–499.

Swets, J. A. Signal detection theory applied to vigilance. In R. R. Mackie (Ed.) *Vigilance: Theory, operational performance, and physiological correlates*. New York: Plenum Press, 1977.

Taft, R. Extraversion, neuroticism, and expressive behavior: An application of Wallach's moderator effect to handwriting analysis. *Journal of Personality*, 1967, *35*, 570–584.

Taine, H. *De L'intelligence*. Paris: Hachette, 1878.

Tanner, J. M. *The physique of the Olympic athlete*. London: Allen & Unwin. 1964.

Tanner, J. M. *Education and physical growth*. New York: International Universities Press, 1978.

Tansley, A. G. *The new psychology and its relation to life*. New York: Dodd, Mead, 1920.

Taylor, J. A. A Personality Scale of Manifest Anxiety. *Journal of Abnormal and Social Psychology*, 1953, *48*, 285–290.

Taylor, W. B. Kinetic theories of gravitation. *Smithsonian Reports* (Washington), 1876, 205–282.

Teasdale, J. D. Psychological treatment of phobias. In N. S. Sutherland (Ed.), *Tutorial essays in psychology* (Vol. 1). Hillsdale, N.J.: Lawrence Erlbaum, 1977.

Teichman, Y. Emotional arousal and affiliation. *Journal of Experimental Social Psychology*, 1973, *9*, 591–605.

Tellegen, S. *Brief manual for the Differential Personality Questionnaire*. Unpublished.

Tennyson, R. D., & Wooley, F. R. Interaction of anxiety with performance on two levels of task difficulty. *Journal of Educational Psychology*, 1971, *62*, 463–467.

Thackray, R. I., Jones, K. N., & Touchstone, R. M. Personality and physiological correlates of performance decrement on a monotonous task requiring sustained attention. *British Journal of Psychology*, 1974, *65*, 351–358.

Thomas, A., & Chess, S. *Temperament and development*. New York: Brunner/Mazel, 1977.

Thorndike, E. C. *Educational psychology*. New York: Teachers College, 1903.

Thurstone, L. L. *Primary mental abilities*. Chicago: University of Chicago Press, 1938.

Thurstone, L. L. The dimensions of temperament. *Psychometrika*, 1951, *16*, 11–20.

Thurstone, L. L., & Thurstone, T. G. *Factorial studies of intelligence and IQ.*. Chicago: University of Chicago Press, 1941.

Tiggemann, M., Winefield, A. H., & Brebner, J. The role of extraversion in the developmetn of learned helplessness. *Personality and Individual Differences*, 1982, *3*, 27–34.

Timm, V. Reliabilität und Faktorenstruktur von Cattell's IUPF test bei einer deutschen Stichprobe. *Zeitschrift für Experimentelle und Angewandte Psychologie*, 1968, *15*, 354–373.

Timm, V. Eigenschaftsratings als Validitätskriterien eines Persönlichkeitsfragebogens. *Diagnostica*, 1971, *17*, 26–45.

Tobena, S. *Intensitat de l'estimul incondicionat i diferencias individuales*. Unpublished Ph.D. thesis, Universidad Autonoma de Barcelona, 1977.

Tobena, S., Garcia-Sevilla, L., & Garau, S. *Studies on an analogue of extraversion in the rat*. Unpublished manuscript. Universidad Autonoma de Barcelona, 1978.

Tranel, N. Effect of perceptual isolation on introverts and extraverts. *Journal of Psychiatric Research*, 1962, *1*, 185–192.

Travis, R. C. The measurement of fundamental character traits by a new diagnostic test. *Journal of Abnormal Psychology*, 1925, *19*, 400–420.

Trouton, D. S., & Maxwell, A. E. The relation between neurosis and psychosis. *Journal of Mental Science*, 1956, *102*, 1–21.

Trown, E. A., & Leith, G. O. H. Decision rules for teaching strategies in primary schools: Personality treatment interaction. *British Journal of Educational Psychology*, 1975, *45*, 120–140.

Tune, G. S. Errors of omission as a function of age and temperament in a type of vigilance task. *Quarterly Journal of Experimental Psychology*, 1966, *18*, 358–361.

Turnbull, A. A. Selling and the salesman: Prediction of success and personality change. *Psychological Reports*, 1976, *38*, 1175–1180.

Types, E. C., & Christal, R. E. *Recurrent personality factors based on trait ratings*. USAF Tackland Air Force base, Personnel laboratories technical report, 1961, ASD-TR-61-97.

Vale, J. R., Ray, D., & Vale, C. S. The interaction of genotype and exogenous neonatal androgen: Agonistic behavior in female mice. *Behavioral Biology*, 1972, *7*, 321–333.

Vandenberg, S. G., & Price, R. A. Replication of the factor structure of the Comrey Personality Scales. *Psychological Reports*, 1978, *42*, 343–352.

Van Hooff, J. *Aspects of social behaviour and communication in human and higher non-human primates*. Rotterdam: Bronder, 1971.

Venables, P. H. The relationships between level of skin potential and fusion of paired light flashes in schizophrenics and normal subjects. *Journal of Psychiatric Research*, 1963, *1*, 279–287.

Venturini, R., de Pascalis, V., Imperiali, M. G., & San Martini, P. EEG alpha reactivity and extraversion–introversion. *Personality and Individual Differences*, 1981, *2*, 215–220.

Verghese, A., Large, P., & Chin, E. Relationship between body build and mental illness. *British Journal of Psychiatry*, 1978, *132*, 12–15.

Vernon, P. E. *The assessment of psychological qualities by verbal methods.* London: H.M.S.O., 1938.

Verschuer, O. *Wirksame Faktoren im Leben der Menschen.* Wiesbaden: Steiner, 1954.

Wachtel, P. L. Conceptions of broad and narrow attention. *Psychological Bulleting*, 1967, *68*, 417–429.

Wakefield, J. A., & Doughtie, E. B. The geometric relationship between Holland's person-altiy model and the Vocational Preference Inventory. *Journal of Counselling Psychology*, 1973, *20*, 513–518.

Wakefield, J. A., Yom, B. H. L., Bradley, P. E., Doughtie, E. B., Cox, J. A., & Kraft, I. A. Eysenck's personality dimensions: A model for the MMPI. *British Journal of Social and Clinical Psychology*, 1974, *13*, 413–420.

Wakefield, J. A., Bradley, P. E., Doughtie, E. B., & Kraft, I. A. Influence of overlapping and non-overlapping items on the theoretical interrelationships of MMPI scales. *Journal of Consulting and Clinical Psycholgoy*, 1975, *43*, 851–857.

Wakefield, J. A., Sasek, J., Brubaker, M. L., & Friedman, A. E. Validity study of the Eysenck Personality Questionnaire. *Psychological Reports*, 1976, *39*, 115–121.

Walker, E. L. Action decrement and its relation to learning. *Psychological Review*, 1958, *65*, 132–142.

Walker, N. K., & Burkhardt, J. F. The combat effectiveness of various human operator controlled systems. *Proceedings of the 17th US Military Operations Research Symposium*, 1965.

Walker, S. F. *Learning and reinforcement.* London: Methuen, 1975.

Walkey, F. H., & Green, D. E. The structure of the Eysenck Personality Inventory: A comparison between simple and more complex analyses of a multiple scale questionnaire. *Multivariate Behavioral Research*, 1981, *16*, 361–372.

Wallach, M. A. Action-analytical v. passive-global cognitive functioning. In S. Messick & J. Ross (Eds.), *Measurement in personality and cognition.* New York: Wiley, 1962.

Wallach, M. A., & Gahm, R. C. Personality functions of graphic constriction and expansiveness. *Journal of Personality*, 1960, *28*, 73–88.

Wambach, R. L., & Panackal, A. A. Age, sex, neuroticism, and traces of control. *Psychological Reports*, 1979, *44*, 1055–1058.

Wankowski, J. A. *Temperament, motivation and academic achievement.* University of Birmingham Educational Survey and Counselling Unit, 1973.

Ward, E. S., & Hemsley, D. R. The stability of personality measures in drug abusers during withdrawal. *International Journal of the Addictives*, 1982, *17*, 575–583.

Ward, W. C. Creativity and environmental cues in nursery school children. *Developmental Psychology*, 1969, *1*, 543–577.

Washburn, M. F., Hughes, G., Steward, C., & Sligh, G. Reaction time, flicker, and affective sensitiveness as tests of extraversion and introversion. *American Journal of Psychology*, 1930, *42*, 412–413.

Watts, F. Desensitization as an habituation phenomenon: Stimulus intensity as determinant of the effect of stimulus length. *Behaviour Research and Therapy*, 1971, *9*, 209–217.

Webb, E. Character and intelligence. *British Journal of Psychology*, Monograph Supplement, 1915.

Weinberg, R. S. The effects of success and failure on the patterning of neuromuscular energy. *Journal of Motor Behavior*, 1978, *10*, 53–61.

Weinberg, R. S., & Hunt, V. The interrelationships between anxiety, motor performance, and electromyography. *Journal of Motor Behavior*, 1976, *8*, 219–224.

Weiner, B., & Schneider, K. Drive versus cognitive theory: A reply to Boor and Harmon. *Journal of Personality and Social Psychology*, 1971, *18*, 258–262.

Weiner, M. J., & Samuel, W. The effect of attributing internal arousal to an external source upon test anxiety and performance. *Journal of Social Psychology*, 1975, *96*, 255–265.

Weisen, A. *Differential reinforcing effects of onset and offset of stimulation on the operant behavior of normals, neurotics, and psychopaths.* Unpublished Ph.D. thesis, University of Florida, 1965.

Weisstein, N. A Rashevsky-Landall neural net: Simulation of metacontrast. *Psychological Review*, 1968, *75*, 494–521.

Weldon, E. An analogue of extraversion as a determinant of individual differences in behaviour in the rat. *British Journal of Psychology*, 1967, *58*, 253–259.

Wells, F. L. *Mental adjustments.* New York: Appleton, 1917.

Weiner, J., & Stromgren, E. Clinical and genetic studies on benign schizophreniform psychoses based on a follow-up. *Acta Psychiatrica Scandinavica*, 1958, *33*, 377–418.

Welsh, G. S. Factor dimensions A and R. In G. S. Welsh & W. G,. Dahstrom (Eds.), *Basic readings in the MMPI in psychology and medicine.* Minneapolis: University of Minnesota Press, 1956.

Werner, H. Studies on contour: I. Qualitative analyses. *American Journal of of Psychology*, 1935, *47*, 40–64.

Wertheimer, M. Figural aftereffects as a measure of metabolic efficiency. *Journal of Personality*, 1955, *24*, 56–73.

Wheelwright, J. B., Wheelwright, J. H., & Buehler, J. A. *Jungian type survey manual: The Gray-Wheelwright Test.* San Francisco: Society of Jungian Analysts, 1964.

Whimbey, E., & Denenberg, V. H. Two independent behavioral dimensions in open-field performance. *Journal of Comparative and Physiological Psychology*, 1967, *63*, 500–504.

White, P. O. Some major components in general intelligence. In H. J. Eysenck (Ed.), *A model for intelligence.* New York: Springer, 1982, pp. 44–90.

White, W. A. *Mechanisms of character formation.* New York: Macmillan, 1916.

Wiersma, H. Die Sekundär funktion bei Psychosen. *Zeitschrift für Psychologie and Neurologie*, 1906, *8*, 1–24.

Wiggins, J. S. *Personality and prediction: Principles of personality assessment.* Reading, Mass: Addison Wesley, 1973.

Wiggins, J. S., & Winder, C. L. The Peer Nomination Inventory: An empirically derived sociometric measure of adjustment in pre-adolescent boys. *Psychological Reports*, 1961, *9*, 643–677.

Wigglesworth, M. J, & Smith, B. D. Habituation and dishabituation of the electrodermal orienting reflex in relation to extraversion and neuroticism. *Journal of Research in Personality*, 1976, *10*, 437–445.

Wilkinson, R. T. The effect of lack of sleep on visual watchkeeping. *Quarterly Journal of Experimental Psychology*, 1960, *12*, 36–40.

Wilkinson, R. T. Interaction of noise with knowledge of results and sleep deprivation. *Journal of Experimental Psychology*, 1963, *66*, 332–337.

Wilkinson, R. T. Effects of up to 60 hours' sleep deprivation on different types of work. *Ergonomics*, 1964, *7*, 175–186.

Willett, R. A. Measures of learning and conditioning. In H. J. Eysenck (Ed.), *Experiments in personality* (Vol. 2). London: Routledge & Kegan Paul, 1960.

Williams, J. L. *Personal space in relation to extraversion-introversion.* Master's thesis, University of Alberta, 1963.

Williams, J. L. An analysis of gaze in schizophrenics. *British Journal of Social and Clinical Psychology*, 1974, *13*, 1–8.

Wilson, G. D., & Maclean, A. Personality, attitudes, and human preferences of prisoners and controls. *Psychological Reports*, 1974, *34*, 847–854.

Wilson, G. D., Tunstall, O. A., & Eysenck, H. J. Individual differences in tapping performance as a function of time on the task. *Perceptual and Motor Skills*, 1971, *33*, 375–378.

Wilson, R. S. Twins: Early mental development. *Science*, 1972, *175*, 914–917.

Wine, J. Test anxiety and direction of attention. *Psychological Bulletin*, 1971, *76*, 92–104.

Winter, K., Broadhurst, A., & Glass, A. Neuroticism, extraversion and EEG amplitude. *Journal of Experimental Research in Personality*, 1972, *6*, 44–51.

Winter, W. D., & Stortroen, M. A. A comparison of several MMPI indices to differentiate psychotics from normals. *Journal of Clinical Psychology*, 1963, *19*, 220–223.

Wissler, C. The correlation of mental and physical tests. *Psychological Revue Monograph*, 1901, No. 3.

Witkin, H. S., Ryk, R. B., Fatterson, H. F., Gardenaugh, P. R., & Pargs, S. A. *Psychological differentiation.* New York: Wiley, 1962.

Worthy, M. *Eye color, sex and race.* Anderson, South Carolina: Drokeltane/Hallus, 1974.

Wundt, W. *Grundzüge der Physiologischen Psychologie* (Vol. 3, 5th Ed.). Leipzig: W. Engelmann, 1903.

Yerkes, R. M., & Dodson, J. D. The relation of strength of stimulus to rapidity of habit-formation. *Journal of Comparative and Neurological Psychology*, 1908, *18*, 459–482.

Young, P. A., Eaves, L. J., & Eysenck, H. J. Inter-generational stability and change in the causes of variation in personality. *Personality and Individual Differences*, 1980, *1*, 35–55.

Yule, W., Gold, R. D., & Busch, C. Long-term predictive validity of the WPPSI: An 11-year follow-up study. *Personality and Individual Differences*, 1982, *3*, 65–72.

Zaffy, D. J., & Bruning, J. L. Drive and the range of cue utilization. *Journal of Experimental Psychology*, 1966, *71*, 382–384.

Zajonc, R. B. Social facilitation. *Science*, 1965, *14*, 269–274.

Zenneck, J. Gravitation. *Encyklopaedie der Mathematischen Wissenschaften.* 1903, *6*, (1), 25–30.

Zerssen, D. V. Konstitution. In K. P. Kisker, J. E. Meyer, C. Miller, & E. Stromgren (Eds.), *Psychiatrie der Gegenwart* (Vol. I/2). Berlin: Springer, 1980.

Zevon, M. A., & Tellegen, A. The structure of mood change: An idiographic/monothetic analysis. *Journal of Personality and Social Psychology*, 1982, *43*, 111–122.

Zuckerman, M. Sensation seeking and risk taking. In C. E. Izard (Ed.), *Emotions in personality and psychopathology.* New York: Plenum Press, 1979. (a)

Zuckerman, M. *Sensation seeking: Beyond the optimal level of arousal.* London: Wiley, 1979. (b)

Zuckerman, M., & Gerbasi, K. C. Dimensions of I-E scale and their relationship to other personality measures. *Emotional and Psychological Measurement*, 1977, *37*, 159–175.

Zuckerman, M., Eysenck, S. B. G., & Eysenck, H. J. Sensation seeking in England and America: Cross-cultural age and sex comparison. *Journal of Consulting and Clinical Psychology*, 1978, *46*, 139–149.

Author Index

Subject Index

Academic achievement, 320–325
Adjustment, 57
Aftereffect. *See* Figural aftereffects;
  Spiral aftereffects
Aggregation. *See* Principle of
  aggregation
American Psychological Association, 77
Animal studies
  conditioning, 240
  personality theory and, 97–102, 210
Antisocial behavior, 329–336
Anxiety
  academic achievement and, 332–324
  attentional mechanisms and, 303–305
  conditioning and, 241–242
  efficiency/effectiveness and, 294–299
  gaze behavior and, 315–316
  habituation and, 232–233
  intimacy and, 316
  learning/memory and, 305–309
  occupational performance and, 326
  performance and, 290–291, 309
  personality theory and, 209–210, 289
  state-trait theory, 18–19, 290–291
  task interactions, 299–303
  theory of, 292
  See also Worry
Appetitive conditioning, 244–245, 246
Approach-avoidance conflicts, 317
Arousal
  attention and, 303
  conditioning and, 241
  electroencephalography and, 223

Arousal (*cont.*)
  intimacy and, 314
  introversion-extraversion, 272–273
  learning and, 270
  measurement of, 217
  memory and, 260–261, 262, 263
  neuroticism and, 233
  perceptual phenomena and, 274–275
  reminiscence and, 271
  schizophrenia, 66–67
  social interaction, 313–314
  stimulation sensitivity, 248–256
  vigilance and, 257–260
Arousal theory, 193, 196–209
  behavioral data and, 284–288
  spiral aftereffect and, 279–280
  support for, 283–284
Ascending reticular activating system
  (ARAS), 197–198
  extraversion and, 218
Ascending reticular formation, 47
Attention, 303–305
Auditory evoked potentials
  intelligence and, 174–175
  sensitivity and, 255, 256
  See also Evoked potentials
Aversive conditioning, 244

Behavior, 311–343
  animal studies, 97–102
  arousal theory and, 284–288
  crime and, 329–336
  educational achievement and, 320–325